Soil and Aquifer Pollution

Springer

Berlin
Heidelberg
New York
Barcelona
Budapest
Hong Kong
London
Milan
Paris
Santa Clara
Singapore
Tokyo

Hillel Rubin · Nava Narkis · Judith Carberry (Eds.)

Soil and Aquifer Pollution

Non-Aqueous Phase Liquids – Contamination and Reclamation

With 181 Figures and 40 Tables

Springer

Prof. Hillel Rubin
Prof. Nava Narkis
Dept. of Civil Engineering
Technion - Israel Institute of Technology
Haifa 32000, Israel

Dr. Judith Carberry
359 N Street, SW
Washington, DC 20024
USA

ISBN 3-540-62586-0 Springer-Verlag Berlin Heidelberg New York

Library of Congress Cataloging-in-Publication Data
Soil and aquifer pollution: non-aqueous phase liquids - contamination and reclamation / Hillel Rubin, Nava Narkis, Judith Carberry. eds. p. cm. Includes bibliographical references and index. ISBN 3-540-62586-0 (hardcover) 1. Oil pollution of soils. 2. Oil pollution of water. 3. Dense nonaqueous phase liquids - Environmental aspects. 4. Soil remediation. 5. Groundwater - Purification. I. Rubin, Hillel. II. Narkis, Nava. III. Carberry, Judith B. TD879.P4S57 1998 628.5'2 - dc21

Printed in Germany

Cover design: *design & production* GmbH, Heidelberg
Typesetting: Best-set Typesetter Ltd., Hong Kong

SPIN 10540264 31/3137 5 4 3 2 1 0 - Printed on acid-free paper

Foreword

The Water Research Institute at the Technion (Israel Institute of Technology) is proud to have initiated and sponsored the International Workshop "Soil and Aquifer Pollution: Non-Aqueous Phase Liquids – Contamination and Reclamation", held May 13th–15th, 1996, on the Technion campus in Haifa.

Groundwater contamination is one of the pressing issues facing Israel and other countries which depend on groundwater for water supply. In Israel, 60% of the water supply comes from groundwater, most of it from two large aquifers. The Coastal Aquifer underlies the area where the largest concentration of human activity already takes place, and where much of future development is expected to occur. It is a phreatic sandstone aquifer, vulnerable to pollution from activities at the surface. The Mountain Aquifer is recharged in the higher terrain to the east, and flows, first in a phreatic zone, then confined, westward and underneath the Coastal Aquifer. This limestone aquifer has higher permeabilities and flow velocities, so pollution can reach the groundwater quite readily. Smaller local aquifers are also important components in the national water system.

While measures are taken to protect these aquifers from pollution, there are locations where contamination has already occurred. Furthermore, accidental pollution may not be totally avoided in the future. Therefore, understanding the processes of groundwater contamination, recommending the proper measures for preventing it, and determining the best means for reclamation once pollution has occurred, are of great practical importance. Non-aqueous phase liquids (NAPLs) are among the most significant contaminants.

Past experience with remediation of polluted groundwater has been less than successful, and has led to the realization that we do not yet understand sufficiently the physical, chemical and biological processes and their interactions, which take place in an aquifer during contamination events and remediation programs. Very large sums of money have been spent in recent years on groundwater remediation, only to find out that, after the project was completed, the quality of groundwater still does not meet the desired and expected standards. This has sent scientists and practitioners "back to the drawing board", to reconsider existing knowledge and fill gaps by basic and applied research.

The Workshop is a contribution to this effort. Experts form Israel and around the world convened to present research experience and results, exchange information, discuss theories and approaches, recommend where research efforts should be placed, and suggest relevant water resources management policies.

The Water Research Institute is grateful to all experts who participated in the Workshop, to the editorial board of this book, and especially to Professor Hillel Rubin, who so ably undertook the organization and management of the Workshop.

This book is a tribute to their efforts, a contribution to the understanding of groundwater contamination by NAPLs, and to the identification of best reclamation and management options.

Uri Shamir, Director, Water Research Institute
Haifa, Fall 1997 Technion - Israel Institute of Technology

Preface

In May 1996, a distinguished group of scientists and engineers gathered at the Technion in Haifa, Israel, to conduct a workshop entitled, Soil and Aquifer Pollution: Non-Aqueous Phase Liquids – Contamination and Reclamation. The purpose of this meeting was to engage interdisciplinary experts in discussion of various aspects of this complex subject. Of all the disciplines represented at the meeting, the scientists and engineers could be characterized as "wet" or "dry." That is, these experts were nominally experimentalists or modelers. The discussion was very beneficial, therefore, since the exchange of information between these two types of researchers is so rare and appreciated. Soil scientists discussed biological topics with microbiologists. Porous media modelers exchanged information on surfactant character with chemists. Much of the interchange took place following the formal presentations which lasted 3 days. Each day was devoted to one or two topics, and each topic was addressed by experts from different disciplines. At the end of the day, the daily chairperson summarized the proceedings of the day and integrated the different perspectives presented on each topic. These summaries provoked further interchange and discussion.

This book is a collection of manuscripts written to describe each presentation at the workshop. The five sections contained herein attempt to cover the topics in depth and from the perspectives of the scientists and engineers contributing their work. The introductory section is followed by sections on physical–chemical aspects of soil and aquifer pollution. A short, but definitive section on bioremediation follows, and then an extensive section on modeling of processes in saturated and unsaturated soils and in fractured rock. The final section presents field applications and case histories of remediation technologies. The breadth and depth of these topics should be of interest to both theoretical researchers and remediation managers and practitioners. This book should indicate both the current directions of work on NAPL contamination, its behavior and remediation, and the unanswered questions which need to be addressed. The editors hope it will be of interest worldwide.

The editors of this volume wish to thank the authors of each chapter and the publishing editor, Ms. Isolde Tegtmeier of Springer-Verlag for her coordination and cooperation in producing this book.

H. Rubin
N. Narkis
J. B. Carberry

Fall 1997

Contents

List of Contributors

Dr. Linda M. Abriola
Dept. of Civil & Environmental Engineering
The University of Michigan
1351 Beal Ave.
Ann Arbor, Michigan 48109-2125, USA

Dr. Robert Armon
Environmental & Water Resources Engineering
Faculty of Civil Engineering
Technion - Israel Institute of Technology
Haifa 32000, ISRAEL

Dr. Jacob Bear
Environmental & Water Resources Engineering
Faculty of Civil Engineering
Technion - Israel Institute of Technology
Haifa 32000, ISRAEL

Dr. Edward J. Bouwer
Dept. of Geography and Environmental Engineering
The Johns Hopkins University
Baltimore, Maryland 21218, USA

Dr. Carol Braester
Technion - Israel Institute of Technology
Faculty of Civil Engineering
Haifa 32000, ISRAEL

Dr. Judith B. Carberry
359 N Street, SW
Washington, DC 20024, USA

Dr. M. Yavuz Corapcioglu
Dept. of Civil Engineering
Texas A&M University
College Station, Texas 77843-3136, USA

Dr. N. I. Galil
Dept. of Civil Engineering
Environmental and Water Resources Engineering
Technion - Israel Institute of Technology
Haifa 32000, ISRAEL

Dr. Ellen R. Graber
Institute of Soils and Water
ARO
The Volcani Center
P.O. Box 6
Bet Dagan 50250, ISRAEL

Dr. Linda S. Lee
Dept. of Agronomy
Purdur University
1150 Lilly Hall
West Lafayette, Indiana 47907, USA

Dr. Vincent Lagendijk
Ist. of Hydraulic Engineering and
Water Resources Management
Aachen University of Technology
Mies-van-der-Rohe Str. 1
52056 Aachen, GERMANY

Dr. Svetlana Lumelsky
Dept. of Civil Engineering
Environmental & Water Resources Engineering
Technion - Israel Institute of Technology
Haifa 32000, ISRAEL

Dr. S. Lumelsky
Ha-Arugot Str., 6/7
Pardes Hana - Karkur
37000, ISRAEL

Dr. Nava Narkis
Technion - Israel Institute of Technology
Dept. of Civil Engineering
Environmental and Water Resources Engineering
Technion City, Haifa 32000, ISRAEL

Dr. J. T. Novak
Dept. of Civil Engineering
Virginia Polytechnic Inst. and
State University
Blacksburg, Virginia 24061, USA

Dr. Jack C. Parker
Environmental Systems & Technologies, Inc.
2608 Sheffield Drive,
Blacksburg, Virginia 24060, USA

Dr. George F. Pinder
College of Engineering and Mathematics
University of Vermont
Burlington, Vermont 05405, USA

Dr. Arieh Pistiner
Ministry of Environment
Water Quality Division
2 Chury St.
Haifa 33045, ISRAEL

Dr. Manahem Rebhun
Division of Environmental & Water Resources Engineering
Dept. of Civil Engineering
Technion - Israel Institute of Technology
Haifa 32000, ISRAEL

Dr. E. Rosenberg
Dept. of Molecular Microbiology and Biotechnology
George S. Wise Faculty of Life Sciences
Tel Aviv University
Ramat Aviv 69978, ISRAEL

Dr. Hillel Rubin
Division of Environmental & Water Resources Engineering
Dept. of Civil Engineering
Technion - Israel Inst. of Technology
Haifa 32000; ISRAEL

Dr. Victor C. Ryzhik
Division of Environmental & Water Resources Engineering
Dept. of Civil Engineering
Technion - Israel Institute of Technology
Haifa 32000, ISRAEL

Dr. D. A. Sabatini
School of Civil Engineering and Environmental Science
The University of Oklahoma
Norman, Oklahoma 73019, USA

Dr. Shaul Sorek
Water Resources Research Center
J. Blustein Institute for Desert Research
Ben-Gurion University of the Negev
Sde Boker Campus 84990, ISRAEL

Dr. Walter J. Weber, Jr.
Environmental and Water Resources Engineering
The University of Michigan
1351 Beal Ave.
Ann Arbor, Michigan 48109-2125, USA

Dr. Bruno Yaron
Institute of Soils and Water
ARO
The Volcani Center
Bet Dagan, ISRAEL

Dr. Uri Zoller
Dept. of Science Education - Chemistry
Haifa University - Oranim
Kiryat Tivon 36006, ISRAEL

Part I

Introduction and General Considerations

1 Overview of NAPL Contamination and Reclamation

H. Rubin[1], N. Narkis[1], and J. B. Carberry[2]

Contents

1 Introduction

Groundwater represents the major source of drinking water in most parts of the world. In Israel about 60% of the water supply comes from aquifers. For many years the public's interest with regard to water needs was concerned with the availability of sufficient quantities of groundwater. In arid and semiarid regions there is an increasing need for adequate storage in which water surplus in wet seasons can be diverted for use in dry seasons. It has been shown that in most cases groundwater aquifers represent the best option for water storage. Therefore, references to groundwater resources include the quantity of groundwater as well as the capacity of the aquifers to store water.

In the past, Israel was a very important producer of oranges. Citrus trees are extremely sensitive to chloride content in water and suffer when chloride concentration is higher than 150 mg/l. Therefore, the major concern regarding groundwater quality was for many years limited to the content of chlorides and salinity in the water.

Scientists directed the interest of water authorities and the public to a detailed awareness of water quality. They showed that many inorganic and organic components may degrade the quality of groundwater and create risks to public health. At

[1]Division of Environmental & Water Resources Engineering, Department of Civil Engineering, Technion – Israel Institute of Technology, Haifa 32000, Israel
[2]JBC, Inc., 359 N Street, SW, Washington, D.C. 20024, USA

present, there are global problems with increasing nitrate concentrations in groundwater due to the use of fertilizers and irrigation with treated effluents and with an increasing concentration of heavy metals and organic micropollutants due to vast industrial activity, etc. The increasing number of cases of soil and aquifer contamination by nonaqueous phase liquids (NAPLs) are complicated and very risky. Such fluids are mainly immiscible fuels, oil distillates, and oil products. Minute quantities of such fluids dissolved in the aqueous phase have a very significant effect on the taste and odor of the water. Human beings are sensitive to the presence of fuel in water even when its concentration is smaller than 1 mg/l. Some of the hydrocarbon fractions in the fuel are poisonous at specific threshold levels. Various organic liquid compounds which incorporate chlorine are suspected to be carcinogenic substances. In every industrial country NAPL contamination of soils and aquifers is very common. However, such a contamination is not always reported, and its effect is not recognized for years after the contamination has occurred. Therefore, the extent of NAPL contamination is not always defined and adequately reported by the water authorities. Statistics have shown that, in Israel, there have been many instances of fuel disposal and leakage which led to soil and aquifer contamination. In Israel risks of NAPL contamination are very high, and damages from such events may be very significant. In the narrow strip of land constituting the coastal zone of Israel there are many operations associated with transport and storage of crude oil, fuels, and other oil products. This piece of land is subject to very intensive urbanization and industrialization. There have been many cases of physical damage to oil pipelines due to construction operations. Corrosion of pipes and storage tanks is also an important source of oil leaks into the environment.

According to the Israeli Ministry of the Environment, during 1995 18 cases of fuel leaks were reported, not including unreported leaks. The total volume of fuel in the reported leaks was about 630 m^3. From the reported leaks, 12 were caused by physical damage and corrosion; 4 leaks occurred in storage tanks; 1 spill occurred in a truck-tanker accident; and 1 leak took place in an industrial plant. Most of these oil leaks took place in the coastal zone, on top of the Coastal Plain Aquifer of Israel. This aquifer constitutes the major source of drinking water in Israel. Such oil leaks continued to occur during the last year, even though the public is aware of the risks to the environment and groundwater. However, in the past, many events of NAPL disposal and contamination of soil and groundwater were not reported or were even known to the authorities and the responsible individuals. Information as to the scale of such contamination, its potential risk to groundwater quality, and how its negative effects should be avoided has not yet been completely understood and evaluated.

In previous references to contamination, fuel and hydrocarbon constituents caused contamination of soil and groundwater. However, there are many other types of organic liquids with very high contamination potential which should also be considered. Some of them, like chlorinated hydrocarbons, are much more toxic and hazardous than fuels. Differences in risks caused by various types of fuel and fuel fractions should also be taken into account.

In the US, there are thousands of recognized contaminated sites. The United States Environmental Protection Agency (USEPA) maintains a national priority list of the most dangerously contaminated sites, and the reclamation of these sites must be addressed (USEPA 1996). Responsible parties or government agencies must develop an approved procedure to restore the site to a negotiated level of contamination, usually based on risks associated with expected or past uses of the site. Each state manages the reclamation of these sites. For example, Pennsylvania has 96 active sites which are in some stage of cleanup. Florida has 50 sites to which a dollar value of cleanup costs has been assigned. Usually the NAPL contaminated sites contain mixtures of constituents originating from industrial dumping on site or at a previous landfill. Some of these sites have contaminants only in the unsaturated soil zone and the most difficult sites contain contaminants which have penetrated an underground aquifer. The latter condition has resulted in very expensive, and in some cases futile, efforts to remediate.

2 Classification and Risk of Nonaqueous Phase Liquids (NAPLs)

NAPLs, like fuels, are identified as liquids lighter than water, called light NAPLs (LNAPLs). NAPLs whose density is higher than that of water are termed dense NAPLs (DNAPLs). LNAPLs are generally petroleum hydrocarbons which can be classified according to the following four basic types: paraffins (alkanes), olefins, naphthenes, and aromatics. Some examples of such compounds and their structure are given by Rubin and Mechrez (1989).

Petroleum hydrocarbon compounds are products obtained in the petroleum industry. In the refinery, crude oil is separated by distillation into various fractions (Wittcoff and Reuben 1980). The first and most volatile fractions are gases, which are not considered NAPLs. The second and third fractions are mixtures of petroleum hydrocarbons, which are the major constituents of gasoline. They are often termed "light distillates". Groundwater contamination by light distillates such as gasoline also incorporates the negative effect of fuel additives, which increase the octane number of the gasoline. Light distillates have comparatively low viscosity and a high volatility potential. Because of the large number of car engines and gasoline stations, soil and groundwater contamination due to light distillates is wide spread. Studies of this type have been reviewed by Corapcioglu and Baehr (1985). Because of the low viscosity, the residual immobile saturation of light distillates is comparatively low. Due to their high volatility, their persistence is low, and treatment of the water contaminated by light distillates is comparatively simple.

The fourth and fifth fractions, termed "medium distillates", are kerosene and diesel oil. These fuels are used for heavy trucks, tractors, jet aircraft and domestic heating. Usually the contamination potential of such distillates is assumed to be the highest among petroleum hydrocarbon fractions. Their average viscosity and low volatility lead to high persistence of these distillates. Groundwater and soil contamination by medium distillates are typical of phreatic aquifers located underneath airports and airbases.

The sixth and seventh fractions of the distillation process are termed "heavy distillates." Although such distillates may sometimes include poisonous compounds, their contamination potential is usually assumed to be comparatively small. Due to their high viscosity, they migrate very slowly through the unsaturated zone, and their immobile degree of saturation is comparatively high (Dietz 1979).

Medium and heavy distillates contain aromatic hydrocarbons. Some of them have proved to be hazardous. For example, benzene has been determined a human carcinogen by the American Cancer Society of the United States. Brumaster and Harris (1982) reported increases in the number of spontaneous abortions and prolonged menstrual bleeding attributed to occupational exposure to benzene, toluene, and xylene. These compounds are also designated as hazardous by the Toxic Substance Control Act of the United States. Furthermore, according to McKee et al. (1972), taste and odor of petroleum hydrocarbons can be detected in concentrations as low as 0.005 mg/l.

The major group of DNAPLs includes chlorinated solvents, which represent the most prevalent organic groundwater contaminants (McCarty and Semprini 1994). These solvents consist primarily of chlorinated aliphatic hydrocarbons. They are usually used for degreasing aircraft engines automobile parts, electronic components, and clothing. Their water solubility is in the range of 1000 mg/l, which is several orders of magnitude higher than the drinking water standards.

The major chlorinated solvents are: carbon tetrachloride (CTC), tetrachloroethene (PCE), trichloroethene (TCE), and 1,1,1-trichloroethane (TCA). These compounds can be transformed by chemical and biological processes in soils to form a variety of other chlorinated aliphatic hydrocarbons. The chlorinated solvents are considered toxic.

3 Phenomenological Characteristics of NAPL Contamination

NAPL quantities which are released at the ground surface percolate almost vertically through the unsaturated zone, with some migration in the horizontal direction. Some of the NAPL is entrapped in that zone. If the released NAPL quantity is large, or the water table is close to the surface, then the NAPL spill may arrive at the capillary fringe, where it migrates horizontally. LNAPL forms a lens of NAPL on top of the groundwater surface, as shown in Fig. 1, which refers to the distribution of the contaminant after the source of NAPL spill has vanished. This figure is typical of many publications concerning the contamination of groundwater by LNAPLs. This description, originally presented by Schwille (1975), eventually became a "classic description" of LNAPL spill migration in the subsurface.

The classic description of LNAPL spill migration may prevail under field conditions only under particular circumstances. If the NAPL spill volume is smaller than

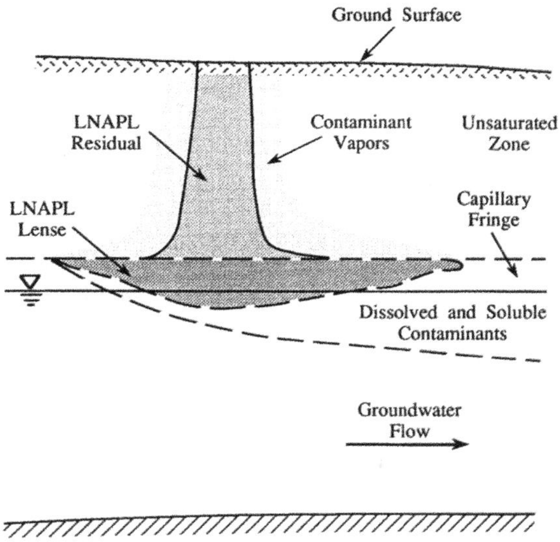

Fig. 1. Phenomenological characteristics of LNAPL spill contamination

the residual NAPL saturation of the soil layer, representing the unsaturated zone, then the NAPL will not arrive at the aquifer. If the spill volume is larger, then the LNAPL creates a lens floating on top of the groundwater table, as shown in Fig. 1. This example refers to a stationary water table, and no groundwater pumping is performed. However, the water table is subject to annual and seasonal fluctuations, which may lead to significant entrapment of the LNAPL in the aquifer and the capillary fringe. Such effects are also associated with pumping of groundwater. The LNAPL in the form of a free-floating lens and the entrapped NAPL are in direct contact with the flowing groundwater. They are subject to continuous dissolution and supply of dissolved and solubilized organic contaminant into the aquifer. It should be noted that the percolation of NAPL in the unsaturated zone may be considerably different from the percolation of water. Loam becomes very quickly impermeable to water after its contact with water. The permeability of this type of soil is almost unaffected by its contact with fuel. During diesel oil spill in the Negev Desert (Perle et al. this Vol.), the fuel percolated to a depth of 5 m in a few hours. Water in a similar quantity as the diesel spill would not be able to penetrate to such a depth.

As shown in Fig. 2, the vertical migration of a DNAPL spill, similar to the LNAPL spill, continues through the unsaturated zone under the forces of gravity and capillary attraction until it reaches the capillary fringe. There, the DNAPL continues to sink within the aquifer until an impermeable layer is reached. In depressions, the DNAPL creates pools or ponds on top of the impermeable layer. Figure 2 refers to the distribution of the contaminant after its source has vanished.

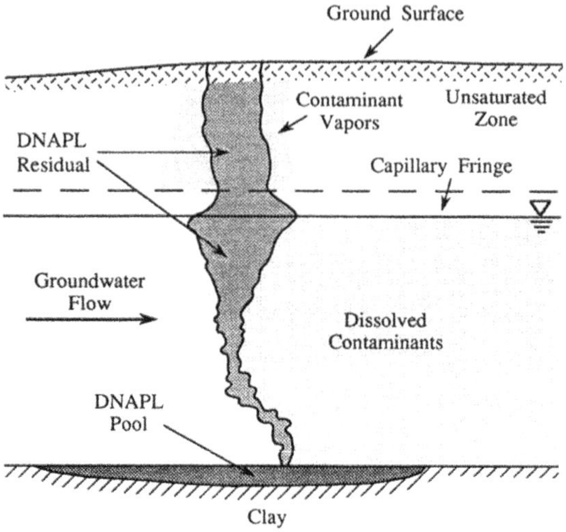

Fig. 2. Phenomenological characteristics of DNAPL spill contamination

Entrapment mechanisms of NAPL within porous media were reviewed by Powers et al. (1991). Blobs or ganglia of NAPL are trapped when interfacial forces are sufficiently strong to overcome viscous and gravitational forces (Handy and Datta 1966; Wardlaw 1982; Mohanty et al. 1987). In the saturated zone, the NAPL usually behaves as the nonwetting phase. Therefore, the entrapped blobs have concave interfaces and are formed within a single pore or several adjacent pores and throats (Schiegg 1980; Chatzis et al. 1983; Conard et al. 1987, 1989; Schwille 1988; Wilson et al. 1988). According to studies with micromodels made from etched glass, it seems that the shape and size of the entrapped blobs are a function of the pore geometry, interfacial tension, wettability, viscosity, and density of the NAPL (Chatzis et al. 1983; Lam et al. 1983; Chen 1986; Li and Wardlaw 1986a,b; Conard et al. 1987, 1989).

4 Physicochemical Processes

The previous section refers to various physicochemical processes associated with phenomenological characteristics of NAPL contamination in soil and aquifers. This section concerns the various physicochemical processes associated with soil and aquifer contamination and reclamation.

Contamination of the unsaturated zone by NAPLs may incorporate the following physicochemical processes: percolation, volatilization of flowing NAPLs, entrapment of the NAPL phase, multiphase flow of NAPL–water–air, dissolution, solubilization, adsorption–desorption, chemical degradation, and volatilization of entrapped NAPLs. In the saturated zone, the following phenomena may take place: multiphase flow of NAPL water, NAPL entrapment, dissolution, solubilization, and adsorption–desorption, as well as chemical degradation of entrapped NAPLs.

Some of the physicochemical processes are classified as flow and transport phenomena, which depend on physical properties of the NAPL, e.g., density and viscosity, as well as physical properties of the soil, such as porosity and permeability. Percolation of NAPLs in the unsaturated zone is in some cases very similar to the percolation of water (Kessler and Rubin 1987). The presence of clay particles in the soil may affect this similarity, as clay expands after contact with water, but does not expand after contact with NAPLs. Therefore, fuel percolation through loam is faster than water percolation (Perle et al. this Vol.). Multiphase flow of water–NAPL–air or water–NAPL can be analyzed by using the experience gained by the oil industry (Peaceman 1977; Faust 1985). The transport of dissolved NAPL constituents with the water phase in the unsaturated zone, as well as in the saturated zone, is basically identical to the transport of neutrally buoyant tracers (Bear 1972). Dissolution and solubility of the NAPL constituents in the water phase are associated with the multiphase flows as well as with the direct contact between the flowing water phase and the NAPL entrapped in the porous medium. Characteristics of the dissolution and solubility of NAPL constituents are reviewed by Powers et al. (1991). Heterogeneity of the porous medium as well as fractures in the aquifers may significantly reduce the efficiency of aquifer remediation methods, based on solubilization of the entrapped NAPL by injection and pumping systems (Rubin et al. 1997a, this Vol.). Enhancement of the entrapped NAPL dissolution and solubility by surfactants are the subject of various studies (Jafvert et al. 1994; Sabatini, this Vol.) concerning the reclamation of contaminated aquifers. Surfactants may increase the equilibrium concentration of dissolved and solubilized NAPLs with the water phase. They allow better exposure of the contaminant to biodegradation. Surfactants also modify coefficients of mass transfer between the aqueous and NAPL phases. Dissolved NAPL constituents may be subject to adsorption by soil minerals. Studies have shown that the adsorption of nonionic organic compounds by soil is often directly related to the organic carbon content of the adsorbing phase (Chiou et al. 1979; Briggs 1974; Karickhoff 1981). This information provides the basis for estimates of the mobility of toxic organic chemicals, including pesticides and herbicides (Gerstl 1989). Experience with such materials is also applicable to the evaluation of contamination and reclamation processes for other types of NAPL.

Highly volatile fractions of NAPL spills are transported into the gaseous phase of the porous medium and may subsequently be lost into the atmosphere, adsorbed into the solid phase, or dissolved in the groundwater. By implementing the soil vapor extraction (SVE) method, the transport of highly volatile entrapped NAPL in the unsaturated zone into the gaseous soil phase is enhanced. Limitations of such an enhancement are given by Hinchee (1994). The SVE method is based on the incorporation of NAPL mass transfer from the liquid phase to the gaseous phase and the transport of the gaseous phase in the porous medium (Baehr et al. 1989). The transport of the entrapped NAPL into the gaseous phase can be enhanced by the injection of steam if the higher temperature yields a more favorable partitioning of the entrapped NAPL. The bioventing method (Rathfelder et al. 1995) causes the flow of the gaseous phase in the unsaturated zone, but its major objective is to enhance bioremediation of the contaminated unsaturated zone. The method of air sparging (Brown 1991) enhances the transport of highly volatile entrapped NAPL in the saturated zone into the air phase and also enhances physical release of entrapped NAPL and bioremediation. The method induces the flow of air bubbles in groundwater. The study of the basic characteristics of such a flow are applicable for the improvement of this method of aquifer cleanup.

5 Biological Processes

Biological processes are involved in the degradation of NAPL contaminants in the unsaturated as well as saturated zones. Bioremediation involves the enhancement of microbial activity to convert contaminants to less harmful species. Microbial activity may create natural surfactants and emulsifiers (Rosenberg and Ron, this Vol.), which better release the entrapped NAPL to biodegradation. Technologies for the bioremediation of excavated soil, unsaturated soil or groundwater, are presently in use and seem to be feasible and economical. Bioremediation offers the most realistic and economical approach for coping with general problems of waste treatment. During the last decade, more and more examples regarding the practical use of bioremediation methods for the cleanup of soil and groundwater have been reported. In situ bioremediation usually involves the enhancement of indigenous soil microflora (Armon and Arbel, this Vol.), whereas introduced microflora is often applied in cases of on-site remediation of excavated soil (Perle et al., this Vol.). Criteria used to determine whether microbial augmentation is necessary have been proposed by Piotrowski and Cunningham (1996). Recently, demonstrations have been conducted, indicating that only oxygen need be introduced to the petroleum plume in order to provide proper conditions for bioremediation (Vogel 1996).

Bioremediation systems insure optimal metabolic activity of biomass, which utilizes the NAPL as a primary or secondary substrate (McCarty and Semprini 1994). Such systems should supply nitrogen, phosphorus, and oxygen to bacteria that are present in the contaminated soil and groundwater. If the site is contaminated by chlorinated organic solvents, then very often a primary substrate should also be supplied. It has been indicated that halogenated aliphatic hydrocarbons can be consumed by methanotrophs (Roberts et al. 1990; Semprini and McCarty 1992) and phenol utilizers (Hopkins et al. 1992). Various technologies applied to enhance bioremediation are reviewed in the following section. Such technologies emphasize the supply of oxygen for aerobic biomass activity. However, bioremediation can also take place with electron acceptors other than oxygen. Such compounds, e.g., nitrates, are more soluble in water than oxygen. Some compounds are only transformed under aerobic conditions, some require reducing conditions, and others are transformed under both aerobic and anaerobic conditions. Combinations of anaerobic remediation followed by aerobic remediation seem promising in some cases of highly chlorinated organic contaminants, since anaerobic degradation alone can produce more toxic degradation products than the original chlorinated contaminant. Recent work by Carberry et al. (Carberry and Benzig 1991; Carberry and Lee 1991, 1992, 1994, 1995; Carberry 1993; Carberry and Kovach 1996) using chemical and biological preoxidation with Fenton's reagent and white rot fungus is very promising for enhancing the biodegradation of recalcitrant chlorinated hydrocarbons.

The objectives of bioremediation should be reasonable and associated with acceptable criteria based on environmental risks. A review by Perle et al. (this Vol.) has indicated that codes for the achievement of soil reclamation vary from country to country and, in the US, even from state to state. The development of codes relevant to possible and acceptable goals of bioremediation seems to be very appropriate. For example, the California State Water Resources Control Board has decided to recommend closing all low-risk contaminated sites and monitoring the results to underground drinking water supplies. This action was taken as a result of the report written by the Lawrence Livermore Laboratory and the University of California, which indicated that petroleum plumes are limited in their migration and do not impose a threat to deep underground aquifers used for drinking water supplies (Rice 1995). New York State, which takes much of its drinking water supply from surface waters, is proposing this action as a model for policies there (Rivkin 1996).

6 Modeling

The modeling of all processes involved in the contamination and reclamation of soil and groundwater is based on the appropriate quantification of the processes mentioned in previous sections. The incorporation of all phenomena and processes associated with NAPL contamination and reclamation is an extremely complicated task. Furthermore, studies of contamination and reclamation usually have quite limited scope and objectives. Therefore, a hierarchy of modeling approaches can be applied to provide coverage of all physicochemical and biological processes, and the practical aspects associated with NAPL contamination and reclamation.

Historically, the cleanup of aquifers contaminated with petroleum hydrocarbons began with the remediation of aquifers located underneath refineries. The pumping of petroleum hydrocarbons in those sites was justified economically: quite simply, the value of the pumped NAPL would cover the expenses associated with the aquifer cleanup. Under such circumstances the cleanup could not be complete, and basic modeling approaches, applied by the oil industry, could be used to provide some estimates of the multiphase flow induced in the aquifer during the cleanup process. The modeling approach developed by Faust (1985) and GeoTrans (1989) exemplifies the use of the experience gained in the oil industry. This approach is based on the reference to the equation of flow of each fluid phase and some constitutive equations representing values of parameters, such as relative permeabilities and capillary pressure. No mass transfer between the various fluid phases is considered. Computer codes with no mass transfer between phases can be useful for the reconstruction of the history of LNAPL contamination in aquifers, buildup of the LNAPL lens on top of the water table, and scenarios of simultaneous pumping of the free-floating LNAPL and groundwater (Rubin et al. 1994). Abriola (1987) reviewed the basic approaches of the modeling of multiphase flow associated with NAPL contamination in soil and aquifer. Generally, NAPLs are subject to dissolution and solubility in concentrations much higher than the allowed concentration according to the drinking water standard. Owing to this property, they may damage the groundwater quality by migrating very large distances within the water phase. Therefore, models incorporating multiphase flow with mass transfer between the fluid phases and dispersion of the dissolved components are presently under development and use (Lagendijk et al. this Vol.). Most of these models assume local equilibrium for component partitioning among the fluid phases. Such an assumption has also led to the development of various analytical methods for the simulation of multiphase flow with mass transfer between the fluid phases (Pistiner et al. 1990; Pistiner and Shapiro, this Vol.). Russell (1995) reports that the suitability of the assumption of local equilibrium between the fluid phases has the greatest impact on the practice of multiphase multicomponent modeling. His review indicates that the local equilibrium assumptions were found to be true if the Damkohler number is sufficiently large. The local equilibrium assumption is relevant in cases in which all fluid phases are subject to flow, as well as to cases of dissolution or volatilization of the entrapped NAPL. If the total quantity of NAPL is entrapped, then only a single fluid phase, aqueous or gaseous, is subject to flow. In such cases, modeling incorporates a single fluid flow with mass transfer from the entrapped NAPL into the flowing fluid phase (Powers et al. 1991, 1992). Such modeling is simpler than multiphase flow modeling, but is applicable after vanishing of the NAPL spill source and pumping of free LNAPL quantities floating on top of the groundwater table. It has been shown in various studies (Rathfelder et al. 1991; Yeh and Tripathi 1991; Reeves and Abriola 1994) that the assumption of local equilibrium is not applicable in many cases of multiphase flow or entrapped NAPL dissolution, associated with the flow of a single phase. The nonequilibrium significantly alters model formulation.

Bioremediation increases the complexity of modeling objectives. Biomass development, decay, and limited transport are additional components that should be incorporated into the modeling procedure. Linear modeling, based on Monod kinetics, has been considered in various studies (Molz et al. 1986; Chen et al. 1992). However, the advection of biomass in soil is usually very limited, and major processes associated with bioremediation involve mainly growth and decay. Therefore, nonlinear modeling of mass growth should probably be applied to the simulation of soil and aquifer bioremediation. Various studies refer to optimization modeling of soil and aquifer reclamation (Gorelick et al. 1984; Ahlfeld et al. 1986, 1988a,b; Sawyer et al. 1995; Xiang et al. 1995).

7 Technology

Reclamation of contaminated soils and aquifers is an emerging industry in industrial countries. Though many failures have been reported with regard to the achievement of reclamation goals, there have also been many successes. Experience gained through demonstration studies and pilot plants has indicated that careful formulation of reclamation goals may help to avoid disappointment and is essential.

The technology of NAPL-contaminated soil and aquifer reclamation involves the provision of adequate monitoring equipment and equipment involved in pumping of free NAPLs, the development of techniques and equipment for the enhancement of chemical and biological processes, and the development of measuring methods by which contamination and reclamation can be assessed.

With regard to the pumping of free LNAPL floating on top of the groundwater table, the development of membranes which separate floating hydrocarbons from water exemplifies the involvement of new technologies in the removal of floating LNAPL (SITE REMEDIATION 1996). Aquifer reclamation has triggered the development of new types of pumps able to pump floating LNAPL with minimum quantities of water (Breslin et al. 1996). Pumping of floating LNAPL lenses is probably the optimal method for the removal of free NAPL from a contaminated aquifer. Following the complete removal of the free NAPL, other technologies should be used to avoid further contamination of the groundwater originating from entrapped NAPL, or nonpoint source of NAPL, which cannot be identified and stopped. A complete description of the technology involved in a project of continuous reclamation of groundwater contaminated by NAPL originating from such a nonpoint source was given by Sontheimer et al. (1989). The remediation of groundwater was achieved in that project by the injection of water enriched with ozone.

Technologies for the pumping and treatment of contaminated water above the surface are gradually being replaced by in-situ remediation of soil and aquifers. Various techniques of air sparging (Brown et al. 1991) are currently being developed and used for the remediation of NAPL-polluted aquifers. This technology enhances oxygenation, dissolution of entrapped NAPL, volatilization, groundwater stripping, and physical displacement of entrapped NAPL. Increasing the efficiency of air-sparging techniques is the objective of various technology developers (Marly and Droste 1965). Enhancement of entrapped NAPL release by the use of various types of surfactants (Zoller, this Vol.) provides some hope for the development of a surfactant industry targeted at soil and aquifer reclamation. Such materials are extensively used in the remediation of excavated, contaminated soil (Perle et al. this Vol.).

The unsaturated zone is the subject of reclamation technologies based on the induced flow of the gaseous phase. Soil vapor extraction and bioventing (Rathfelder et al. 1995) technologies are being developed and used in many remediation projects.

Contemporary remediation in most cases involves the enhancement of the biological activity of microorganisms, which consume NAPL constituents as primary or secondary substrate. Techniques of applying biological emulsifiers are also being developed and used (Rosenberg and Ron, this Vol.). The application of technologies using indigenous microflora as well as introduced microorganisms (Armon and Arbel, this Vol.) are currently being considered in various reclamation projects.

In cases of halogenated organics contamination, bioremediation technologies involving the development of mixed microflora, supply of primary substrate, and optimization of bioremediation processes are the current challenge of soil and aquifer reclamation (Bouwer 1994; McCarty and Semprini 1994; Wilson and Kampbell 1994).

8 Summary and Conclusions

The contamination and reclamation of soils and aquifers polluted by NAPLs present interdisciplinary research and development challenges. Reference is made to the phenomenological characteristics of NAPL spill migration in the unsaturated and saturated zones. The classification of NAPLs with regard to their typical characteristics and contamination potential are reviewed. It is considered that NAPL contamination research is involved with studies of the physicochemical and biological processes of contamination and reclamation. Such studies should be quantified by adequate modeling approaches. The incorporation of theoretical research activities and laboratory studies enables upscaling to pilot plants and field studies. The latter provide guidelines for technological developments and practical environmental reclamation.

References

Abriola LM (1987) Modeling contaminant transport in the subsurface: an interdisciplinary challenge. Rev Geophys 25:125–134

Ahlfeld DP, Mulvey JM, Pinder GF (1986) Designing optimal strategies for contaminated groundwater remediation. Adv Water Resour 9:77–83

Ahlfeld DP, Mulvey JM, Pinder GF, Woo E (1988a) Contaminated groundwater remediation design using simulation optimization and sensitivity theory 1. Model development. Water Resour Res 24:431–441

Ahlfeld DP, Mulvey JM, Pinder GF, Woo E (1988b) Contaminated groundwater remediation design using simulation optimization and sensitivity theory 2. Analysis of field site. Water Resour Res 24:443–452

Baehr AL, Hoag GE, Marley MC (1989) Removing volatile contaminants from the unsaturated zone by inducing advective air-phase transport. J Contam Hydrol 4:1–26

Bouwer EJ (1994) Bioremediation of chlorinated solvents using alternate electron acceptors. In: Norris RD et al. (eds) Handbook of bioremediation. Lewis, Ann Arbor, Michigan, pp 149–175

Bear J (1972) Dynamics of fluids in porous media. American Elsevier, New York

Breslin M, Feola S, Lui R (1996) Pumps: quality reliability key. Int Groundwater Techn 1(7):13–16

Briggs GG (1974) A simple relationship between soil adsorption of organic chemicals and their octanol/ water partition coefficients. Proc 7th Br Insectic Fungic Conf 1973, Brighton, pp 83–86. The Boots, Nottingham
Brown RA, Herman C, Henry E (1991) The use of aeration in environmental clean ups. Proc Haztech Int Pittsburgh Waste Conf, Pittsburgh, Pennsylvania
Brumaster DE, Harris RH (1982) Groundwater contamination an emerging threat. Technol Rev 84:50–62
Carberry JB (1993) Enhacement of bioremediation by partial pre-oxidation. In: Wise DL, Trantollo DH (eds) Process engineering for pollution control and waste minimization. Marcel Dekker, New York
Carberry JB, Benzig TM (1991) Peroxide pre-oxidation of recalicitrant toxic waste to enhance biodegradation. Water Sci Technol 23:367
Carberry JB, Kovach LB (1997) Influence of phanerochaete chrysosporium pretreatment on the biological degradation rate of pentachlorophenol. J Biotechnol Bioeng (in press)
Carberry JB, Lee SH (1991) Aerobic biodegradation of trichloroethylene enhanced by chemical oxidants. Hazardous and Industrial Wastes, Proc 23rd Mid-Atlantic Industrial Waste Conf, Technomic Publ, Lancaster, Pennsylvania
Carberry JB, Lee SH (1992) Biological degradation of pentachlorophenol enhanced by chemical oxidation pretreatment. J Water Environ Res 64:682
Carberry JB, Lee SH (1994) Enhancement of PCP bioremediation by Fenton's reagent partial oxidation. In: ACS Symp Ser on Emerging Technologies for Toxic Waste Treatment IV, p 197
Carberry JB, Lee SH (1995) Enhancement of PCB congener biodegradation following the partial pre-oxidation with Fenton's reagent. Water Sci Technol 30:105
Chatzis I, Morrow NR, Lim HT (1983) Magnitude and detailed structure of residual oil saturations. Soc Petrol Eng J 22:311–326
Chen JD (1986) Some mechanisms of immiscible fluid displacement in small networks. J Colloid Interface Sci 110:488–503
Chen YM, Abriola LM, Alvarez PJJ, Anid PJ, Vogel TM (1992) Modeling transport and biodegradation of benzene and toluene in sandy aquifer material: Comparison with experimental measurements. Water Resour Res 28:1833–1847
Chiou CT, Peters LJ, Freed VH (1979) A physical concept of soil-water equilibria for nonionic organic compounds. Science 206:831–832
Conard SH, Hagan EF, Wilson JL (1987) Why are residual saturation of organic liquids different above and below the water table? In: Proc NWWA Conf on Organic chemicals and petroleum hydrocarbons. National Well Water Association, Dublin, Ohio
Conard SH, Wilson JL, Mason W, Peplinski W (1989) Observing the transport and fate of petroleum hydrocarbons in soils and groundwater using flow visualization techniques. In: Symp on Environmental concerns in the petroleum industry. American Association of Petroleum Geologists, Palm Springs, California
Corapcioglu NY, Baehr AL (1985) The transport and fate of petroleum products in soils and groundwater. Final Project Rep No CEE-8401438. Civil Engineering, City College of New York, New York
Dietz DN (1979) Pollution of permeable strata by oil components. In: Water pollution by oil. Applied Science, pp 127–139
Faust CR (1985) Transport of immiscible fluids within and below the unsaturated zone: numerical model. Water Resour Res 21:587–596
GeoTrans Inc (1989) SWANFLOW Simultaneous water air and nonaqueous phase flow, Computer code documentation. Herndon, Virginia
Gerstl Z (1989) Predicting the mobility and availability of toxic organic chemical. In: Gerstl Z, Chen Y, Mingelgrin U, Yaron B (eds) Toxic organic chemicals in porous media. Springer, Berlin Heidelberg New York, pp 151–162
Gorelick SM, Voss CI, Gill PE, Murray W, Sounders A, Wright MH (1984) Aquifer reclamation design – the use of contaminant transport simulation combined with nonlinear programing. Water Resour Res 20:415–427
Handy LL, Datta P (1966) Fluid distribution during immiscible displacement in porous media. Soc Petrol Eng J 6:261–266

Hinchee RE (1994) Bioventing of petroleum hydrocarbons. In: Norris RD et al. (eds) Handbook of bioremediation. Lewis, Ann Arbor, Michigan, pp 39–59

Hopkins GD, Semprini L, McCarty PL (1992) Evaluation of enhanced in-situ aerobic biodegdradation of trichloroethylene and *cis*-and-*trans*-1,2-dichloroethylene by phenol utilizing bacteria. Symp on Bioremediation of hazardous wastes, EPA's Biosystems Technology Development Program. Chicago, Illinois, Abstract

Jafvert CT, Va-Hoof PL, Heath JK (1994) Sulibilization of non-polar compounds by non-ionic surfactant micelles. Water Res 28:1009–1017

Karickhoff SW (1981) Semi-empirical estimation of sorption of hydrophobic pollutants on natural sediments and soils. Chemosphere 10:833–846

Kessler A, Rubin H (1987) Relationships between water infiltration and oil spill migration in sandy soils. J Hyrol 91:187–204

Lam AC, Schechter RS, Wade WH (1983) Mobilization of residual oil under equilibrium and non-equilibrium conditions. Soc Petrol Eng J 23:781–790

Li Y, Wardlaw NC (1986a) The influence of wettability and critical pore-throat size ratio on snap-off. J Colloid Interface Sci 109:461–472

Li Y, Wardlaw NC (1986b) Mechanisms of non-wetting phase trapping during imbibition at slow rates. J Colloid Interface Sci 109:473–486

Marly MC, Droste EX (1995) Sparging strips trapped contaminants. Int Groundwater Technol 1(11):18–23

McCarty PL, Semprini L (1994) Ground-water treatment for chlorinated solvents. In: Norris RD et al. (eds) Handbook of bioremediation. Lewis, Ann Arbor, Michigan, pp 87–116

McKee JF, Laverty FB, Hertel RM (1972) Gasoline in groundwater. J Water Pollut Fed 44:293–302

Mohanty KK, Davis HT, Scriven LE (1987) Physics of oil entrapment in water wet rocks. SPE Reservoir Eng 2:243–261

Molz FJ, Widdowson MA, Benefield LD (1986) Simulation of microbial growth dynamics coupled to nutrient and oxygen transport in porous media. Water Resour Res 22:1207–1216

Peaceman DW (1977) Fundamentals of numerical reservoir simulation. Elsevier, New York

Piotrowski M, Cunnigham J (1996) Factors to consider before adding microbes and nutrients in "Soil and Groundwater Cleanup"

Pistiner A, Shapiro M, Rubin H (1990) Similarity solutions for immiscible phase migration in porous media: An analysis of free boundaries. Trans Porous Media 5:491–515

Powers SE, Loureiro CO, Abriola LM, Weber WJ Jr (1991) Theoretical study of the significance of nonequilibrium dissolution of nonaqueous phase liquids in subsurface systems. Water Resour Res 27:463–477

Powers SE, Abriola LM, Weber WJ Jr (1992) An experimental investigation of nonaqueous phase liquid dissolution in saturated subsurface systems: Steady state mass transfer rates. Water Resour Res 28:2691–2705

Rathfelder K, Yeh WWG, Mackay D (1991) Mathematical simulation of soil vapor extraction systems: Model development and numerical examples. J Contam Hydrol 8:263–297

Rathfelder K, Lang JR, Abriola LM (1995) Soil vapor extraction and bioventing applications limitations and future research directions. Reviews of Geophysics Supplement, US National Report to International Union of Geodesy and Geophysics 1991–1994, pp 1067–1081

Reeves HW, Abriola LM (1994) An iterative compositional model for subsurface multiphase flow. J Contam Hydrol 15:249–276

Rice D (1995) California leaking underground fuel tank historical analysis. Lawrence Livermore National Laboratory Report

Rivkin R (1996) California Water Board Guidelines on Petrol Cleanups may impact remediation efforts. Report to the New York State Department of Environmental Resources

Roberts PV, Hopkins GD, Mackay DM, Semprini L (1990) A field evaluation of in-situ biodegradation of chlorinated ethenes: Part 1 Methodology and field site characterization. Ground Water 28:591–604

Rubin H, Mechrez E (1989) Transport of organic pollutants in a multiphase system. In: Gerstl Z, Chen Y, Mingelgrin U, Yaron B (eds) Toxic organic chemicals in porous media. Springer, Berlin Heidelberg New York, pp 231–250

Rubin H, Braxein A, Daniels H, Rouvè G (1994) Migration of oil and petroleum pollutants in soils and groundwater. In: Zoller U (ed) Groundwater contamination and control. Marcel Dekker, New York, pp 355–375

Rubin H, Rathfelder K, Abriola LM (1997) Modeling quasi-steady NAPL dissolution in fractured permeable media. ASCE J Environ Eng 123:205–216

Russell TF (1995) Modeling of multiphase multicontaminant transport in the subsurface. Reviews of Geophysics Supplement, US National Report to International Union of Geodesy and Geophysics 1991–1994, pp 1035–1047

Sawyer CS, Ahlfeld DP, King AJ (1995) Groundwater remediation design using a three-dimensional simulation model and mixed-integer programming. Water Resour Res 31:1373–1385

Schiegg HO (1980) Field infiltration as a method for the disposal of oil-in-water emulsions from the restoration of oil-polluted aquifers. Water Resour Res 14:1011–1016

Schwille F (1975) Groundwater pollution by mineral oil product. IAHS-AISH 103:226–240

Schwille F (1988) Dense chlorinated solvents in porous and fractured media. Lewis, Chelsea, Michigan

Semprini L, McCarty PL (1992) Comparison between model simulations and field results of in-situ biorestoration of chlorinated aliphatics: Part 2. Cometabolic transformations. Ground Water 300:37–44

SITE REMEDIATION Inc (1996) KLEER Remedial support skimmer for floating hydrocarbons. Int Groundwater Technol 2(1):33

Sontheimer H, Nagel G, Werner P (1989) Restoration of aquifers polluted with hydrocarbons. In: Gerstl Z, Chen Y, Mingelgrin U, Yaron B (eds) Toxic organic chemicals in porous media. Springer, Berlin Heidelberg, New York, pp 320–334

USEPA (1996) Superfund National Priority List, Washington, DC

Vogel TM (1996) Bioventing initiative at Air Force Bases, Report to Air Force Center for Environmental Excellence, Univ Michigan, Ann Arbor

Wardlaw NC (1982) The effects of geometry wettability viscosity and interfacial tension on trapping in single pore-throat pairs. J Can Petrol Technol 21:21–27

Wilson JT, Kampbell DH (1994) Bioventing of chlorinated solvents for ground-water cleanup through bioremediation. In: Norris RD et al. (eds) Handbook of bioremediation. Lewis, Ann Arbor, Michigan, pp 117–129

Wilson JL, Conard SH, Hagan E, Peplinski WR (1988) The pore level spatial distribution and saturation of organic liquids in porous media. In: Proc NWWA Conf on Petroleum hydrocarbons and organic chemicals in the subsurface. National Well Water Association, Dublin, Ohio, pp 107–133

Wittcoff HA, Reuben BG (1980) Industrial organics chemical in perspective. Wiley, New York, pp 41–45

Xiang Y, Sykes J, Thompson N (1995) Alternative formulations for optimal groundwater remediation design. J Water Res Plan Manage 121:171–181

Yeh GT, Tripathi VS (1991) A model for simulating transport of reactive multispecies components: model development and demonstration. Water Resour Res 27:3075–3094

2 Solubilization of Dense NAPLs: Exploring the Implications of Experimental Observations on Field-Scale Surfactant-Enhanced Remediation

T. J. Dekker[1,2] and L. M. Abriola[1]

Contents

1 Introduction

The widespread presence of entrapped dense nonaqueous phase liquids (DNAPLs) in the subsurface has been identified as one of the most intractible problems in the remediation of contaminated aquifers (National Research Council 1994). Over the past 40 years, industrial and commerical operations in the U.S. have consumed large quantities of chlorinated solvents, mostly for degreasing manufactured parts and in dry cleaning operations. The occurrence of organic contamination at numerous hazardous waste sites in North America associated with the use of these solvents is well documented (e.g., Schwille 1988; Mercer and Cohen 1990), as is the minimal success of conventional pump-and-treat remediation methods at such sites (Mercer and Cohen 1990; Haley et al. 1991). Because of their high densities relative to water, DNAPLs tend to migrate deep into aquifer formations under gravity forces, leaving a residual liquid entrapped in aquifer pores under capillary forces. The remediation problem posed by these highly inaccessible entrapped organics is compounded by their typically low aqueous phase solubilities, which enhances their persistence as a long-term source of organic pollution. A clear need exists for innovative remediation techniques which address the accessibility and low solubility problems currently associated with NAPL remediation.

The use of aqueous surfactant solutions to enhance the performance of pump-and-treat remediation has received much attention in recent years (e.g., West and

[1]Environmental and Water Resources Engineering, The University of Michigan, 1351 Beal Ave., Ann Arbor, Michigan 48109-2125, USA
[2]Present address: Camp Dresser & McKee, One Woodward Avenue, Suite 1500, Detroit, Michigan 48226, USA

H. Rubin/N. Narkis/ J. Carberry
Soil and Aquifer Pollution

Harwell 1992). Surfactant enhanced aquifer remediation, or SEAR, is based upon the ability of surfactant solutions to mobilize and/or solubilize an entrapped organic phase. Aquifer remediation by mobilization of DNAPLs, however, carries the implicit risk of exacerbating the contamination problem by potentially allowing dense NAPLs to migrate to deeper and less accessible regions of the formation (see Pennell et al. 1996a for further disscussion). For this reason, micellar solubilization is favored as a more controllable method of aquifer remediation. Micellar solubilization utilizes the tendency of surfactants in solution to form spherical aggregates, or micelles, above the critical micelle concentration (CMC). These micelles contain a hydrophobic core into which nonpolar organics will readily partition. Micellar surfactant solutions may be used to greatly increase the aqueous phase solubility of typical DNAPLs, facilitating higher NAPL recovery under pumping conditions. Experimental studies of micellar solubilization show that relatively dilute solutions of surfactant can increase aqueous solubility of organics by several orders of magnitude (Gannon et al. 1989; Edwards, et al. 1991; Pennell et al. 1993). Recently, surfactants have been shown to effectively remove entrapped residual organics in column studies (Fountain et al. 1991; Pennell et al. 1993, 1994).

Modeling of the SEAR process has its origin in the modeling of tertiary oil recovery, where surfactants are used to enhance oil recovery through mobilization (Thomas et al. 1984; Camilleri et al. 1987; Scott et al. 1987). Following the lead of the oil industry work, early modeling of the micellar solubilization process typically assumped local equilibrium partitioning between the organic and aqueous phases (Wilson 1989; Clarke et al. 1991; Brown et al. 1995). However, recent studies of column scale solubilization give evidence of strong local nonequilibrium (Pennell et al. 1993, 1994; Abriola et al. 1995). In an attempt to address this issue, Abriola et al. (1993a) presented a model which incorporates a rate-limited form for mass transfer of organic to the aqueous phase. While this modeling approach has been shown to effectively reproduce the behavior of column-scale NAPL solubilization, little research has been performed to date to investigate the influence of column scale and/or heterogeneity-induced solubilization rate limitations on the field-scale performance of SEAR.

In this chapter, mathematical modeling is used to examine the potential performance of SEAR under field conditions representative of natural sandy aquifers. The first portion of this chapter reviews the conceptual and mathematical framework for SEAR modeling within the context of available experimental observations. Numerical simulations are then employed in the second portion of the chapter to illustrate significant processes influencing field scale recovery of DNAPLs during enhanced solubilization operations.

2 Mathematical Model

A DNAPL-contaminated aquifer represents a three-phase system, composed of organic (o), aqueous (w), and solid matrix phases (m). Modeling of SEAR requires the tracking of a minimum of three chemical constituents: DNAPL, surfactant, and water. The mathematical framework is rooted in constituent mass balance equations in a phase α (Abriola 1989):

$$\frac{\partial}{\partial t}\left(\varepsilon_\alpha \rho^\alpha \omega_i^\alpha\right)+\nabla\cdot\left(\rho^\alpha \omega_i^\alpha \boldsymbol{q}^\alpha\right)-\nabla\cdot\left(\varepsilon_\alpha \rho^\alpha \mathbf{D}_{h_i}^\alpha \cdot \nabla\omega_i^\alpha\right)=\sum_{\beta\neq\alpha} E_i^{\alpha\beta}, \tag{1}$$

where ε_α is the volume fraction of the α phase, ω_i^α is the mass fraction of component i in phase α, $\mathbf{D}_{h_i}^\alpha$ is the hydrodynamic dispersion tensor, $\boldsymbol{q}^\alpha$ is the Darcy velocity, ρ^α is the density of the α phase, and $E_i^{\alpha\beta}$ represents exchange of mass of component i between the α and β phases. Equation (1) is defined for components $i=1, \ldots, n_c$ where n_c is the total number of components, and α and β can represent the organic, water and solid phases. A phase balance equation can be obtained by summing (1) over all species within each phase, yielding

$$\frac{\partial}{\partial t}\left(\varepsilon_\alpha \rho^\alpha\right)+\nabla\cdot\left(\rho^\alpha \boldsymbol{q}^\alpha\right)=E^\alpha. \tag{2}$$

Here, E^α is the sum of the right hand side of Eq. (1) over all species. A conservative form of the component balance equation is obtained by substitution of (2) into (1), yielding

$$\varepsilon_\alpha \rho^\alpha \frac{\partial \omega_i^\alpha}{\partial t}+\rho^\alpha \boldsymbol{q}^\alpha \nabla\cdot\omega_i^\alpha-\nabla\cdot\left(\varepsilon_\alpha \rho^\alpha \mathbf{D}_{h_i}^\alpha \cdot \nabla\omega_i^\alpha\right)=\sum_{\beta\neq\alpha} E_i^{\alpha\beta}-\omega_i^\alpha E^\alpha, \tag{3}$$

which is subject to the constraints:

$$\sum_i \omega_i^\alpha = 1$$
$$\sum_\alpha \varepsilon_\alpha = 1. \tag{4}$$

The above expressions are for the very general case in which mass transfer of all species is considered between all phases. These equations can be simplified by considering the particular characteristics of the contaminated aquifer system and making appropriate assumptions. Assuming that the DNAPL has reached residual (immobile) levels in the formation subsequent to infiltration and redistribution, Eq. (2) for the organic phase becomes:

$$n\frac{\partial}{\partial t}\left(s_o \rho^o\right)=E^{ow}, \tag{5}$$

where s_o is the saturation of the organic phase, n is the matrix porosity, and E^{ow} represents all mass exchange between the organic and aqueous phases ($E^{ow}=\sum_i E_i^{ow}$). Here, it is assumed that the organic is nonwetting with respect to the solid; mass exchange between the solid and organic phases is neglected (i.e., $E^o \equiv E^{ow}$). The influence of matrix compressibility has also been neglected.

If the organic liquid is at immobile levels, only the aqueous phase flow need be modeled. An aqueous phase flow equation can be obtained by incorporating a modified Darcy law expression into Eq. (2) to yield:

$$\frac{\partial}{\partial t}\left[ns_w\rho^w\right] = \nabla\cdot\left[\rho^w\frac{\boldsymbol{k}k_{rw}}{\mu_w}\cdot\left(\nabla P_w - \gamma_w\nabla z\right)\right] \tag{6}$$

where $\boldsymbol{k}$ is the permeability tensor, k_{rw} is the relative permeability of the medium to the aqueous phase, μ_w is the dynamic viscosity, P_w is the pressure, and γ_w is the specific weight of the aqueous phase. Here, the direct influence of interphase partitioning on aqueous phase flow has been neglected in comparison to pressure and gravity terms. Note, however, that aqueous phase density will, in general, be a function of phase composition. In the simulations reported here, ρ^w was calculated with a weighted mass fraction expression.

Because only the water contacts the solid, the solid phase balance equation for a constituent i can be written directly from Eq. (3) as:

$$(1-n)\frac{\partial}{\partial t}\left(\rho^m\omega_i^m\right) = E_i^{mw}. \tag{7}$$

This expression neglects the bulk movement of the solid phase and surface diffusion of solute i. If Eq. (7) is incorporated into the aqueous phase balance equation of a constituent [Eq. (3)], a more recognizable form of the aqueous phase transport equation results:

$$nS_w\frac{\partial C_i}{\partial t} + \rho_b\frac{\partial Q_i}{\partial t} + \boldsymbol{q}^w\cdot\nabla C_i - \nabla\cdot\left(ns_w\mathbf{D}_{h_i}\cdot\nabla C_i\right) = E_i^{wo}, \tag{8}$$

in which C_i is the aqueous phase concentration of species i ($= \rho^i\omega_i$), Q_i is the sorbed phase concentration of species i ($=\omega_i^m$), ρ_b is bulk density ($= (1-n)\rho^m$), and E_i^{wo} represent mass exchange of species i between the organic and aqueous phases. The hydrodynamic dispersion tensor, $\mathbf{D}_{h_i}$, represents dispersive effects due to Fickian diffusion and to differential local advection.

Critical to the modeling of the surfactant-enhanced NAPL solubilization problem is an appropriate description of the mass exchange terms which appear in equations (5), (7), and (8). While a number of recent studies have investigated the dissolution of NAPLs in packed columns (Powers et al. 1991; Powers 1992; Miller et al. 1990, Mackay et al. 1991), relatively little work has been directed toward the quantification of the rates of micellar solubilization. Abriola et al. (1993a) modeled a series of experiments exploring the surfactant-enhanced solubilization of entrapped dodecane in a sandy soil. In this study, as in the aforementioned dissolution studies, column data were fit by assuming a linear driving force expression for mass exchange. The linear driving force model is based on the assumption that mass transfer is controlled by a rate-limiting diffusion process across a liquid barrier, usually a thin film at the interface between phases. Flux of a solute due to diffusion across the interface can be represented by (Weber and DiGiano 1996):

$$E_i^{wo} = k_f' a_o \left(C_{Ii} - C_i \right), \tag{9}$$

where C_{Ii} represents concentration of the solute at the interface, and C_i is concentration in the bulk phase. The k_f' parameter is a mass transfer coefficient (L/t), which is proportional to the diffusion coefficient of species i divided by the boundary layer thickness. This quantity is multiplied by the specific interfacial area for mass transfer, a_o (1/L), which represents contact area between the two phases. A modified form of Eq. (9) is usually posed in terms of measurable parameters as

$$E_i^{wo} = k_f a_o \left(C_{ei} - C_i \right), \tag{10}$$

where C_{ei} represents the equilibrium saturated concentration of orgainc. In column-based mass exchange studies, an expression of the type (10) is frequently cast in terms of a lumped mass exchange coefficient due to the difficulty of quantifying the interfacial contact area between phases.

Evaluation of effluent concentration measurements from column studies reveals that lumped mass exchange coefficients for micellar solubilization (Pennell et al. 1993; Luning Prak et al. 1996) may be as much as two orders of magnitude smaller than those predicted from correlations developed from simple dissolution measurements under similar hydrodynamic conditions (Powers et al. 1991). This suggests that micellar solubilization is a more complex process than that modeled conceptually by orgainc diffusion across a boundary layer. Limited attention, however, has been directed towards the evaluation of the processes limiting micellar solubilization. The available mechanistic models for surfactant solubilization suggest that it may involve a number of sequential steps, which include micelle formation, diffusion of the micelle to and from the phase interface, and incorporation of organic into the micelle (Chan et al. 1976; Carrol 1981). Recent measurements suggest that solubilization will depend on the selected surfactant-organic system and may also depend to a lesser extent on surfactant concentration (Abriola et al. 1995; Luning Prak et al. 1996). While the linear driving force model reduces the solubilization process to a single diffusion-controlled step, it can reproduce the behavior of surfactant flushing at the column scale (Abriola et al. 1993a) and will be employed here.

Following Abriola et al. (1993a), the lumped mass transfer coefficient, $k_{eff} = k_f a_o$, is expressed as a function of the hydrodynamics of the system:

$$k_{eff} = a q^b + \hat{k}_o, \tag{11}$$

where q is the local Darcy velocity, $\hat{k}_o$ is the mass transfer coefficient under conditions of no flow, and a and b are empirical fitting parameters. For the simulations presented in this chapter, the specific interfacial area for mass transfer, a_o, has been described in terms of spherical entrapped NAPL ganglia as presented in Abriola et al. (1993a). This approach is simlilar to that of Powers et al. (1991), and can calibrate well with column dissolution data (Powers et al. 1994).

The equilibrium saturated concentration of organic, C_{ei}, used in Eq. (10) is typically observed to be a linear function of the surfactant concentration, C_s, above the CMC. Thus, the transport equations for surfactant and organic [Eq. (8)] are implicitly coupled through the organic-aqueous mass exchange term. In this chapter it is assumed that the surfactant does not partition significantly into the organic phase. This is a reasonable assumption for the selected ethoxylated nonionic surfactant employed here [POE(20) sorbitan monooleate], due to its high HLB (hydrophilic-lipophilic balance) value (HLB = 15). Batch solubilization experiments exploring the solubilization of tetrachloroethylene (PCE) by POE(20) sorbitan monooleate produced the following relation:

$$C_{eo} = 2.4 \times 10^{-4} + 0.891 C_s, \tag{12}$$

in which concentrations of the organic and surfactant are given in g/g (Pennell et al. 1996b).

Also critical to modeling of the SEAR process is an appropriate representation of sorption of the organic and surfactant components. Sorption of POE (20) sorbitan monooleate to clean Ottawa sand has been shown to conform to a Langmuir isotherm (Pennell et al. 1993):

$$Q_s = \frac{Q_{ms} b_s C_s}{1 + b_s C_s}, \tag{13}$$

where Q_{ms} represent a maximum sorption capacity, and b_s is a constant characteristic of the steepness of the sorption isotherm. Recent column studies of surfactant sorption to clean sands suggest that equilibrium may not be attained at that scale (Pennell et al. 1996c). Accordingly, a nonequilibrium sorption expression based on a kinetic derivation of the Langmuir isotherm (Dekker and Abriola 1996b) was employed in this work:

$$\frac{\partial Q_s}{\partial t} = k_d \left(1 + b_s C_s\right) \left[\frac{Q_{ms} b_s C_s}{1 + b_s C_s} - Q_s \right]. \tag{14}$$

Organic component sorption was modeled with a linear partitioning relation, in which the partition coefficient was scaled as a function of surfactant concentration and sorbed surfactant. Further information on this scaling procedure may be found in Edwards et al. (1994) and Dekker and Abriola (1996b). Note that sorption processes play a negligible role in the simulations presented here, due to the large surfactant concentration used and the low organic carbon content of the sandy aquifer material.

The above expressions form a closed system of coupled equations in the unknows P_w, C_s, C_o, and s_o. The resulting problem is slightly nonlinear, due to the dependence of the interfacial area for mass transfer on s_0 and the use of the Langmuir equation, in which the sorbed phase concentration is dependent on the surfactant concentration, C_s. This nonlinearity makes solution by analytical means infeasible. Thus, a two-dimensional surfactant solubilization simulator was developed by extensive modification of the US Geological Survey (USGS) Saturated-Unsaturated transport (SUTRA) code (Voss 1984). The resulting simulator, MSURF2D, can be used for areal or cross-sectional simulation of surfactant-enhanced aquifer remediation processes. The simulator allows for anisotropy and heterogeneity in permeability and dispersivity values, and permits spatial variation in most other aquifer properties, including capillary properties, sorption coefficients, and NAPL ganglia geometry characteristics. This two-dimensional simulator, an extension of that presented in Abriola et al. (1993b), was verified through comparisons with one-dimensional solubilization simulations (Abriola et al. 1993a) and analytical solutions for simple transport problems. As a further measure of solution quality, an internal mass balance check is employed for fluid phases and transported constituents.

3 Numerical Simulations

To examine the factors influencing SEAR performance at the field scale, the simulator descirbed above was used to predict the recovery of a 75-l spill of PCE in a saturated sandy aquifer. A formation having a statistical distribution and spatial correlation structure similar to the well-characterized Borden sand (Woodbury and Sudicky 1991) was chosen for the simulation domain. The multiphase flow

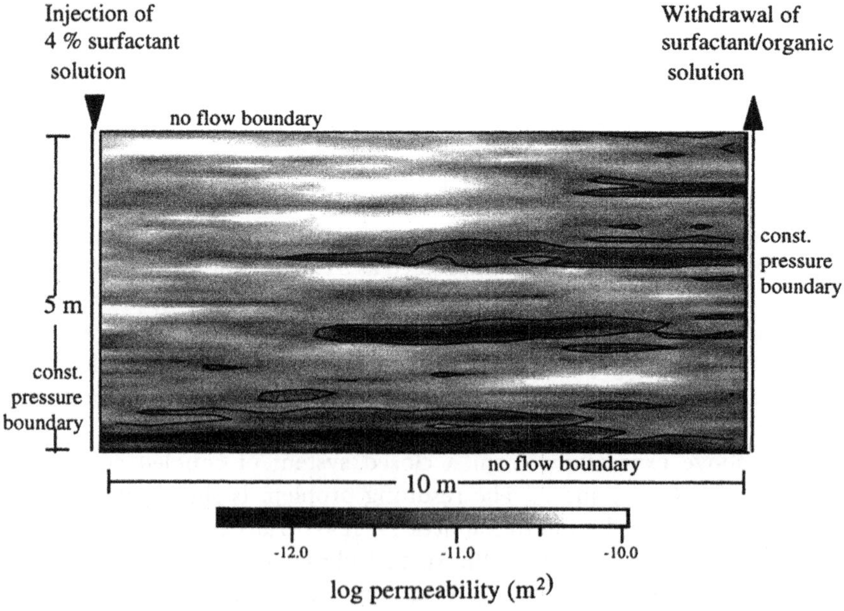

Fig. 1. Simulation scenario and permeability distribution (contour indicates $k = 5 \times 10^{-12}$ m^2)

Table 1. Simulation input data

Parameter		Units	References
Water:			
Density	0.999	g/cm^3	Lide (1994)
Viscosity	0.0114	g/cm s	Lide (1994)
Organic:			
Density	1.625	g/cm^3	Verschueren (1977)
Diffusion coeff.	6.56×10^{-6}	cm^2/s	Hayduk and Laudie (1974)
Solubility	240	mg/l	Verschueren (1977)
Linear adsorp. coeff.	0.1541	cm^3/g	Ptacek and Gillham (1992)
Surfactant:			
Density:	1.08	g/cm^3	Pennell et al. (1993)
Diffusion coeff.	1.48×10^{-6}	cm^2/s	Hayduk and Laudie (1974)
Langmuir Q_{ms}	0.152	mg/g	Pennell et al. (1993)
Langmuir b_s	0.026	l/mg	Pennell et al. (1993)
Noneq. sorp. rate	0.01	1/h	Laboratory Studies of POE (20) (1995)
Solubilization rate parameters:		units for q: cm/h, k_{eff}: 1/h	
a	0.0250		Pennell et al. (1993)
b	0.192		Pennell et al. (1993)
$\hat{k}_o$	0.0109		Pennell et al. (1993)
Porous medium:			
Porosity	0.34		Sudicky (1986)
Hyd. conductivity:		cm/s	
variance($\ln(K)$)	0.24		Woodbury and Sudicky (1991)
$\ln(K)_{mean}$	−4.62		Woodbury and Sudicky (1991)
horiz. corr. length	5.1	m	Woodbury and Sudicky (1991)
vert. corr length	0.21	m	Woodbury and Sudicky (1991)
Dispersivity (h, v)	61.0, 1.0	cm	Sudicky (1986)
Anisotropy factor	0.5		Brown et al. (1995)
Grid spacing: Δx	0.5	m	
Δz	0.1	m	

model MVALOR, an updated version of the VALOR code (Abriola et al. 1992), was employed to develop the initial DNAPL residual distribution conditions for the simulations. This code solves coupled two- or three-phase immiscible flow equations, incorporating hysteresis in capillary pressure relations and organic entrapment. For the simulations, capillary pressure parameters were scaled to the permeability field by the use of a modified Leverett function (see Dekker and Abriola, 1996a for further detail). Figure 1 presents the formation permeability distribution used in the simulations. This spatially correlated permeability field was generated using the turning bands method of Tompson et al. (1989). Statistical parameters, including the mean and variance of the log hydraulic conductivity and the vertical and horizontal correlation lengths, are given in Table 1. Note that the variance represents a moderate level of heterogeneity, similar to that reported for a sand and gravel formation in Switzerland (Jussel et al. 1994) and larger than that reported for the relatively homogeneous Borden formation.

To establish initial conditions for the MSURF2D simulations, MVALOR was used to simulate a spill event. The PCE was released at a constant rate under natural aquifer gradient conditions for a period of 5 days, after which redistribution was allowed to occur over a period of time sufficient for the organic to reach an immobile state. This saturation distribution was then used as input to MSURF2D. Because infiltration and entrapment of the organic mass was expected to occur rapidly relative to the slow rate of organic dissolution in the absence of surfactant, this sequential modeling approach was judged appropriate for the scenarios examined here.

A schematic of the cleanup scenario is presented in Fig. 1. A 4% by weight solution of POE(20) sorbitan monooleate is introduced at the left side of the model domain and effluent is recovered at the right boundary. This nonionic surfactant was selected based on its reported efficiency in the recovery of organic residuals from soil columns (Pennell et al. 1993, 1994) and its environmental acceptability. The imposed 2% hydraulic gradient is typical of pump-and-

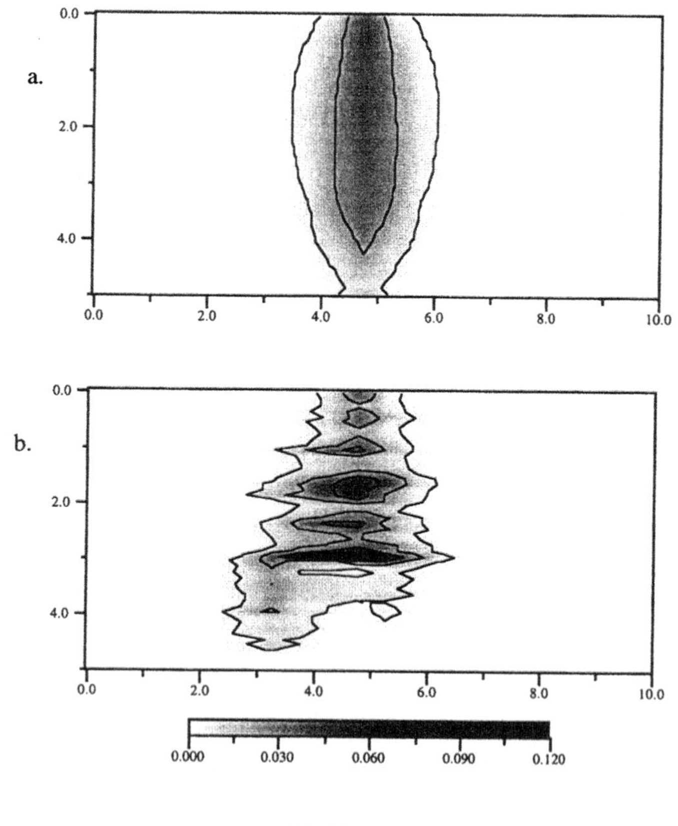

Fig. 2a–b. Distribution of entrapped PCE. **a** Homogeneous formation; **b** Jussel formation

treat remediation scenarios. The domain is saturated, with zero-flow boundary conditions enforced at the top and bottom of the model domain, and pressure-type boundary conditions at the left and right. Model simulations were continued until all of the organic liquid was removed from the domain and the organic plume had exited the downstream end. All model input parameters are summarized in Table 1.

Figure 2 illustrates the influence of formation heterogeneity on the distribution of entrapped PCE. Here two MVALOR simulations are presented under identical release conditions, one for a homogeneous aquifer and another for the aquifer depicted in Fig. 1. All other simulation parameters, including the average

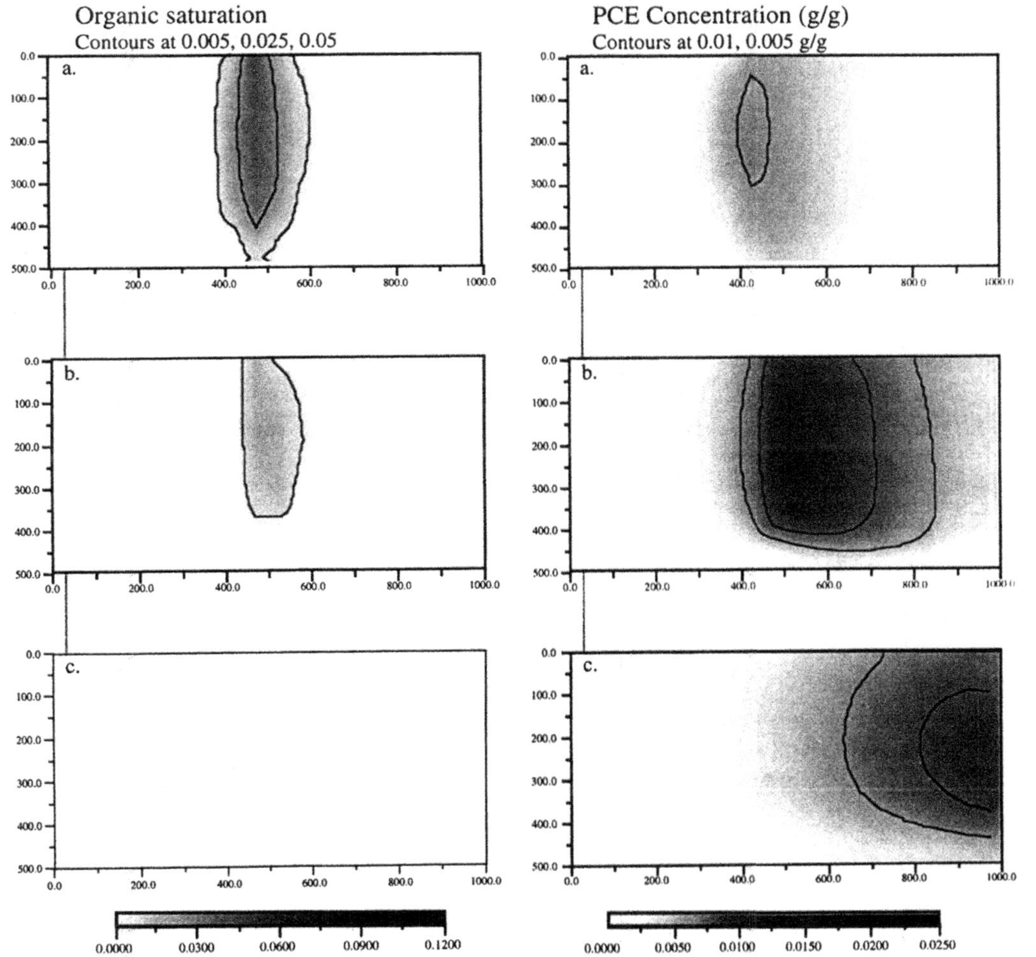

Fig. 3a–c. Homogeneous formation simulation results showing organic saturation and PCE concentration after flushing with: **a** 5000 l; **b** 10000 l; **c** 18000 l

permeability, were the same for both scenarios. Comparison of the figures reveals that the presence of heterogeneity tends to increase the lateral spreading of the organic and to decrease spill penetration depth. These effects are due to the tendency of the PCE to pool on lower permeability layers before penetration. An increase in heterogeneity also results in an increase in the maximum entrapped saturation in the domain.

Figures 3 and 4 graphically depict the surfactant-enhanced removal of the entrapped PCE distributions illustrated in Fig. 2a,b, respectively. Here, saturations and aqueous phase concentrations of organic are plotted as grayscale maps, with organic saturation contours at 0.005, 0.025, and 0.05; and concentration

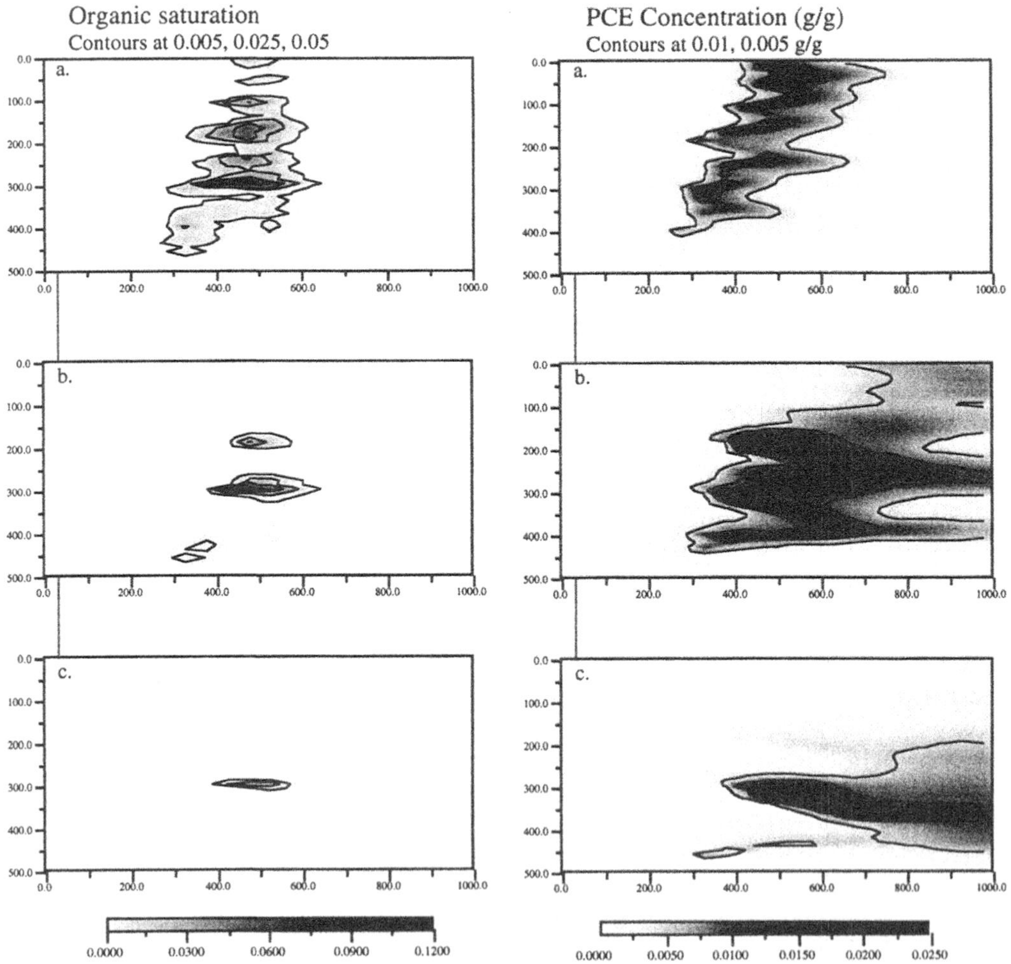

Fig. 4a–c. Jussel formation simulation results showing organic saturation and PCE concentration after flushing with: **a** 5000 l; **b** 10 000 l; **c** 18 000 l

contours at 0.01, and 0.005 g/g. In the homogeneous domain, the surfactant front just reaches the region of entrapped NAPL after 5000 l of flushing, resulting in increased solubilized concentration in the aqueous phase. By contrast, in the heterogeneous formation, a significant portion of the NAPL in the higher permeability regions at the top of the formation has been removed after flushing with the same volume. After 10 000 l have been injected, only isolated regions of organic remain in the heterogeneous formation, in zones of relatively low permeability. In the homogeneous case, removal is somewhat more uniform, due to the more regular saturation and permeablity distributions. After 12 400 l (15.4 days), the NAPL has been completely removed from the homogeneous domain. In the heterogeneous aquifer, however, a portion of the liquid PCE remains within the lower permeability regions, acting as a significant and persistent source of organic to the aqueous phase. For this scenario, 25 800 l (24.3 days) were required for the complete removal of the organic liquid. To appreciate the efficiency of SEAR for NAPL recovery, these injection volumes and remediation times should be compared with those required for NAPL dissolution during a traditional pump-and-treat operation at the same injection rate [1 180 000 l (1410 days) for the heterogeneous case].

Inspection of the aqueous phase contours in Figs. 3 and 4 reveals interesting behavior of the solubilized PCE plume. Note that in the areas of highest organic concentration, the trajectory of the plume has a downward component, due to the relatively high density of the plume in comparison with the injected surfactant solution. This gravity plunging effect appears more pronouncedly in the heterogeneous case, probably due to the lower flow rate and corresponding high concentrations present in the low permeability regions. The solubilized organic plume is observed to migrate downward, out of the low permeability region, and then to the right, as a region of higher flow velocity is encountered.

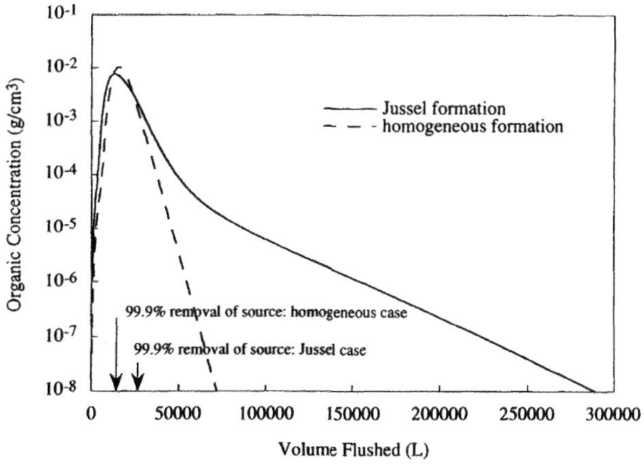

Fig. 5. Effect of heterogeneity on removal of the organic plume

Figure 5 plots flux-averaged organic concentration of the recovered solution as a function of flushing volume. Here the influence of heterogeneity is very marked in the long tailing behavior of the concentration profile. In comparison with the homogeneous formation, approximately four times the flushing volume is required to reduce concentrations to regulatory levels in the heterogeneous case. Surprisingly, for the heterogeneous case, an additional 329000 l (305 days) of surfactant flushing is required to remediate the formation, subsequent to complete recovery of the NAPL. This is compared with an additional 67960 l (81 days) for remediation of the homogeneous formation. Also note the decrease in concentration peak for the heterogeneous case. This decreased peak value is caused by NAPL entrapped in less accessible regions and by increased dilution effects which result from the smaller penetration depth of the spill. The observed long tailing behavior is probably due to the slow, diffusion-limited transport of organic from regions of low permeability (see Fig. 4). Transport of solubilized PCE into such low permeability zones is facilitated by the density plunging of the plume.

The potential influence of local mass transfer limitations on NAPL recovery is explored in Fig. 6. Here, model predictions using the mass transfer expression developed for surfactant-enhanced solubilization of dodecane (Abriola et al. 1993a) are compared with predictions under conditions of local equilibrium. These two representations of mass transfer are assumed to cover the range of expected behavior under field conditions. Mass transfer coefficients computed by these correlations differ by three orders of magnitude. Note that, although average ganglia diameter was correlated with grain size distribution in these simulations, no attempt was made to correlate mass transfer coefficients with permeability, due to the lack of information on surfactant solubilization kinetics. Examination of Fig. 6 reveals that use of equilibrium mass exchange rate has the effect of sharp-

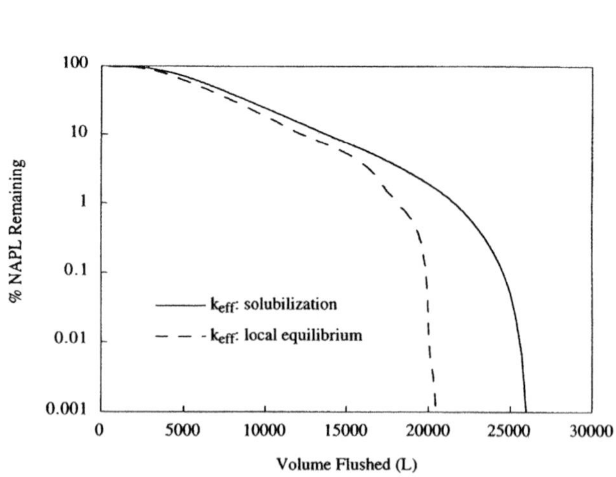

Fig. 6. Effect of mass-transfer coefficient on removal of the organic phase

ening the NAPL removal curve, indicating more rapid removal of the most persistent fraction of NAPL. A decrease in the mass transfer coefficient results in a 27% increase in the required flushing volume. This dependence of flushing volume on k_{eff} appears to be relatively weak, however, given the wide range of values explored.

4 Conclusions

This chapter has presented a conceptual and mathematical framework for the modeling of surfactant-enhanced solubilization in saturated groundwater systems. Based upon available laboratory observations, the model incorporates rate-limited mass transfer to the flowing aqueous phase and to the aquifer solids. To facilitate the field-scale application of this model, spatial variations in system properties are accommodated, including aquifer permeabilities and scaled capillary parameters as well as composition-dependent phase densities.

Model simulations were used to illustrate the potential advantage of SEAR over traditional pump-and-treat technologies and to explore the potential influence of heterogeneity on the recovery of an entrapped DNAPL. Simulations revealed that spatial variability in conductivity and capillary properties will act to enhance the lateral spread of the organic and to increase maximum saturation levels attained in the formation. Entrapment in layers of lower permeability within the heterogeneous formation was observed to more than double the injection volume required for NAPL recovery. For the scenario examined, spatial variability also resulted in a more than fourfold increase in the required injection volume for complete plume recovery. In addition, model simulations suggested that density plunging of the solubilized plume could hinder organic recovery and must be considered in remedial design.

The simulations presented in this chapter were relatively insensitive to local mass transfer coefficients, implying that heterogeneity-induced (macroscopic) mass transfer limitations may dominate, even for formations with only moderate variances in log permeability. More work is needed to systematically explore the influence of heterogeneity on SEAR performance over a wider range of contamination and recovery scenarios. Experience gained in such studies should prove invaluable for the identification of information critical to the successful implementation of SEAR and for the optimization of recovery schemes.

Acknowledgments. This research was supported by the United States Environmental Protection Agency through Grant R-815750 to the Great Lakes and Mid-Atlantic Hazardous Substance Research Center. Additional funding was provided by the National Science Foundation under Grant EID-9023090. This work has not been subject to Agency review and therefore does not necessarily reflect the views of the Agency; no official endorsement should be inferred.

References

Abriola LM (1989) Modeling multiphase migration of organic chemicals in groundwater system – a review and assessment. Environ Health Perspect 83:117–143

Abriola LM, Rathfelder K, Yadav S, Maiza M (1992) VALOR Code Version 0.0: A PC Code for Simulating Immiscible Contaminant Transport, Final Report RP2879-08. Electric Power Research Institute, Palo Alto, California

Abriola LM, Dekker TJ, Pennell KD (1993a) Surfactant-enhanced solubilization of residual dodecane in soil columns. 2. Mathematical modeling. Environ Sci Technol 27(12):2341–2351

Abriola LM, Dekker TJ, Rathfelder KM (1993b) Recent advances in the modeling of organic liquid contaminant migration and persistence in aquifer systems. In: Eckstein Y, Zaporozec A (eds) Hydrogeologic investigation, evaluation, and ground water modeling. Proc Industrial and Agricultural impacts on the hydrologic environment, Water Environment Federation, Alexandria, Virgimia, pp 303–316

Abriola LM, Pennell KD, Pope GA, Dekker TJ, Luning-Prak DJ (1995) Impact of surfactant flushing on the solubilization and mobilization of dense nonaqueous-phase liquids. In: Surfactant-enhanced aquifer remediation – emerging technologies. ACS Symp Ser 594:11–23

Brown CL, Pope GA, Abriola LM, Sepehnoori K (1995) Simulation of surfactant enhanced aquifer remediation. Water Resour Res 30(11):2959–2977

Camilleri D, Fil A, Pope GA, Rouse BA, Sepehrnoori K (1987) Improvements in physical-property models used in micellar/polymer flooding. SPE Reservoir Eng 2(4):433–440

Carroll BJ (1981) The kinetics of solubilization of nonpolar oils by nonionic surfactant solutions. J Colloid Interface Sci 79(1):126–135

Chan AF, Evans DF, Cussler EL (1976) Explaining solubilization kinetics. AIChE 22(6):1006–1012

Clarke AN, Plumb PD, Subramanyan TK, Wilson DJ (1991) Soil clean-up by surfactant washing. I. Laboratory results and mathematical modeling. Separation Sci Technol 26(3):301–343

Cvetkovic VD, Shapiro AM (1990) Mass arrival of sorptive solute in heterogeneous porous media. Water Resour Res 26(9):2057

Dekker TJ, Abriola LM (1996a) The influence of formation heterogeneity on DNAPL migration and entrapment. Water Resour Res (submitted)

Dekker TJ, Abriola LM (1996b) An assessment of the effects of field-scale formation heterogeneity on surfactant-enhanced aquifer remediation (in preparation)

Edwards DA, Luthy RG, Liu Z (1991) Solubilization of polycyclic aromatic hydrocarbons in micellar nonionic surfactant solutions. Environ Sci Technol 25(1):127–133

Edwards DA, Liu Z, Luthy RG (1994) Surfactant solubilization of organic compounds in soil/aqueous systems. J Environ Eng 120(1):5–22

Fountain JC, Klimek A, Beikirch MG, Middleton TM (1991) The use of surfactants for in situ extraction of organic pollutants from a contaminated aquifer. J Haz and Mater 28:295–311

Gannon OK, Bibring P, Raney K, Ward JA, Wilson DJ, Underwood JL, Debelak KA (1989) Soil clean up by in-situ surfactant flushing. III. Laboratory results. Separation Sci Technol 24(14):1073–1094

Haley JL, Hanson B, Enfield C, Glass J (1991) Evaluating the effectiveness of groundwater extraction systems. Ground Water Mon Rev 11(1):119–124

Hayduk W, Laudie H (1974) Prediction of diffusion coefficients for nonelectrolytes in dilute aqueous solution. AlchE J 20:611–615

Jussel P, Stauffer F, Dracos T (1994) Transport modeling in heterogeneous aquifers. 1. Statistical description and numerical generation of gravel deposits. Water Resour Res 30(6):1803–1817

Laboratory studies of POE(20) Monooleate sorption. University of Michigan, Ann Arbor

Lide DR (1994) CRC handbook of environmental data on organic chemicals. CRC Press, Ann Arbor

Liu Z, Edwards DA, Luthy RG (1992) Sorption of non-ionic surfactant onto soil. Water Res 26(10):1337–1345

Luning Prak DJ, Pennell KD, Abriola LM, Weber WJ Jr (1996) Rate controlled micellar solubilization of an LNAPL in aquifer materials. Proc ASCE Assessment and Remediation Conf, Non-aqueous phase liquids (NAPLs) in the subsurface environment, Nov 1996, Washington, DC (in press)

Mackay D, Shiu WY, Maijanen A, Feenstra S (1991) Dissolution of non-aqueous phase liquids in groundwater. J Contam Hydrol 8:23–42

Mercer JW, Cohen RM (1990) A review of immiscible fluids in the subsurface: properties, models, characterization, and remediation. J Contam Hydrol 6:107–163

Miller CT, Poirer-McNeill MM, Mayer AS (1990) Dissolution of trapped nonaqueous phase liquids: mass transfer characteristics. Water Resour Res 26(11):2783–2796

National Research Council (1994) Alternatives for groundwater cleanup. National Academy Press, Washington, DC

Pennell KD, Abriola LM, Weber WJ (1993) Surfactant-enhanced solubilization of residual dodecane in soil columns. 1. Experimental investigation. Environ Sci Technol 27(12):2232–2340

Pennell KD, Jin M, Abriola LM, Pope GA (1994) Surfactant enhanced remediation of soil columns contaminated by residual tetrachloroethylene. J Contam Hydrol 16:35–53

Pennell KD, Pope GA, Abriola LM (1996a) Influence of viscous and bouyancy forces on the mobilization of residual tetrachloroethylene during surfactant flushing. Environ Sci Technol 30(4):1328–1335

Pennell KD, Adinolfi A, Abriola LM, Diallo M (1996b) Solubilization of dodecane, tetrachloroethylene and 1,2-dichlorobenzene in micellar solutions of ethoxylated nonionic surfactants. Environ Sci Technol (submitted)

Pennell KD, Wigder DE, Dekker TJ, Abriola LM (1996c) Influence of rate-limited sorption on the transport of an ethoxylated nonionic surfactant. (in preparation)

Powers SE (1992) Dissolution of nonaqueous phase liquids in saturated subsurface systems. Doctoral Thesis, The University of Michigan, Ann Arbor

Powers SE, Loureiro CO, Abriola LM, Weber WJ (1991) Theoretical study of nonequilibrium dissolution of nonaqueous phase liquids in subsurface systems. Water Resour Res 27(4):463–477

Powers SE, Abriola LM, Dunkin JS, Weber WJ (1994) Phenomenological models for transient NAPL-water mass transfer processes. J Contam Hydrol 16:1–33

Ptacek CJ, Gillham RW (1992) Laboratory and field measurements of nonequilibrium transport in the Borden aquifer, Ontario, Canada. J Contam Hydrol 10:119–158

Schwille F (1988) Dense chlorinated solvents in porous and fractured media. Translated by JF Pankow. Lewis, Chelsea, Michigan

Scott T, Sharpe SR, Sorbie KS, Clifford PJ, Roberts LJ, Foulser RWS, Oakes JA (1987) A general purpose chemical flood simulator. Proc 9th SPE Symp on Reservoir Simulation, Feb 1–4, 1987, San Antonio, Texas

Sudicky EA (1986) A natural gradient experiment on solute transport in a sand aquifer: spatial variability of hydraulic conductivity and its role in the dispersion process. Water Resour Res 22(13):2069–2082

Thomas CP, Fleming III PD, Winter WK (1984) A ternary, two-phase, mathematical model of oil recovery with surfactant systems. Soc Pet Eng J 24(6):606–616

Tompson AFB, Ababou R, Gelhar LW (1989) Implementation of the three-dimensional turning bands random field generator. Water Resour Res 25(10):2227–2243

Voss CI (1984) USGS SUTRA manual. US Geological Survey National Center, Reston, Virgimia

Weber WJ Jr, DiGiano F (1996) Process dynamics in environmental systems. Wiley Interscience, New York

West CC, Harwell HJ (1992) Surfactants and subsurface remediation. Environ Sci Technol 26(12):2324–2330

Wilson DJ (1989) Soil clean up by in-situ surfactant flushing. I. Mathematical modeling. Separation Sci Technol 24(11):863–892

Woodbury AD, Sudicky EA (1991) The geostatistical characteristics of the Borden aquifer. Water Resour Res 27(4):533–546

Part II

Physical–Chemical Considerations

3 Behavior of Volatile Organic Liquid Mixtures in the Soil Environment

B. Yaron[1], I. Dror[1], E. Graber[1], J. Jarsjo[2], P. Fine[1], and Z. Gerstl[1]

Contents

1 Introduction

Complex mixtures of volatile organic liquid compounds (VOLM), such as petroleum hydrocarbons, contain components whose vapor pressures range across several orders of magnitude. During their transport through soils or while floating on the groundwater, parts of these mixtures will volatilize, leaving a liquid enriched by the less volatile, components.

This Chapter provides a conceptual approach for understanding the behavior of volatile organic liquid mixtures in the soil environment, based on the results of our research group (Acher et al. 1989; Yaron 1989; Yaron et al. 1989; Galin et al. 1990a,b; Fine and Yaron 1993; Gerstl et al. 1994; Jarsjo et al. 1994, 1997; Nye et al. 1994; Fine et al. 1996; Dror et al. 1996), and will also include new, unpublished data. Our conceptual approach provides experimental evidence of abiotic processes governing soil-VOLM interactions, as well as parameters to be used in modeling their behavior in the unsaturated zone. We will take into consideration that volatilization, dissolution, sorption, and transport form a continuous abiotic process which controls the transport of VOLM in porous media from the surface to

[1] Institute of Soils and Water, ARO, The Volcani Center, Bet Dagan, Israel
[2] Department of Civil and Environmental Engineering, Royal Institute of Technology, Stockholm, Sweden

H. Rubin/N. Narkis/ J. Carberry
Soil and Aquifer Pollution

groundwater. After a VOLM spill on the surface, its components will be redistributed in the soil. Retention by the solid phase will initially be via entrapment in the soil pores. The highly volatile fraction will volatilize in the gas phase of the porous medium. From there it may be lost to the atmosphere, sorbed onto soil solids, or dissolved into soil water or groundwater. Less volatile components will be transported in the nonaqueous phase through the porous medium. Differing aqueous solubilities for the disparate components will also result in differential dissolution and transport as solutes in soil water. If the amount of VOLM in the soil's upper layer is greater than the soil retention capacity, the hydrocarbon mixture will travel into the vadose zone as a separate, nonaqueous phase. Abiotic VOLM-soil interactions will be discussed in the light of this approach.

2 Experiments

The experiments summarized below were performed on a broad spectrum of soils whose characteristics are presented in Table 1. Five petroleum hydrocarbons [m-xylene, isopropylbenzene (ps-cumene), t-butyl benzene, n-decane, and n-dodecane] and commercial 'kerosene' (constituting a mixture of more than 100 different hydrocarbons) were used in our experiments. Details on experimental procedure have been published elsewhere (cited in the Sect. 1).

Table 1. Characteristics of the soils used in experiments

Soil	Water retention capacity (WRC) (wt%)	Bulk density (g cm^{-3})	Total porosity (%)	Water content in air-dry soil (wt%)	Organic matter (%)	Texture[a]		
						Clay (%)	(Silt) (%)	Sand (%)
Coarse sand 1	1.0	1.82	30	0.35	0.1	–	–	100
Coarse sand 2	10	1.80	30	0.28	0.4	0	14	86
Medium sand	2.4	1.84	29	0.40	0.1	–	–	100
Fine sand	2.8	1.73	33	0.48	0.1	–	–	100
Sandy loam 1	47	1.42	44	2.46	5.4	19	14	67
Sandy loam 2	46	1.38	46	2.29	4.4	15	18	67
M.[b] loam	19	1.47	42	2.5	0.5	16	36	48
M.[b] clay	28	1.39	54	4.9	1.2	40	17	43
M.[b] Evesham clay	nd	nd	nd	nd	3.7	35	nd	nd
Glacial clay	44	1.37	44	3.19	1.3	51	21	28
Post gl. clay	50	1.44	40	3.36	2.6	44	25	31
Peat	74	0.68	63	5.03	27.8	–	–	–

nd, not determined.
[a] According to the USDA system of classification.
[b] Montmorillonitic.

3 Results and Discussion

3.1 Volatilization Process

The mass transfer of spilled VOLM from natural bodies of water directly into the atmosphere, and the effect of this process on their distribution between the gaseous and liquid phases, were the subject of a series of studies (Drivas 1982; Mackay and Yeun 1983; Roberts and Dandiker 1983; Stiver and Mackay 1984; Burris and MacIntyre 1986).

If the mass transfer of these same VOLM components occurs through a layer of soil, their rate of volatilization is affected by temperature, soil constituents, and soil water content. There is much literature on volatilization from the soil by compounds of intermediate vapor pressure (e.g., Spencer and Farmer 1980; Jury et al. 1984), but little on volatile organic compounds of low molecular weight. A study by Karimi et al. (1987) showed that diffusion of benzene through unsaturated soil is greatly reduced as soil bulk density and water content increase. The studies cited addressed the volatilization of only a single compound from soil.

Baehr and Corapcioglu (1987) modeled the movement of a hydrocarbon plume in moist soil both into groundwater and into the atmosphere. Baehr (1987) emphasized the importance of the distribution of volatile solutes between soil water and soil atmosphere. If the various petroleum components reach the groundwater as a water-immiscible phase, volatilization continues through the soil. Acher et al. (1989) and Yaron et al. (1989) showed that the volatilization of each hydrocarbon added to the soil surface in a mixture simulating a spilled petroleum product (kerosene) occurred at different rates. The same behavior was observed by Galin et al. (1990a,b) and Fine and Yaron (1993) for the case of industrial kerosene.

Nye et al. (1994) developed and tested a model for volatilization of a multicomponent liquid through dry soil by comparing the evaporation of m-xylene and n-dodecane through soils with very low sorbing capacity (sand) and very high sorbing capacity (Evesham clay soil). In the case of a pure liquid, three stages were identified (Fig. 1):

1. A soil-accumulation stage, in which the soil sorbs the vapors diffusing up the column.
2. A steady-state stage, in which the vapor evaporates at a constant rate into the atmosphere after diffusing up the column under a constant concentration gradient. If the rate of sorption is slow, the time taken to arrive at a steady state would increase, and at the limit no steady state would be reached before the supply was exhausted.
3. A soil-depletion stage, during which gas molecules sorbed by the soil diffuse into the atmosphere.

The same stages were recognized when a mixture of volatile components evaporated – with important differences. In the accumulation stage, the components were sorbed according to their volatility, diffusivity, and competitive isotherm (Fig. 2a). No true steady-state stage was reached because the composition of the

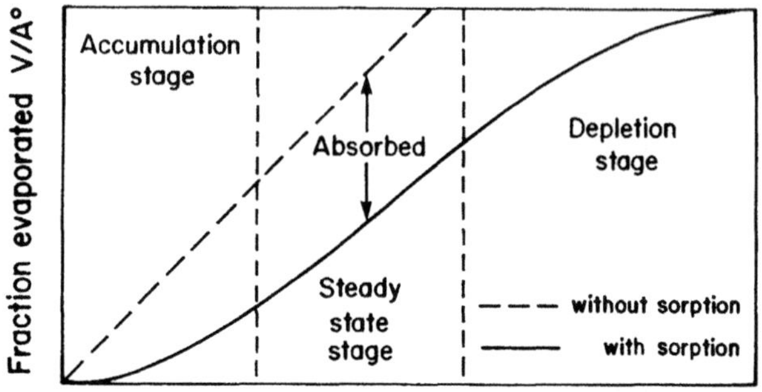

Fig. 1. Effect of soil sorption on fraction evaporated in dimensionless time T : τ = sc°d θft/aol. c° is equilibrium gas phase concentration; s is cross-sectional area of the column (m^2); d is the diffusion coefficient; *f* is the impedence factor; θ is air-filled pore volume; a_o is the initial amount of liquid added to the system (mol); l is the length of the soil columns and t is the time. (Nye et al. 1994)

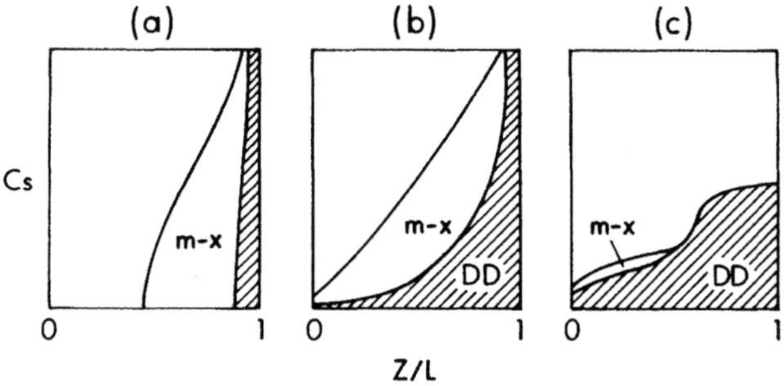

Fig. 2a–c. Profiles of adsorbed m-xylene and n-dodecane evaporating from 1 : 1 mixture. **a** Accumulation stage; **b** quasi-steady-state stage; **c** depletion stage. (Nye et al. 1994)

source liquid was continuously changing, and the soil composition adjusted correspondingly. When the supply of source liquid was large compared with the soil's total sorption capacity, adjustment was slow and evaporation approached a steady state (Fig. 2b). In the depletion stage, the soil, already enriched by the less volatile components, lost these slowly. The soil still contained a small proportion of the more volatile components and did not lose them completely until all the components disappeared (Fig. 2c).

The effect of component properties on their volatilization from a mixture of petroleum hydrocarbons adsorbed on sandy clay soil was discussed by Acher et al. (1989), who observed that more than 99% of the adsorbed aromatic compounds (m-xylene, pseudo-cumene, and t-butylbenzene) volatilized in 2 h. The aliphatic compounds had a lower rate of loss, especially n-dodecane, of which 40% remained after 16 h (see Table 2).

Nye et al. (1994) studied the relative volatilization from solution and soil of m-xylene and n-dodecane separately and mixed. Figure 3 compares observed weight losses by evaporation for the different mixtures and soils with model-simulated

Table 2. Relative volatilization of VOLM components from an air-dried silty clay soil. (After Acher et al. 1989)

VOLM components	Losses[a] after	
	16 h	2 h
m-Xylene	99.3	99.0
ps-Cumene	99.7	94.0
t-Butylbenzene	99.7	95.2
n-Decane	89.5	71.5
n-Dodecane	61.3	14.3

[a] At 22 °C, as percentage of the amount adsorbed.

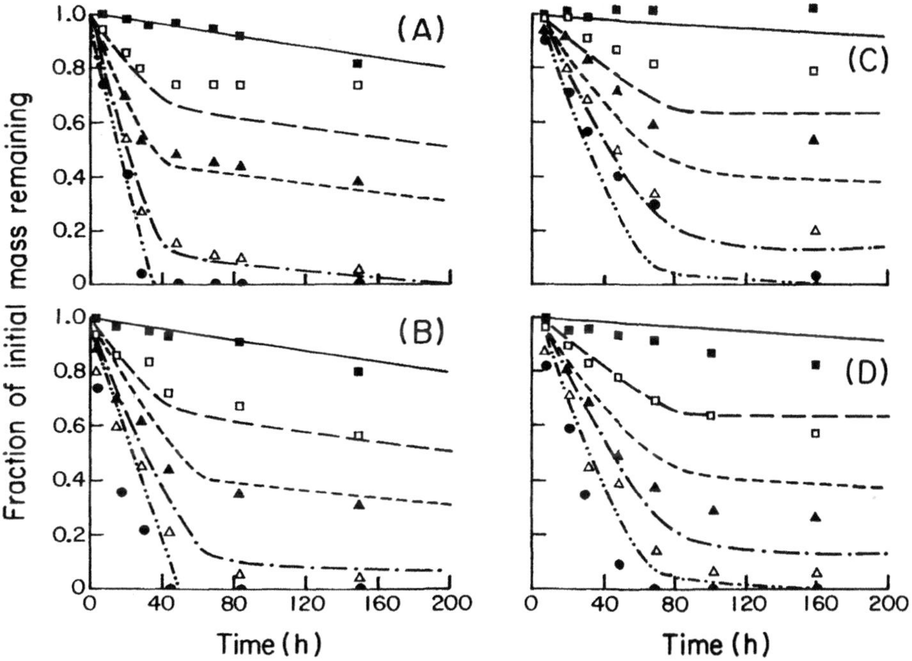

Fig. 3A–D. Evaporation of m-xylene (*solid squares*), n-dodecane (*solid circles*), and m-xylene-dodecane mixtures (32:68 v:v *open squares*; 54:46 *solid triangles*: 84:16 *open triangles*). Lines are model predictions, symbols are actual data. The results are plotted as fraction of initial mass remaining after time-dependent evaporation from **A** liquid; **B** medium sand; **C** oven-dry Evesham clay; **D** air-dry Evesham clay. (Nye et al. 1994)

weight losses. The m-xylene:n-dodecane ratios (w/w) in the mixture were: M_1, 32.3:67.7; M_2, 53.9:46.1; M_3, 83.4:16.6. The fraction of the initial mass of liquid remaining is plotted against time in Fig. 3. It may be observed that:

1. The amounts evaporated for a given time and soil diminished according to the order: m-xylene > M_3 > M_2 > M_1 > n-dodecane.

2. The amounts evaporated from a given liquid mixture diminished in the order: solution > sand > Evesham clay (air- and oven-dry). Differences between air- and oven-dry soils are thought to be due to changes in soil moisture content during the runs.

The rate of volatilization of the mixture is controlled by the ratio between its two components. As volatilization proceeded, the decrease in the m-xylene-dodecane ratio in the mixture led to a decrease in the rate of volatilization. Interactions between the two components and sorption at soil surfaces temporarily reduced vapor phase concentration and diffusion into the gas phase. Once the system reached a steady state, however, the final apparent steady-state vapor diffusion coefficient was a function of air-filled porosity and not related to properties of the soil matrix. As in the case of individual petroleum hydrocarbon mass transfer into the atmosphere, the greatest remaining portion of initial mass occurred in oven-dry Evesham clay soil and the highest volatilization rate occurred in coarse sand.

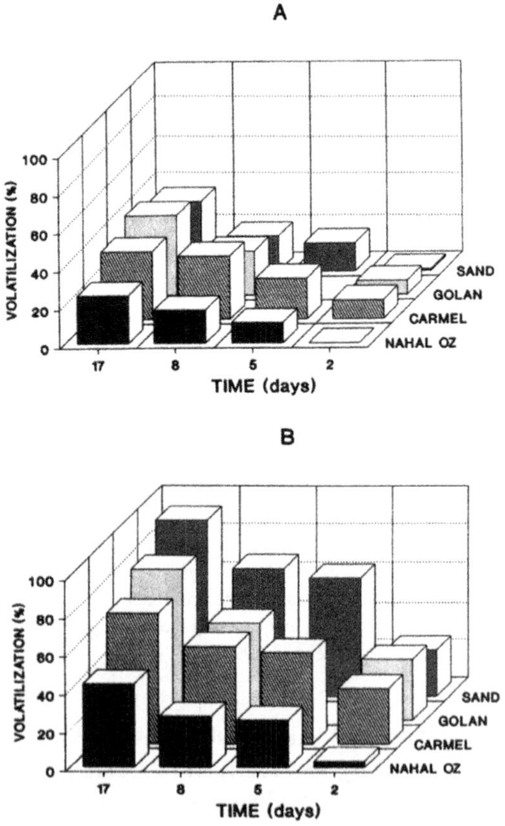

Fig. 4A,B. Volatilization of kerosene from the 5–15-cm soil layer (in percent of initial content). A Free; B enhanced venting. (Fine and Yaron 1993)

The effects of soil properties and soil environment on VOLM volatilization was tested in experiments with kerosene. Fine and Yaron (1993) reported the results of a semi-pilot study in which the volatilization of kerosene from four soils was examined as a function of natural or enhanced venting. The soils were characterized according to clay, moisture, and organic matter (OM) content: dune sand, very fine montmorillonitic clay soil with 1.6% organic matter OM (Golan), very fine montmorillonitic clay soil with 5.1% OM (Carmel), and fine loam soil with 2% OM (Nahal Oz). From Fig. 4 it may be observed that VOLM volatilization was strongly affected by soil properties. Induction of vapor phase transport by enhanced venting significantly increased the loss of kerosene through volatilization, but in both cases the impact of soil properties is evident.

Soil aggregation status, air permeability, and soil moisture content also affect VOLM volatilization from soil. The effect of soil aggregation on kerosene volatilization was examined by comparing volatilization from two fractions of Golan vertisol, the <2 mm grain size fraction and 2–5 mm aggregates. The soil fractions differed also in air permeability, their respective K_a-values being 0.082 ± 0.009 and $0.145 \pm 0.011\,cm^2$. Figure 5 shows that the rate of kerosene removal from the two soil fractions was related to their air permeability and to the specific venting treatment. The greater the air permeability, the more readily the soil released kerosene under both enhanced and free-venting treatments. The most efficient

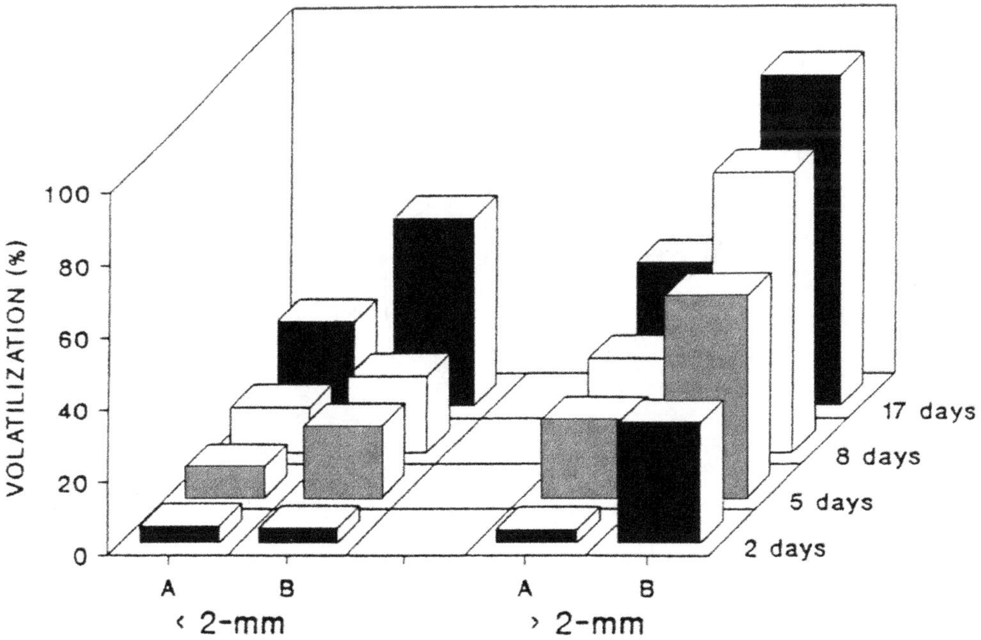

Fig. 5. Volatilization of kerosene from Golan vertisol as affected by aggregate size and venting method. **A** Free; **B** enhanced venting. (Fine and Yaron 1993)

removal of kerosene from soil was obtained with enhanced venting of coarse aggregates, which facilitated the removal of four to seven times more kerosene than the finely aggregated soil under free venting.

3.2 Volatilization-VOLM Composition Relationship

Volatilization of volatile organic liquid mixtures brings about changes in both the physical and chemical properties of the residual liquid. This process is exemplified by the behavior of kerosene. The differential volatilization of selected kerosene compponents during volatilization up to 50% (w/w) is presented in Fig. 6. It may be seen that the lighter fractions evaporate at the beginning of the volatilization process and with an increasing percentage of evaporation, additional components volatilize. This leads to a relative increase of the heavier fractions of kerosene in the remaining liquid.

Differential volatilization through soils leads to changes in residual kerosene composition. Figure 7 shows gas chromatograms of kerosene recovered from sand,

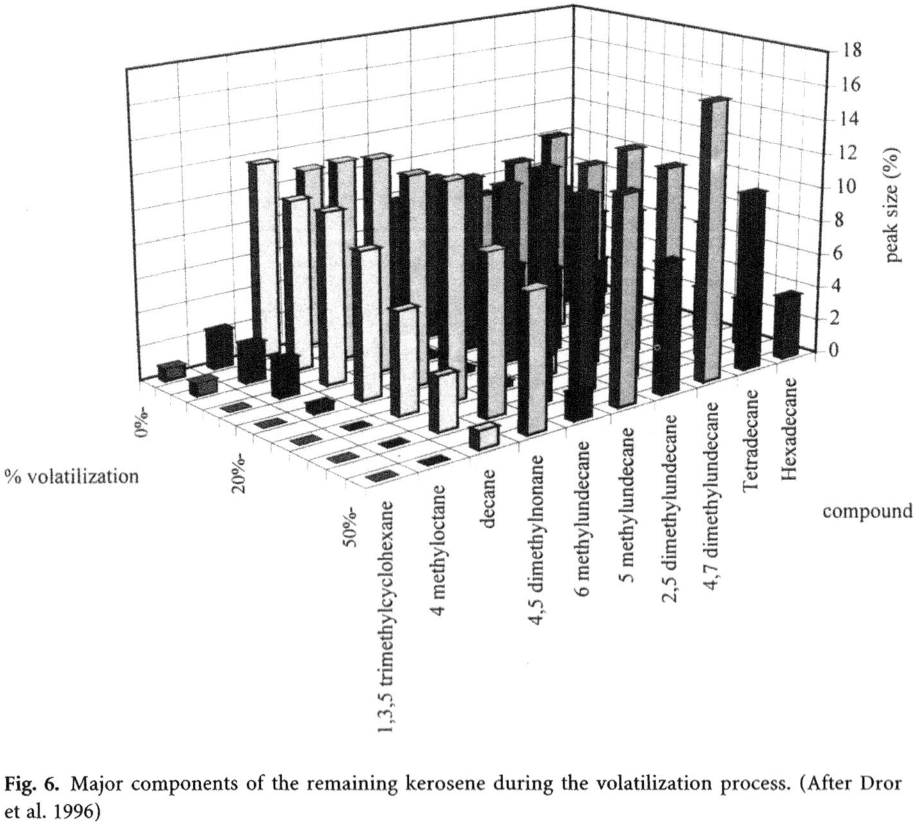

Fig. 6. Major components of the remaining kerosene during the volatilization process. (After Dror et al. 1996)

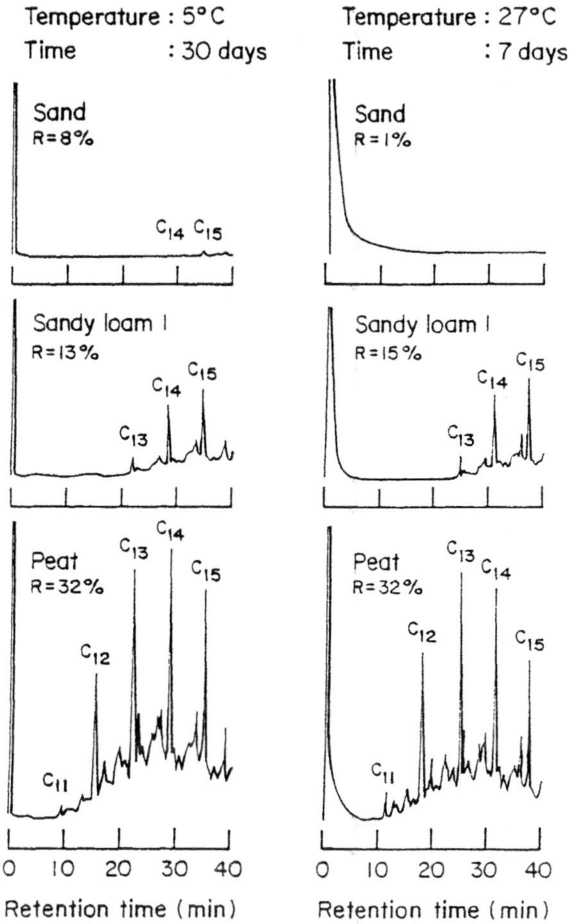

Fig. 7. Gas chromatograms of kerosene recovered from sand, sandy loam 1 and peat, after volatilization at 5 and 27 °C for 30 and 7 days, respectively. *R* Remaining kerosene (% of initial amount). (Jarsjo et al. 1994)

sandy loam 1 and peat after volatilization at 5 and 27 °C, for 30 and 7 days, respectively. All the lighter components have disappeared from the sand and the sandy loam 1, and the residual kerosene is similar regardless of temperature. The relative concentration of the light fractions C_{10}–C_{12} diminished with time in all soils, and those of the heavy fractions C_{14} and C_{15} increased.

Changes in the chemical properties of the residual kerosene in porous material as a result of volatilization of light fractions also cause changes in the physical properties of the remaining petroleum product: Table 3 summarizes the effect of differential volatilization on kerosene viscosity, surface tension, and density. When 20, 40, and 60% of the initial amount had been removed by the transfer of light hydrocarbon fractions to the atmosphere, the viscosity of the remaining

Table 3. Effect of volatilization on physical properties of residual kerosene. (Galin et al. 1990b)

Amount volatilized (%)	Density (g cm^{-3})	Surface tension (N m^{-1})	Viscosity (Pa s $\times 10^{-3}$)
0	0.805	2.75	1.32
20	0.810	2.78	1.48
40	0.818	2.80	1.78
60	0.819	2.78	1.96

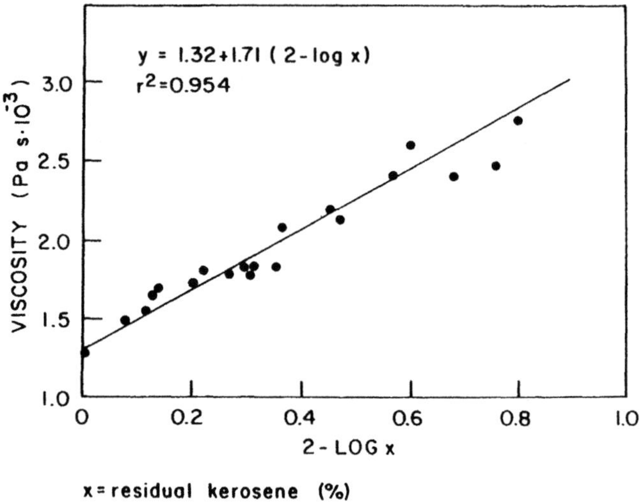

Fig. 8. The relationship, due to losses occurring by volatilization, between *y* kerosene viscosity and *x* the amount of remaining kerosene (%). (After Galin et al. 1990b)

kerosene was strongly affected. Negligible effects for liquid density and no effect for the surface tension were observed.

Figure 8 shows the relationship between the viscosity of the liquid and the percent of kerosene remaining following volatilization from coarse, fine, and medium sands. The highest increase in viscosity was obtained for coarse sand with an initial kerosene concentration equivalent to the residual retention capacity and the lowest increase for fine sand with initial kerosene concentration equivalent to the saturated retention capacity. Viscosity increased with time of volatilization in the order coarse > medium > fine sand for each initial concentration.

3.3 VOLM Dissolution in Soil Water

Excess aqueous solubility of aromatic constituents from petroleum products (e.g., gasoline, diesel fuel, etc.) was recently described by Cline et al. (1991) and Lea et al. (1992a,b) and was explained by co-solvency due to the presence of polar compounds in the mixture. This co-solvent effect potentially increases aqueous concentrations of hydrocarbons. To examine the effect of volatilization on the dissolution of VOLM in the aqueous phase, we will present recent results from Dror et al. (1996) on kerosene solubility. Figure 9 compares a chromatogram of kerosene (Fig. 9a) with a chromatogram of the fraction of kerosene soluble in an

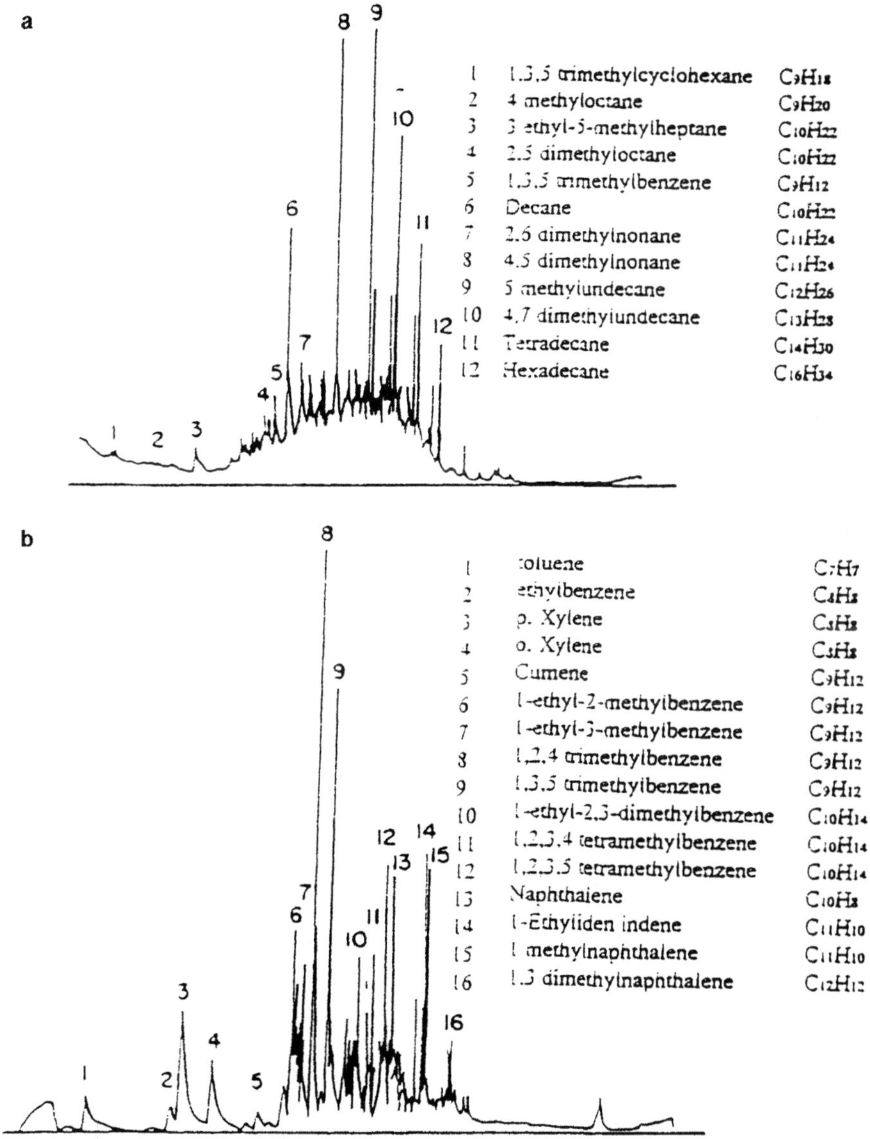

Fig. 9a,b. Gas chromatography of **a** kerosene neat and **b** kerosene dissolved in 0.01 n NaCl aqueous solution. (Dror et al. 1996)

aqueous electrolyte solution of 0.01 N NaCl (Fig. 9b) at ambient temperature (22 °C). Aliphatic and branched aliphatic hydrocarbons in the range of C_9 to C_{16} make up the major group of components in neat kerosene with only a minor set of aromatic components. In the aqueous electrolyte solution, on the other hand, aromatic compounds, especially the branched benzenes and naphthalenes, make up the majority of the compounds. The C_9 to C_{16} aliphatic and branched aliphatic components do not appear in the chromatogram of the aqueous electrolyte solution. The reason for the difference between the neat and aqueous solutions is the wide range of aqueous solubilities typical of the different components of kerosene. Aromatic components are several orders of magnitude more soluble than aliphatic components. Therefore, aromatic compounds, which were minor constituents of neat kerosene, dissolve in the aqueous solution much more than the major group of aliphatic compounds.

The effect of volatilization of neat kerosene on dissolution into the aqueous phase (0.01 N NaCl water solution) is depicted in Fig. 10. It was observed that an increase in volatilization led to a change in the composition of dissolved components and to a change in their relative abundance. Volatilization-induced changes in aqueous phase composition can be subdivided into three distinct sets (Fig. 11). One set of compounds is characterized by a decreasing relative peak size down to total disappearance of the components. The second set is characterized by increasing relative component peak size at the beginning of volatilization followed by decreasing relative component peak size as volatilization proceeds. The third set is characterized by increasing relative peak size of the components with continued volatilization.

3.4 VOLM Adsorption on Soil Solid Phase

Sorption of petroleum hydrocarbons by the soil solid phase and on various soil constituents has been the subject of numerous studies over the years. Nathwani and Philips (1977) studied the sorption of selected hydrocarbons from the vapor phase (benzene, o-xylene, toluene, and n-hexadodecane) on a number of Canadian soils and found that for concentrations between 1 and 100 ppm, hydrocarbon sorption was modeled by the Freundlich isotherm. Soils with a higher organic matter content exhibited higher values of the distribution coefficient. Many other studies found that sorption of organic compounds from vapor, aqueous, and organic solvent phases onto clay surfaces could be described by the Langmuir isotherm or by more complex, composite isotherms (Green-Kelly 1954a; German and Harding 1969; Olejnik et al. 1974).

Nye et al. (1994) studied the kinetics of vapor sorption of m-xylene and n-dodecane on oven- and air-dried Evesham clay soil and on a medium sand. Sorption equilibrium for both hydrocarbons was reached in less than 2 h on the sand and in 40 to 80 h on the clay soil (Fig. 11). M-xylene reached sorption equilibrium faster on oven-dry clay soil than on air-dried clay soil. The total sorption capacity for m-xylene on air-dried clay was nearly twice that of the oven-dried clay, while

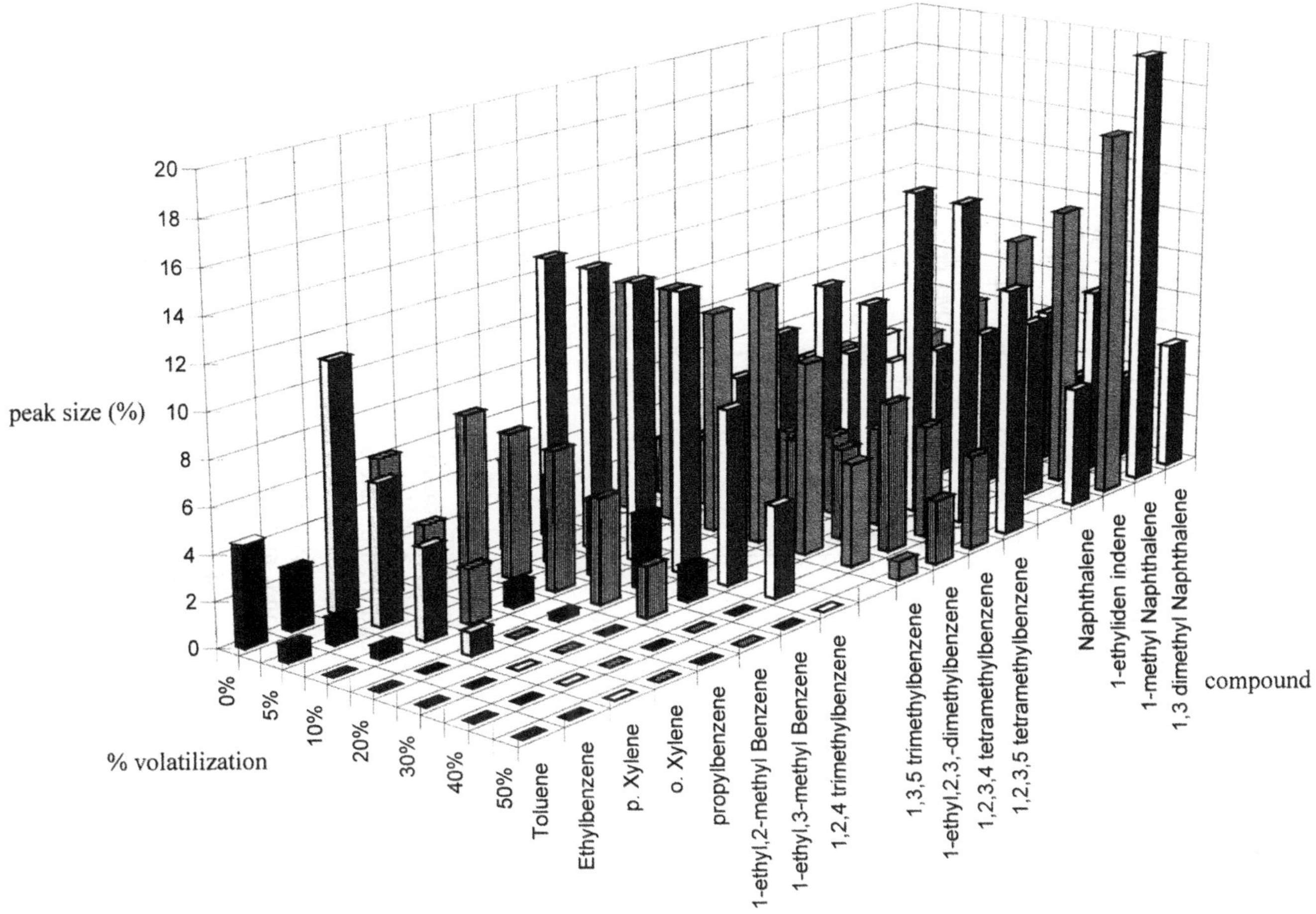

Fig. 10. Major components of kerosene dissolved in 0.01 *n* NaCl aqueous solution as observed after an initial volatilization of 0–50% (w/w). (After Dror et al. 1996)

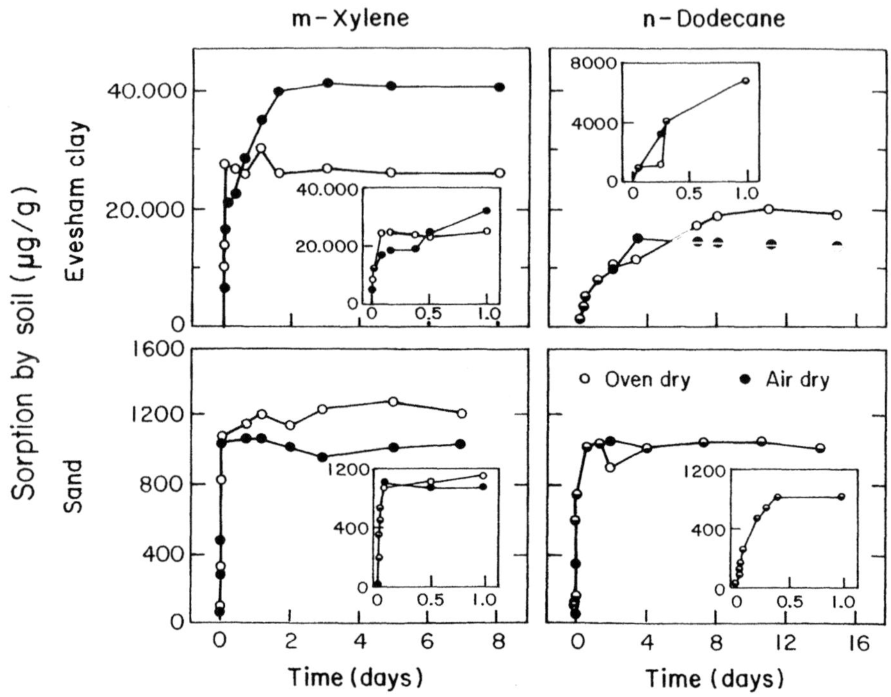

Fig. 11. Kinetics of m-xylene and n-dodecane vapor sorption by medium sand and Evesham clay soil. (Nye et al. 1994)

the sorption capacity of the sand was relatively unaffected by its moisture content. The total sorption capacity of the clay for m-xylene was 20 to 40 times that of the sand, and its sorption capacity for n-dodecane was approximately 15 times that of the sand. The uptake rate and total uptake of n-dodecane were equal in both the air- and oven-dried sand, but the total sorbed amount in the air-dried clay was less than in the oven-dried clay. In comparison, m-xylene sorption was greater in oven-dried sand than air-dried sand, and far higher in air-dried clay than that in oven-dried clay (Fig. 11). This behavior can be partly explained on the basis of structure-related size exclusion phenomena. The n-dodecane molecule is too large to penetrate clay interlayer spaces and can sorb only to the external surface area. In the air-dried clay, water successfully competes for sorption sites, resulting in a reduction of n-dodecane sorption compared with that in oven-dried clay. In the sand, the moisture content for air-dried and oven-dried sands are so similar that there is little or no competition for external sorption sites between water and n-dodecane. The smaller, planar m-xylene molecule can, by comparison, penetrate interlayer spaces in the expanded air-dried clay, while in the oven-dried clay these interlayer spaces are not accessible.

Soil composition has a strong impact on the kinetics of hydrocarbon sorption from gaseous hydrocarbon mixtures (Yaron et al. 1989; Nye et al. 1994). Vapor phase sorption of five compounds (m-xylene, ps-cumene, n-decane, n-but.-benzene, and n-dodecane) from a mixture of three soils with differing clay and organic matter contents (Bet Dagan: clay 12%, OM 0.5%; Gilat: clay 16%, OM 0.6%; Oxford: clay 35%, OM 7.2%) was studied (Yaron et al. 1989). For oven-dry soils, sorption of the studied compounds was greatest on the Gilat soil and was essentially equal in the Bet Dagan and Oxford soils. Despite the fact that the Oxford soil has twice the clay content of the Gilat soil, its sorptive capacity is significantly lower. This may be because sorption sites on mineral surfaces were occluded by soil organic matter. Complementary studies of sorption from the same hydrocarbon mixture on peat showed very low sorption capacity. In sorption from the vapor phase, free mineral surfaces rather than organic matter control hydrocarbon sorption. This is in contrast to sorption from the aqueous phase, where soil organic matter content is strongly correlated with sorptive capacity (Karickhoff 1981; Chiou 1989; Chin et al. 1991). In a related study it was shown that the sorption order from the mixture (m-xylene > n-decane > ps-cumene > t-butylbenzene > n-dodecane) corresponds to increasing boiling point and molar weight, rather than to vapor pressure (Acher et al. 1989).

An increase in soil moisture content led to a decrease in soil sorptive capacity in a silt loamy soil for all the hydrocarbons studied (Yaron et al. 1989). At a soil moisture content equivalent to 70% of field capacity (11%), the adsorption of hydrocarbons was almost negligible. This behavior was also observed for the same hydrocarbon mixture by Acher et al. (1989), and by Chio and Shoup (1985) and Barbee and Brown (1986) for benzene and xylene, respectively.

3.5 Volatilization-Transport Relationship

Multiphase transport of petroleum hydrocarbons in porous media is affected by the hydraulic and physical properties of the soil as well as by its hydration status. Acher et al. (1989) showed that during VOLM infiltration into soil, there is simultaneous but faster vapor transport, resulting in a penetration front in advance of the liquid front.

When a liquid kerosene applied at the surface of the soil column was depleted and the kerosene wetting front ceased to advance, there was a redistribution of the different hydrocarbon components according to a dynamic soil-vapor adsorption-desorption equilibrium (Acher et al. 1989). This vapor movement resulted in a redistribution of the more volatile components and hence to a change in residual kerosene composition. The initial soil moisture content affected the penetration of the kerosene both as an immiscible liquid and as vapor. The liquid hydrocarbon infiltration rate and extent in a sandy loam increased with increasing soil moisture content (Acher et al. 1989). Infiltration rate and extent for the vapor, in comparison, was inversely related to soil moisture content. The highest rate and greatest depth of m-xylene vapor penetration were observed in oven-dry soil (moisture content 0%), with decreasing penetration rate and depth as the moisture content increased. The distribution of the less volatile n-dodecane showed no dependence on soil moisture content. In another series of column experiments, a soil moisture content of more than 40% prevented the development of a separate vapor phase. These results are similar to those of Barbee and Brown (1986), who reported that xylenes applied to a sandy loam with a moisture content close to field capacity were retained by the soil and their movement significantly attenuated by the presence of water in soil pores.

As previously shown, volatilization-induced chemical changes in a VOLM (kerosene) lead to changes in viscosity. Since the rate of transport of VOLM is affected not only by the properties of the porous media but also by fluid density and viscosity, the conductivity of kerosene in soils was drastically changed as a result of volatilization.

Figure 12 shows the effects of changes in kerosene viscosity on its conductivity in sand, loam, and clay soils. An increase in viscosity from 1.32 to 1.96×10^{-3} Pa s led to a 28% decrease in kerosene conductivity in the sand, a 76% decrease in the loam soil, and an 83% decrease in the clay soil. Gerstl et al. (1994) studied conductivity (relative to the kerosene conductivity in the respective oven-dry media) in sand, loam, and clay soil columns with the initial moisture contents varying from zero to field capacity (Fig. 13). Kerosene conductivity in sand did not vary with moisture content, whereas its conductivity in the clay soil increased as the moisture content increased from oven-dry to 70% of field capacity. With the soil moisture content at 70% field capacity, kerosene conductivity reached a maximum and at field capacity, kerosene conductivity decreased to zero. In the loam soil,

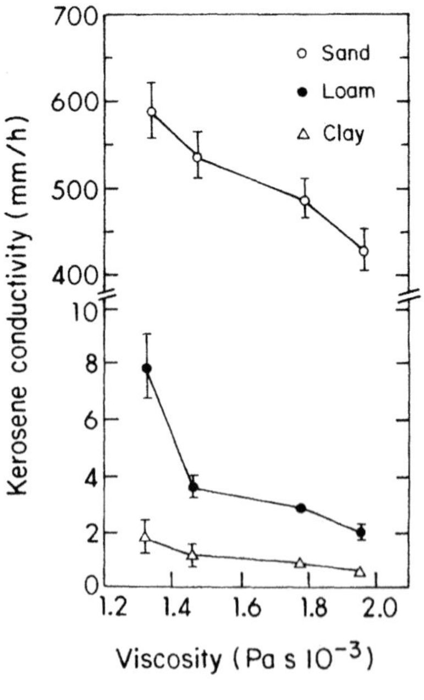

Fig. 12. Kerosene conductivity in soils as affected by type of soil and liquid viscosity. (Gerstl et al. 1994)

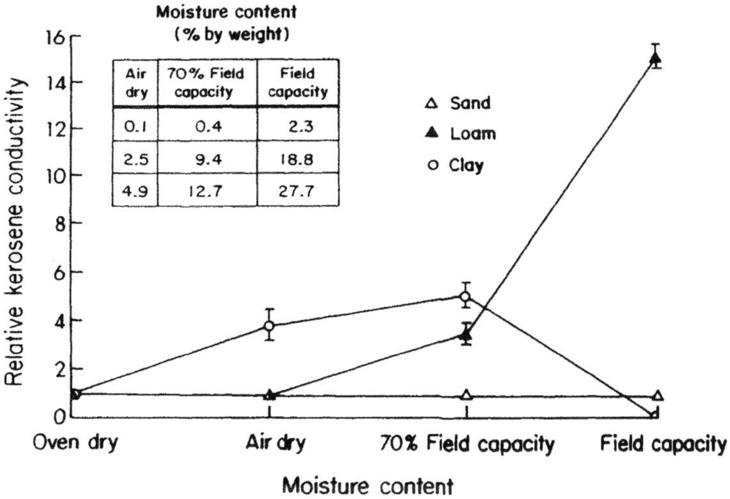

Fig. 13. Kerosene conductivity relative to that for oven-dry media as affected by increasing water content (by weight). (Gerstl et al. 1994)

kerosene conductivity increased as the moisture content varied from air-dry to field capacity. Gerstl et al. (1994) explain this process as follows: In the first stage of hydration (from oven- to air-dry) the water molecules partially cover the sorption sites in the soil. Since water sorbed on the soil particles prevents interaction (sorption) between the solid phase and hydrocarbon molecules, a kerosene layer is not formed. In the oven-dry clay soil the kerosene infiltrating the soil is adsorbed on the pore surfaces creating a wall effect, which retards movement because of the formation of a viscous layer. However, in the air-dry soil the pore walls are partly hydrated, the wall effect is reduced, and the kerosene conductivity increases relative to the oven-dry soil. The kerosene conductivity was not affected by partial hydration for either the loam soil or sand, both of which are characterized by larger pores than those in the clay soil. The difference between the viscosities of water and kerosene and their immiscibility, result in a well-defined interface. The shearing effect between the two liquids is smaller than within each liquid, which explains the increase in kerosene conductivity in the hydrated (70% field capacity) soils. This explanation should be examined in additional studies.

The conductivity relationships for water and VOLM phases are commonly assumed to be related to an intnirsic permeability that is a characteristic property of the porous medium and is independent of the fluid flowing through the medium (e.g., Abriola and Pinder 1985; Osborne and Sykes 1986; Pinder and Abriola 1986; Corapcioglu and Baehr 1987; Kuppusamy et al. 1987). The saturated fluid conductivity (K_s) is related to the intrinsic permeability (k) as

$$K_s = \frac{K\rho g}{\mu}, \tag{1}$$

where ρ and μ are the fluid density and viscosity, respectively, and g is the gravitational constant.

Jarsjo et al. (1997) show that for multiphase conditions, relationships between immiscible fluid conductivities that are based on Eq. (3) are often used in the literature (e.g., Cary et al. 1989a,b) and may be expressed as:

$$K_0 = K_{sw} \frac{\mu_w}{\mu_0} \frac{\rho_0}{\rho_w} f(\theta, \varphi; \varepsilon), \tag{2}$$

in which K_0, μ_0, ρ_0, and φ are the unsaturated conductivity, viscosity, density, and volume fraction of the VOLM, respectively, and K_{sw}, μ_w, ρ_w, and θ are the saturated conductivity, viscosity, density, and volume fraction of water, respectively. Furthermore, the function f in Eq. (2) quantifies the relative VOLM conductivity based on Θ, Ψ, and a soil parameter ε that includes soil porosity and pore-size distribution-related soil coefficients. The function f is often derived from relative water conductivity relationships, where the actual form of f depends on the conceptualization of phase distribution within the pore space (Cary et al. 1989b).

Regardless of the form of f, Eq. (2) is based on the assumption that Eq. (1) is valid, such that an intrinsic permeability k exists that can be used to scale the saturated conductivity K_{s2} of any fluid to a known saturated fluid conductivity K_{s1} as

$$K_{s2} = K_{s1} \frac{\mu_1}{\mu_2} \frac{\rho_2}{\rho_1}. \tag{3}$$

From a practical point of view, it is convenient to express the hydraulic conductivity of a soil by using intrinsic permeability, because it is, in theory, independent of the pore fluid. It is therefore of interest to investigate under which conditions Eq. (3) is applicable.

Figure 14 illustrates the observed and predicted mean kerosene conductivity according to Eq. (3), based on the viscosity ratios and the reference kerosene conductivity $K_{sk}(0\%)$, along with the mean conductivity values for the kerosene mixtures in the soils. For inert sands the difference between observed and predicted conductivity is relatively small. For peat, the prediction of kerossene (60% volatilization) conductivity [Eq. (3)] is considerably higher than the observed kerosene (60% volatilization) conductivity. Furthermore, for the montmorillonitic loam and clay, the continuous increase in viscosity of the kerosene mixtures with increasing volatilized mass percentage seem to explain only a smaller part of the observed changes in hydraulic conductivity.

Jarsjo et al. (1997) show that for temperature-induced changes in kerosene viscosity, the resulting saturated conductivity ratios were consistent with the corresponding viscosity ratios, i.e., with the predictions of Eq. (3), within the range of experimental deviations (see Table 4). For volatilization-induced viscosity changes, which reflect irreversible changes in kerosene chemical composition, the theoretical scaling relationship showed relatively small deviations from corresponding viscosity ratios for inert, sandy soils (coarse, medium and fine sand).

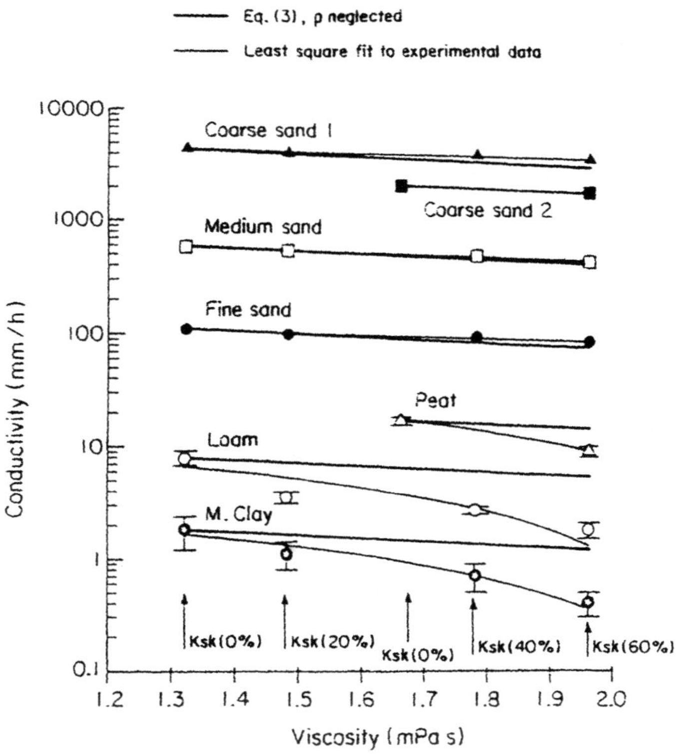

Fig. 14. Observed (*symbols*) and predicted (*solid line*) conductivities for kerosene mixtures with different viscosities and chemical compositions. (Jarsjo et al. 1997)

Table 4. Mean kerosene conductivity ± one experimental standard deviation at 24 °C for kerosenes of different chemical composition. Conductivity values are in mm/h. (After Jarsjo et al. 1997)

Soil	Kerosene mixture			
	Kerosene-60% ($\mu = 1.96$ mPa s)	Kerosene-40% ($\mu = 1.78$ mPa s)	Kerosene-20% ($\mu = 1.48$ mPa s)	Kerosene-9% ($\mu = 1.32$ mPa s)
Coarse sand 1	3480 ± 150	3790 ± 160	3960 ± 130	4350 ± 110
Coarse sand 2	1740 ± 290	–	–	1990 ± 290
Medium sand	427 ± 27	482 ± 28	538 ± 34	591 ± 38
Fine sand	83 ± 6	92 ± 8	98 ± 8	109 ± 14
M.[a] loam	1.8 ± 0.3	2.7 ± 0.2	3.5 ± 0.5	7.9 ± 1.4
M.[a] clay	0.4 ± 0.1	0.7 ± 0.2	1.1 ± 0.3	1.8 ± 0.6
Peat	9.0 ± 1.0	–	–	16.8 ± 1.2

[a] Montmorillonitic.

However, for volatilization-induced viscosity changes in soils containing clay or organic matter, the observed saturated conductivity ratios deviated significantly from the theoretical predictions of Eq. (3). A comparison of four different kerosene mixtures in two montmorillonitic soils further showed that the discrepancies between these two ratios increased with increasing changes in liquid chemical composition. Thus, the results indicate that properties other than fluid viscosity may have an effect on intrinsic permeability of potentially interacting soils with respect to VOLM residues (e.g., Bear 1972).

The laboratory observations on VOLMs support the hypothesis that chemical composition of a liquid mixture may be critical to its permeability value. Furthermore, they demonstrate that volatilization of the liquid is capable of changing the VOLM chemical composition sufficiently to yield considerable conductivity changes, which are not attributed to viscosity changes, in interacting soils. The preferential volatilization of lighter kerosene components changes the composition of the kerosene such that the relative amount of heavy compounds increases with increasing volatilized mass percentage (Galin et al. 1990a; Jarsjo et al. 1994). The remaining liquid kerosene thus becomes more hydrophobic with increasing volatilization, which may, in turn, result in greater adsorption and possibly explain the observed reductions of conductivity of interacting soils.

For swelling clays, another possible explanation for changes in conductivity as a result of changes in chemical composition of the liquid mixture may be given by the diffuse double-layer theory that relates increased swelling with an increased fluid dielectric constant and hence reduces intrinsic permeability (e.g., Yong et al. 1992). In that example, however, the dielectric constant of the fluid did not change appreciably with volatilization. For a detailed discussion of this point, see Graber and Mingelgrin (this vol.).

Acknowledgments. The research reported here was partially supported by a grant from the Water Research Institute of the Technion (WRI) – Israel, Institute of Technology, Haifa, Israel.

References

Abriola LM, Pinder GF (1985) A multiphase approach to the modeling of porous media contamination by organic compounds. 1. Equation development. Water Resour Res 21:11–18

Acher AJ, Broderie B, Yaron B (1989) Soil and groundwater pollution by petroleum products. 1. Multiphase migration of kerosene components in soil columns. J Contam Hydrol 4:333–345

Baehr AL (1987) Selective transport of hydrocarbons in the unsaturated zone due to aqueous and vapor phase partitioning. Water Resour Res 23:1926–1938

Baehr AL, Corapcioglu MY (1987) A compositional multiphase model for groundwater contamination by petroleum products 2. Numerical solution. Water Resour Res 23:201–213

Barbee GC, Brown KW (1986) Movement of xylene through unsaturated soils following simulated spills. Water Air Pollut 29:321–333

Bear J (1972) Hydraulics of groundwater. McGraw Hill, New York

Burris DR, McIntyre G (1986) Solution of hydrocarbons in a hydrocarbon-water system with changing phase composition due to evaporation. Environ Sci Technol 20:296–299

Cary JW, McBride JF, Simmons CS (1989a) Observation of water and oil infiltration into soil: some simulation challenges. Water Resour Res 25:73–80
Cary JW, McBride JF, Simmons CS (1989b) Predicting oil infiltration and redistribution in unsaturated soils. Soil Sci Soc Am J 53:335–342
Chin YP, Weber WJ, Chiou CT (1991) A thermodynamic partition model for binding of nonpolar organic compounds by organic colloids and implications for their sorption to soils and sediments. In: Baker RA (ed) Organic substances and sediments in water vol 1. Humics and soils. Lewis Ann Arbor, Michigan, pp 251–273
Chiou CT (1989) Theoretical considerations of the partition uptake of nonionic organic compounds by soil organic matter. In: Sawheny BL, Brown K (eds) Reaction and movement of organic chemicals in soils. SSSA Spec Publ no 22. Soil Science Society of America, Madison, Wisconsin, pp 1–30
Chiou CT, Shoup TD (1985) Soil sorption of organic vapors and effects of humidity on sorptive mechanism and capacity. Environ Sci Technol 19:1196–1200
Cline VP, Delfino JJ, Rao PSC (1991) Partitioning of aromatic constituents into water from gasoline and other complex solvent mixtures. Environ Sci Technol 25:915–920
Corapcioglu MY, Baehr A (1987) A compositional multiphase model for groundwater contamination by petroleum products. 1. Theoretical consideration. Water Resour Res 23:191–200
Drivas PJ (1982) Calculation of evaporative emissions from multicomponent liquid spills. Environ Sci Technol 16:726–728
Dror I, Gerstl Z, Yaron B (1996) Dissolution of kerosene in water as affected by volatilization. Progress report to Water Research Institute (WRI), Technion, Haifa, Israel
Durner W (1994) Hydraulic conductivity estimation for soils with heterogeneous pore structure. Water Resour Res 30:211–223
Fine P, Yaron B (1993) Outdoor experiments on enhanced volatilization by venting of kerosene component from soil. J Contam Hydrol 12:355–374
Fine P, Graber ER, Yaron B (1997) Soil interaction with petroleum hydrocarbons: abiotic processes. Soil Technol 10:133–153
Galin Ts, Gerstl Z, Yaron B (1990a) Soil pollution bypetroleum products. III. Kerosene stability in soil columns as affected by volatilization. J Contam Hydrol 5:375–385
Galin Ts, McDowell C, Yaron B (1990b) The effect of volatilization on the mass flow of a nonaqueous pollutant liquid mixture in an inert porous medium: experiments with kerosene. J Soil Sci 41:631–641
German WL, Harding DA (1969) The adsorption of aliphatic alcohols by montmorillonite and kaolinite. Clay Miner 8:213–227
Gerstl Z, Galin Ts, Yaron B (1994) Mass flow of a volatile organic liquid mixture (VOLM) in soils. J Environ Qual 23:487–493
Green-Kelly R (1954a) Sorption of aromatic compounds by montmorillonite, part I. Orientation studies. Trans Faraday Soc 51:412–424
Jarsjo J, Destouni G, Yaron B (1994) Retention and volatilization of kerosene: laboratory experiments on glacial and post-glacial soils. J Contam Hydrol 17:167–185
Jarsjo J, Destouni G, Yaron B (1997) On the relation between viscosity and hydraulic conductivity values for volatile organic liquid mixtures in soils. J Contam Hydrol 25:113–127
Jury WA, Spencer WF, Farmer JW (1984) Behavior assessment model for trace organics in soil. IV. Review of experimental evidence. J Environ Qual 13:580–586
Karickhoff SW (1981) Semi-empirical estimation of hydrophobic pollutants on natural sediments and soil. Chemosphere 10:833–846
Karimi AA, Farmer JW, Cliath MM (1987) Vapor phase diffusion of benzene. J Environ Qual 16:38–44
Kuppusamy T, Sheng J, Parker JC, Lenhard RJ (1987) Finite-element analysis of multiphase immiscible flow through soils. Water Resour Res 23:625–631
Lee LS, Hagwall M, Delfino JJ, Rao PSC (1992a) Partitioning of polycyclic aromatic hydrocarbons from diesel fuel into water. Environ Sci Technol 26:2104–2110
Lee LS, Rao PSC, Okuda I (1992b) Equilibrium partitioning of polycyclic aromatic hydrocarbons from coal tar into water. Environ Sci Technol 26:2110–2115

Mackay D, Yeun TK (1983) Mass transfer coefficient correlations for volatilization of organic solutes from water. Environ Sci Technol 13:211–212

Nathwani JS, Philips CR (1977) Adsorption-desorption of selected hydrocarbons in crude oil and soils. Chemosphere 4:157–162

Nye PH, Yaron B, Galin Ts, Gerstl Z (1994) Volatilization of a multicomponent liquid through dry soils: testing a model. Soil Sci Soc Am J 58:269–277

Olejnik S, Posner AM, Quirk JP (1974) Swelling of montmorillonite in polar organic liquids. Clays Clay Miner 22:361–365

Osborne M, Sykes J (1986) Numerical modeling of immiscible organic transport at the Hyde Park landfill. Water Resour Res 22:25–33

Pinder GF, Abriola LM (1986) On the simulation of nonaqueous phase organic compounds in the subsurface. Water Resour Res 22:109–119

Roberts PV, Dandiker GP (1983) Mass transfer of volatile organic contaminants from aqueous solution to the atmosphere during surface aeration. Environ Sci Technol 13:484–489

Spencer WF, Farmer JW (1980) Assessment of the vapor behavior of toxic organic chemicals. In: Haque R (ed) Dynamics exposure and hazard assessment of toxic chemicals. Michigan state University, East Lansing, pp 143–161

Stiver W, Mackay D (1984) Evaporation rate of spills of hydrocarbons and petroleum mixtures. Environ Sci Technol 18:834–840

Yaron B (1989) On the behavior of petroleum hydrocarbons in the unsaturated zone: abiotic aspects. In: Gerstl Z, Chen Y, Mingelgrin U, Yaron B (eds) Toxic organic chemicals in porous media. Springer, Barlin Heidelberg New York, pp 211–230

Yaron B, Sutherland P, Galin Ts, Acher AJ (1989) Soil and groundwater pollution of petroleum products. II. Adsorption-desorption of kerosene vapors as affected by type of soil and its moisture content. J Contam Hydrol 4:347–358

Yong RN, Mohamed AMO, Waretkin BP (1992) Principles of contaminant transport in soils. Elsevier, Amesterdam

4 Permeability of Porous Media as Affected by Shrinkage and Swelling of Clays

E. R. Graber[1] and U. Mingelgrin[1,2]

Contents

1 Introduction

Hydraulic conductivity (K) of a porous medium as defined by Darcy's law

$$K = -\frac{\frac{Q}{A}}{\frac{dh}{dl}} \tag{1}$$

is a transport coefficient which depends both on matrix and pore fluid properties, where Q is discharge, A cross-sectional area, and dh/dl the change in head per unit change in length. The relevant pore fluid properties were determined empirically to be density (ρ) and dynamic viscosity (μ). Intrinsic permeability (k) was defined to describe the conductive properties of a medium independent of the fluid flowing through it (Nutting 1930)

$$K = \frac{K\mu}{\rho g}, \tag{2}$$

with g as the gravitational constant. In a rigid, noninteracting porous medium with an effective pore radius of r, k was shown to be equal to (Bear 1972)

[1] Institute of Soils and Water, The Volcani Center, ARO, P. O. Box 6, Bet Dagan 50250, Israel
[2] Ministry of the Environment, 5 Kanfei Nesharim St., Givat Shaul, Jerusalem 95464, Israel

H. Rubin/N. Narkis/ J. Carberry
Soil and Aquifer Pollution

$$k = f(\sigma)f(\theta)r^2, \tag{3}$$

where $f(\sigma)$ is a function of pore shape and $f(\theta)$ a function of porosity, both dimensionless.

In the more general case of a nonrigid, interacting (e.g., swelling) porous medium in an approximately fixed volume, both $f(\sigma)$ and $f(\theta)$ may be dependent on r. The product $f(\sigma)f(\theta)$ can then be resolved into the product of C (a shape constant whose magnitude and dimensions are system dependent), n (the number of pores in a fixed volume), and an r-dependent term, such that k is approximated by (Graber and Mingelgrin 1994)

$$k = Cnr^i. \tag{4}$$

From Eq. (4) it is evident that for a porous medium comprised in part of expandable elements such as clay minerals, "intrinsic" permeability is a misnomer.

Instrinsic permeability of a porous medium constituted in part by a swelling clay can change, for example, when the chemistry of the aqueous pore fluid is altered. When pore water ionic concentration is altered, clay particles in the matrix may undergo significant volume changes. As a result of such volume changes, pore diameters in the porous medium can increase or decrease. Since the rate of flow through a pore is proportional to the fourth power of its diameter (Pouiselle's law), swelling or shrinkage of clay particles can strongly affect the permeability of porous media.

Organic liquids can also cause clay minerals to undergo volume changes (Bradley 1945; MacEwan 1948; Barshad 1952), which can in turn lead to permeability changes. This was observed, for example, in compacted clay used to line organic solvent waste impoundments (e.g., Green et al. 1981; Acar and Seals 1984; Brown and Thomas 1984; Hettiaratchi and Hrudey 1987). Clay shrinkage and swelling can similarly result in permeability changes in organic-liquid permeated clay-bearing soils (Schramm et al. 1986; Amoozegar et al. 1986; Gerstl et al. 1994; Jarsjo et al. 1997).

Swelling and shrinkage of clays and resultant permeability changes are commonly thought to result from the electrostatic effect of the pore fluid on the thickness of diffuse double layers (DDL) at clay surfaces (Acar and Seals 1984; Brown and Thomas 1987; Hetterarchi and Hrudey 1987). It has been suggested that polar fluids with high dielectric constants (ε), such as water, weaken attractive forces between cations and negatively charged clay surfaces, resulting in swelling and permeability reduction. Nonpolar fluids with low dielectric constants (such as hydrocarbons) have no such ability to counteract attractive electric forces. In such fluids, DDLs are compressed, leading to fabric shrinkage and large increases in permeability.

This review critically examines clay shrinkage and swelling data as a component of our efforts to understand the movement of waste organic liquids through clay liners and in the subsurface. It was recently shown (Demond et al. 1996) that estimated transport parameters for organic liquids can differ significantly from measured transport parameters because of poor estimates for relative permeability. Many estimation and modeling attempts assume that porous media are rigid and noninteracting. This assumption is untenable in media consisting in part of clay minerals. As such, it is neccessary to elucidate the mechanisms controlling clay volume-change behavior in an effort to understand resultant permeability changes in organic fluid-contaminated porous media.

2 DDL Theory Versus Surface Solvation

Two distinct theories have been developed in the literature to explain the swelling of clays: the diffuse double layer theory and the theory of surface solvation.

2.1 DDL Theory of Clay Swelling

The basic theory of diffuse double layers (DDL) at charged colloidal surfaces was developed in the early 1900s by Gouy and Chapman. Cations are attracted towards negatively charged surfaces and anions are repulsed. These interactions obey Coulomb's law

$$\mathrm{F} = \frac{qq'}{\varepsilon\lambda^2}, \tag{5}$$

where F is the force of attraction between an ion and the charged surface, q and q' are the electrical charges, λ is the charge separation, and ε is the dielectric constant of the fluid. From Eq. (5) it is clear that F increases as the charge on the colloid or ion increases, decreases with increasing ε, and decreases as the square root of the distance between the charged colloid and ion.

The tendency of a cation to concentrate at a negatively charged clay surface is balanced by a countertendency for diffusion away from the high concentration region, resulting in a diffuse distribution of charges between the surface and bulk solution. Maximal cation and minimal anion concentrations will occur at the clay surface. A semiquantitative indication of the thickness of the diffuse double layer (H) in angstroms is given by (van Olphen 1963):

$$\mathrm{H} = \frac{\varepsilon\kappa T}{8\pi\eta q^2 \mathrm{Z}^2}, \tag{6}$$

where η is the number of ions per cm^3, κ is the Boltzman constant, Z is valence, and T is absolute temperature. From Eq. (6) it is seen that H is linearly dependent on ε. The thicker the DDL, the greater the repulsion between two approaching clay surfaces.

The distribution of cations, anions, and electrical potential is assumed to be symmetrical between two adjacent surfaces. If there is an excess of cations at the midplane (m) between two surfaces such that $C_m > C_o$, where C_o is the concentra-

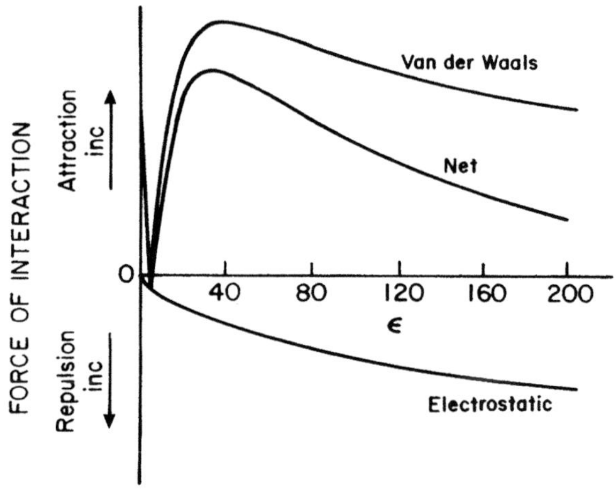

Fig. 1. Forces of interaction at clay surface. Repulsive forces are attributed to electrostatic interactions in the DDL and attractive forces are attributed to van der Waals dispersion forces. The net interaction curve is the sum of repulsive and attractive forces. (Adapted from Moore and Mitchell 1974)

tion in the bulk solution, then an osmotic gradient will exist between the midplane and bulk solution. This leads to fluid imbibing, or swelling, towards the midplane until the tendency to swell is counterbalanced by interparticle attractive forces often attributed to van der Waals dispersion forces (MacEwan 1948; Norrish 1954; van Olphen 1963). The net potential curve of particle interaction can then be constructed by summing repulsive (DDL) and attractive (van der Waals) forces at every point. Moore and Mitchell (1974), for example, developed an analytical expression for the sum of repulsive and attractive forces at clay surfaces as a function of fluid ε (Fig. 1). In a series of experiments they found that shear and axial strengths of consolidated clays permeated with organic fluids could be described by this treatment. Minimal interaction was predicted when the pore fluid dielectric constant equals that of the clay ($\varepsilon \approx 4$). At minimal interaction, swelling should be maximal. As ε becomes larger or smaller than 4, net interactive forces increase, resulting in shrinkage.

2.2 Surface Solvation Theory of Clay Swelling

The DDL theory for interaction between clay surfaces is applicable only to swelling in infinitely dilute clay suspensions (Delville and Laszlo 1990) and clay systems saturated with monovalent cations (Aylmore and Quirk 1962). Low and his coworkers showed that the theory does not hold for real clay-solvent systems at finite concentrations (Low 1980, 1981; Viani et al. 1983, 1985; Chen et al. 1980). It was suggested that clay swelling in the crystalline region (interlayer spacing <20 Å) could be attributed mainly to surface solvation (MacEwan 1948) or complex formation (Bradley 1945; Barshad 1952). Low and others (Low 1980, 1981; Viani et al. 1983; 1985; Chen et al. 1987) have shown that swelling in the noncrystalline region can also be attributed to surface solvation. The possible influence of DDL repulsion on swelling in concentrated solutions was considered negligible by these authors. For example, Chen et al. (1987) showed that organic solvents affect illite swelling by interfering with surface hydration, with little or no relationship to fluid ε.

Besides the evidence that the DDL theory does not successfully describe swelling in concentrated clay-solvent systems, studies have suggested that even in infinitely dilute suspensions, the counterbalancing attractive van der Waals forces (e.g., Moore and Mitchell 1974) are too small to offset repulsive forces in smectites and vermiculite (Norrish and Rausell-Colom 1961; van Olphen 1962). These and other authors suggested that the attractive forces which limit swelling are edge-to-face bonds between sheets (e.g., Emerson 1963; Rowell 1965). Israelachvili and Pashley (1982) measured the strength of interaction between mica surfaces made hydrophobic by adsorption of a monolayer of hexadecyltrimethylammonium bromide (CTAB). The hydrophobic attraction between CTAB-coated surfaces in aqueous solutions was found to be substantially greater than the expected van der Waals attraction. This may be analogous to the behavior of hydrophilic surfaces in apolar organic solvents, lending credence to the asserted importance of surface solvation in clay swelling.

3 Permeability and Conductivity Changes in Porous Media

Many of the studies of permeability and solvent conductivity (K_s, a generalization of hydraulic conductivity K to all fluids) of soils and clays have revealed that permeability many to organic fluids is greater than to water. However, no single relationship between k or K_s and fluid dielectric constant has been developed.

Green et al. (1981) measured the solvent conductivity of remolded clay to benzene, xylene, carbon tetrachloride, trichloroethene, acetone, methanol, glycerol, and water. They found that for CCl_4, solvent breakthrough was so fast (Fig. 2) that it appeared to be caused by flow via preferential pathways. For matrix flow, however, they asserted that conductivity increased as ε increased, although this is not supported by the data (Fig. 2).

Brown and Thomas (1984) measured K_s for three compacted clays to solvents including water, acetone, xylene, gasoline, kerosene, diesel fuel, and motor oil in fixed wall laboratory permeameters. In all cases, conductivity to the organic liquids was higher than to water, in most cases by one to four orders of magnitude (Fig. 3). Variability between replicates was so high, however, that most of the differences were not statistically significant. High variability in such experiments often indicates that flow was through preferential pathways or along the walls of the permeameter. In a subsequent field cell verification study (Brown et al. 1986), xylene was found to traverse compacted liners via shrinkage cracks, whereas acetone was transported mainly through the matrix itself. In both cases, conductivity of the liners to organic solvents was higher than to water. On a different set

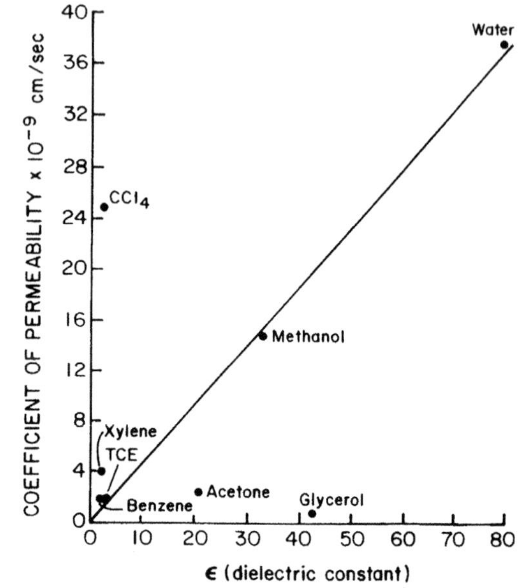

Fig. 2. Solvent conductivity (K_s) of the Ranger Shale to solvents as a function of solvent dielectric constant (ε). (Adapted from Green et al. 1981)

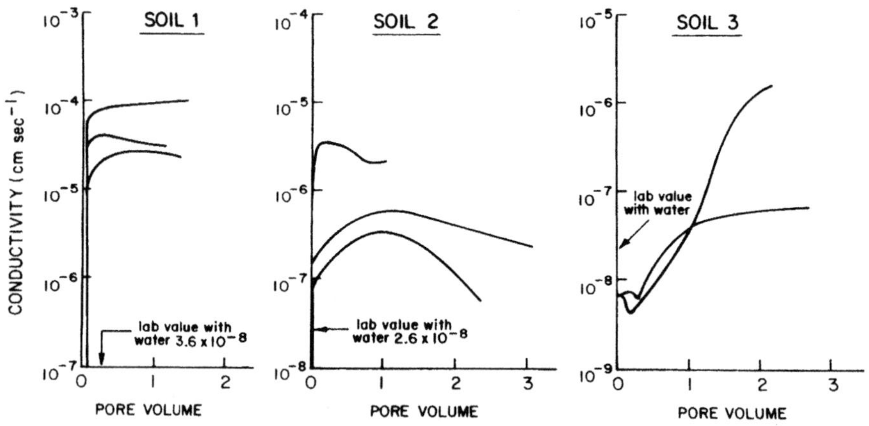

Fig. 3. Solvent conductivity (K_s) of three soils to acetone. Each line represents a single replicate. (Adapted from Brown and Thomas 1984)

of soils Anderson et al. (1985) also showed that compacted clays were more conductive to organic solvents than to water by two to four orders of magnitude. Cracks and voids were observed in the cells permeated with organic liquids. Brown and Thomas (1987) attempted to discern the mechanism by which low ε organic

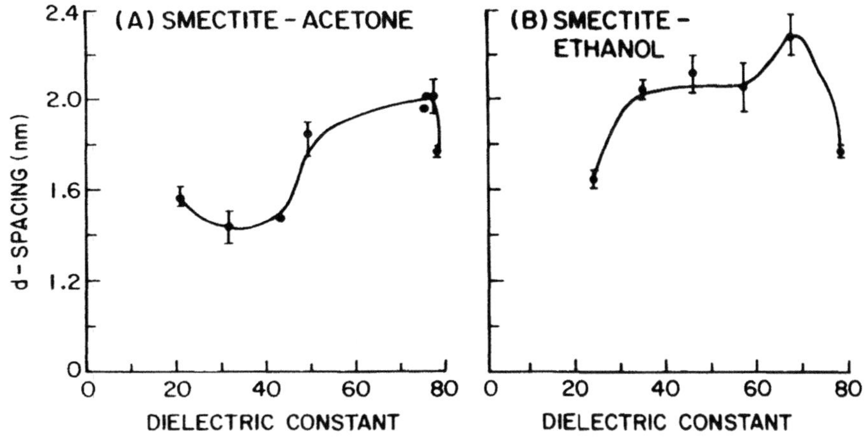

Fig. 4 A,B. Basal d(001) spacing of a smectite equilibrated with organic solvent-water mixtures as a function of mixture dielectric constant. **A** Acetone-water; **B** ethanol-water. (Adapted from Brown and Thomas 1987)

fluids caused flocculation or clay shrinkage by studying K_s, electrophoretic mobility, zeta potential, and swelling of clay soils in solvents. They found that smectites exhibited swelling maxima in organic fluid-water mixtures of intermediate ε values (Fig. 4). This was attributed to changes in surface electrical properties of clays in contact with the different fluids. Hettiaratchi and Hrudey (1987) also observed swelling maxima in fluids of intermediate ε in their study of the free swell of soils in organic solvents.

Intrinsic permeability of eight different soils to xylene, kerosene, isopropanol, ethylene gylcol, and water was found (Schramm et al. 1986) to vary with an inverse log linear relationship to fluid dielectric constant (Fig. 5). They attributed this finding to the expansion and contraction of diffuse double layers at clay surfaces as a function of ε. In a related work, Amoozegar et al. (1986) examined the infiltration of six organic liquids and water through five soils. They found no relationship between solvent properties, including ρ, μ, and ε, and the coefficient of infiltration.

Acar and Olivieri (1989) studied liquid limits, plasticity,and free swell of Na and Ca montmorillonite and kaolinite in various liquids. The free swell of Na and Ca montmorillonites was nearly invariant in the range of fluids studied, while kaolinite displayed a considerable spread in the experimental data (Fig. 6). They suggested that the free swell of kaolinite was reasonably well described by the net interparticle interaction curve determined by Moore and Mitchell (1974; Fig. 1).

In a study of mass flow of kerosene in loam, sand, and clay soils, Gerstl et al. (1994) found that as kerosene volatilized and became enriched in heavier fractions, the intrinsic permeability of the loam and clay soils decreased (Table 1). This

Fig. 5. Log of intrinsic permeability (k) as a function of fluid dielectric constant (ε) for three soils. (Adapted from Schramm et al. 1986)

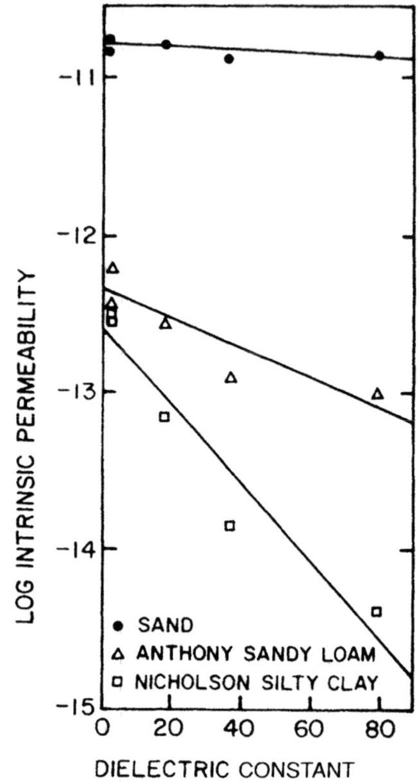

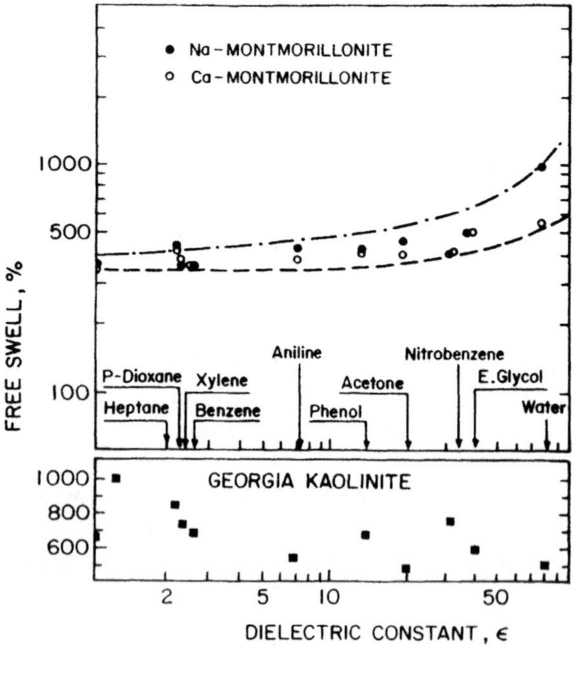

Fig. 6. Log-log plot of percent free swell of clays (Na and Ca montmorillonite, and kaolinite) against solvent dielectric constant (ε). (Adapted from Acar and Seals 1987)

Table 1. Intrinsic permeability (k) as a function of kerosene viscosity. Increasing kerosene viscosity (as a result of volatilization) was the consequence of an increase in heavy hydrocarbon fractions (C_{13}–C_{15}) and a decrease in light hydrocarbon fractions (C_9–C_{11}) in the kerosene mixture. (Gerstl et al. 1994)

Viscosity (Pa s × 10^{-3})	Intrinsic permeability (mm^2)		
	Sand	Loam	Clay
1.32	2.73×10^{-5}	3.65×10^{-7}	8.13×10^{-8}
1.48	2.79×10^{-5}	1.84×10^{-7}	5.84×10^{-8}
1.76	3.00×10^{-5}	1.71×10^{-7}	4.52×10^{-8}
1.96	2.92×10^{-5}	1.27×10^{-7}	2.69×10^{-8}

was attributed either to sorption of viscous hydrocarbon components at soil surfaces creating an immobile flow region and thus narrower effective pores, or to swelling of clay minerals with changing dielectric constant of pore fluid as light fractions volatilized. In this case however there was no apparent change in ε as volatilization proceeded. A similar effect was observed by Jarsjo et al. (1997).

Despite the evidence that no simple relationship between k, K_s and ε exists, such variations in permeability have nevertheless been almost universally attributed to clay shrinkage and swelling as a function of the pore fluid dielectric constant according to the classical DDL theory. Changes in DDL thickness are thought to lead to shrinkage or swelling of clay particles and, consequently, to permeability changes. The measurements made, however, have frequently conflicted with each other and with DDL theory (Figs. 1–6). As pointed out by Graber and Mingelgrin (1994), the conflicting results may be in part a consequence of differences in particle aggregation between the various porous media. Permeability can either decrease or increase upon swelling in an aggregated medium, depending on its structure and composition. For example, in a free-swelling system of uniform spheres of radius R where intrinsic permeability is proportional to R^2, as R increases, intrinsic permeability also increases. If interstices between spheres are filled with cuboids (for example, a mixture of sand and clay) and the volume of each cuboid increases by swelling, the intrinsic permeability of the system will decrease.

4 Swelling and Shrinkage of Clays

In line with the often contradictory evidence from permeability and conductivity studies, many of the results of clay swelling and shrinkage studies have been inconclusive. As a whole, these results do not support the view that the pore fluid dielectric constant controls clay swelling, and there is no clear consensus concerning the mechanisms governing volume changes of clays in organic liquids.

For example, Barshad (1952) studied the interlayer spacing of dehydrated montmorillonite and vermiculite immersed in organic liquids of varying dielectric constants and dipole moments. The extent of interlayer expansion was strongly affected by the size, charge, and amount of interlayer cations, and by the dipole moment and ε of the immersion fluid. He did not find a monotonic relationship between clay expansion and ε.

Swelling in the crystalline as well as the noncrystalline (osmotic, diffuse) regions of Na, Cs, Mg, and Ca montmorillonite were measured in polar organic liquids with similar dipole moments but very different ε values (Olejnik et al. 1974). Na montmorillonite swelled osmotically in both formamide and N-methyl formamide, whereas Cs montmorillonite gave osmotic swelling in formamide only. They concluded that interactions in the interlayer spaces between solvent molecules and the strongly charged clay surfaces or highly concentrated cations affect ε of the solvent at the surfaces. This would obscure any possible relationship between interlayer spacing and bulk fluid ε.

Berkheiser and Mortland (1975) determined the basal spacing of montmorillonite in a range of organic solvents. Full- and reduced-charge montmorillonites saturated with Na, Ca, Cu, or tetramethylammonium ions were immersed in a series of solvents. These authors found little correlation between the degree of

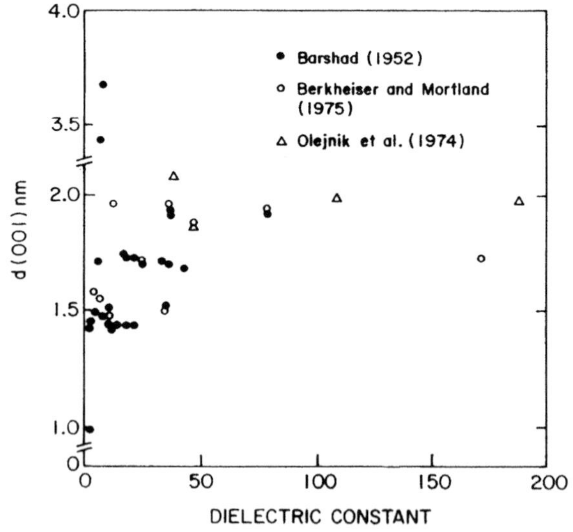

Fig. 7. Summary of results of swelling experiments by several authors. Interlayer d(001) basal spacing of Ca montmorillonite in various organic solvents plotted as a function of the solvent dielectric constant (ε). (Adapted from Murray and Quirk 1982)

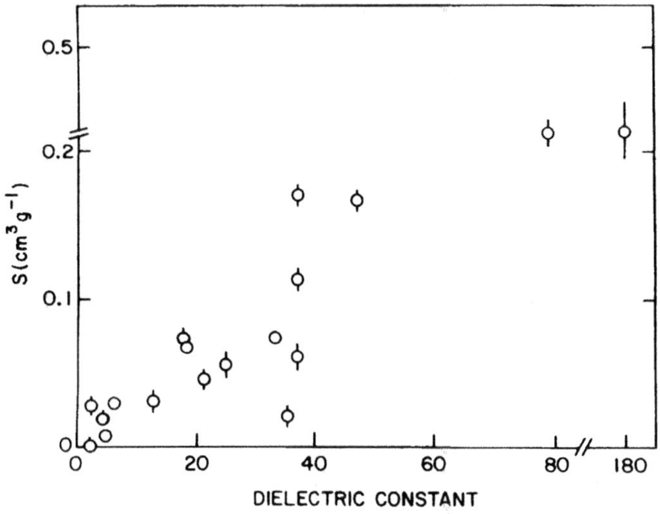

Fig. 8. Bulk swelling (S) of Urrbrae B illite clay in various organic solvents as a function of the solvent dielectric constant (ε).(Adapted from Murray and Quirk 1982)

swelling and such parameters as the Gutman donor number, bulk fluid dielectric constant, bulk surface tension, or solvent dipole moment.

In a review of studies of Ca montmorillonite interlayer basal spacing in organic fluids, Murray and Quirk (1982) recognized that there was little relationship between interlayer spacing and ε (Fig. 7). Bulk swelling did show a linear but poor relationship to ε (Fig. 8). These authors attempted to model bulk swelling as a condensor array, similar to the double-layer treatment of the interlayer region.

Brindley (1980) studied the swelling of montmorillonites in water-dimethylsulfoxide (DMSO) systems. In Ca and Na montmorillonites, the greatest swelling was observed in an intermediate mixture of DMSO (ε = 48) and water (ε= 78), whereas Mg-montmorillonite exhibited the greatest swelling in water. Cs-montmorillonite exhibited little swelling until the mixture contained a threshold percentage of DMSO, after which swelling was uniform with increasing percentage DMSO. Olejnik et al. (1974) as well as Chen et al. (1987) found that maximum swelling for certain clay-organic fluid systems occurred at an intermediate ε.

5 Regular Solution Theory

It is clear from the data above that the DDL theory fails to explain the observed relationship between swelling and solvent properties in concentrated clay-liquid systems. Based on earlier suggestions by MacEwan (1948) and Low and coworkers

(Low and Margheim 1979; Low 1980, 1981; Viani et al. 1983) that solvation of clay surfaces is responsible for swelling in both the crystalline and diffuse regions, Graber and Mingelgrin (1994) examined clay swelling in organic fluids in the context of the theory of regular solution (e.g., Hildebrand et al. 1970). They constructed an alternative to the DDL model by invoking an analogy between the swelling of expandable clays and the swelling of organic polymers in organic fluids.

5.1 Polymer Swelling and Regular Solution Theory

Regular solution theory is a thermodynamic model originally developed to describe miscibility in binary, small-molecule, nonpolar systems (Hildebrand et al. 1970). By assuming that volume change upon mixing (ΔV^M) is zero, Hildebrand developed a simple expression for the molar change in energy upon mixing (ΔE^M), such that

$$\Delta E^M = V_m \phi_1 \phi_2 (\delta_1 - \delta_2)^2, \quad (7)$$

where V_m is the molar volume of the mixture, ϕ_i is the volume fraction of component i, and δ_i its solubility parameter, defined as

$$\delta = \left(\frac{\Delta E^V}{V} \right)^{\frac{1}{2}}, \quad (8)$$

ΔE^V being the molar energy of vaporization of the pure component, and V its molar volume. As the solubility parameters of the two components become closer in value, the interaction energy (e) between the components compensates more fully for the energy loss upon dispersion of the pure components (Barton 1983). Solubility is thus highest when the solubility parameters of the two components are equal.

Despite problems involved with the quantitative application of the regular solution theory (Bagley and Scigliano 1976), it is applied with surprising success, not only to binary mixtures of small molecules, but also to the swelling or dissolution of polymers (Hildebrand et al. 1970; Bagley and Scigliano 1976; Barton 1983). Swelling of macromolecules in non-ionizable solvents results from surface solvation or dissolution of solvent molecules in the solid matrix. The degree of swelling (solubility) depends on the relative magnitudes of the densities of interaction energy (e) between solid-fluid (e_{sf}), solid-solid (e_{ss}), and fluid-fluid (e_{ff}). Equation (7) thus can be formally extended to polymer swelling. Maximum swelling of a polymer should be observed when $\Delta E^M = 0$, namely, the swelling spectrum (ratio of swollen volume to unswollen volume plotted against solvent δ) of a polymer should display a maximum in the solvent which has a δ value most closely matching that of the polymer. Such swelling maxima were indeed observed (Fig. 9; Mark and Tobolsky 1950) and can be used to determine solubility parameters of polymers.

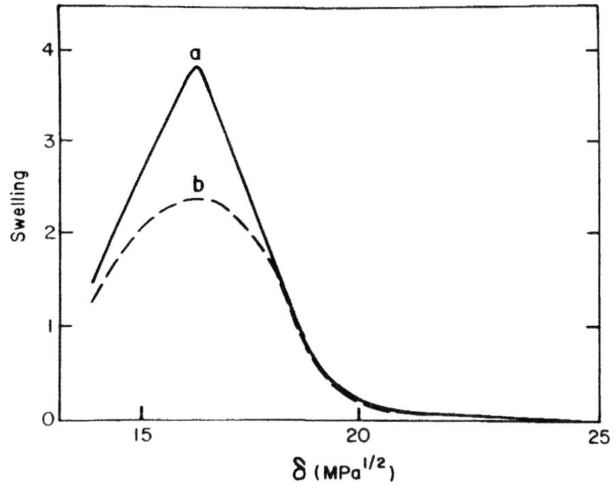

Fig. 9. Swelling (in terms of solvent volume imbibed per unit volume of solid) of natural rubber as a function of solvent solubility parameter (δ). *a* Pure gum; *b* tire tread. (Adapted from Mark and Tobolsky 1950)

5.2 Clay Swelling and Regular Solution Theory

It was suggested by Graber and Mingelgrin (1994) that the swelling of clay minerals can be described by an extension of the regular solution theory as applied to polymers. Clays are composed of repeating aluminosilicate units which commonly form a large, two-dimensional lattice; clay swelling, like that of any polymer, should involve surface solvation or imbibing of the solvent by the polymer. Thus a clay should exhibit a swelling maximum in the solvent whose solubility parameter is nearest to that of the clay. Clays can be expected to differ in their e_{ss} and e_{sf} values and, consequently, in their δ values. A swelling maximum should hence appear in a different solvent for each clay.

To test this theory, swelling and permeability data from the literature were correlated with δ values of the various solvents. For example, Olejnik et al. (1974) reported that dispersion of a Cs montmorillonite occurred in formamide, but displayed only limited crystalline swelling in solvents which retrospectively turned out to have higher or lower δ values (Fig. 10). A maximum in swelling spectrum was also reported by many other authors for different organic liquid-clay systems (e.g., Brindley et al. 1969; Brindley 1980; Brown and Thomas 1987; Figs. 4, 11). Free-swelling volumes from Hettiaratchi and Hrudey (1987) for smectite-bearing soils in organic liquid-water mixtures plotted versus δ also display maxima in mixtures of intermediate δ.

Certain clay systems displayed monotonic increases in swelling with increasing δ, for example, the Na montmorillonite of Olejnik et al. (1974; Fig. 10). If the solubility parameter of the clay is equal to or higher than that of the tested solvent with the highest δ value, a monotonic increase in swelling rather than a maximum will be observed.

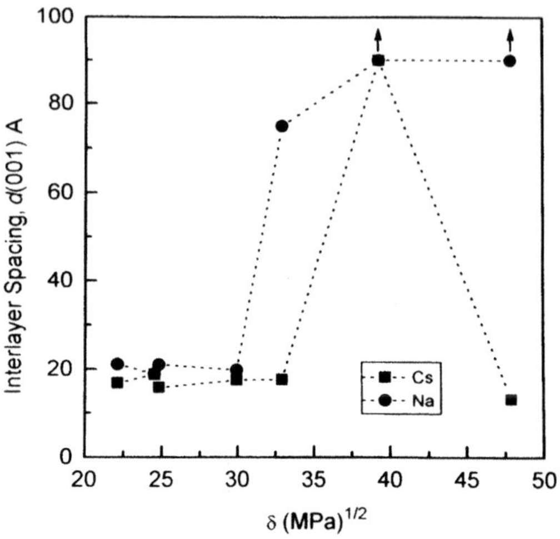

Fig. 10. Interlayer d(001) basal spacing of Cs and Na montmorillonites as a function of solvent solubility parameter (δ). Interlayer basal spacing data from Olejnik et al. (1974)

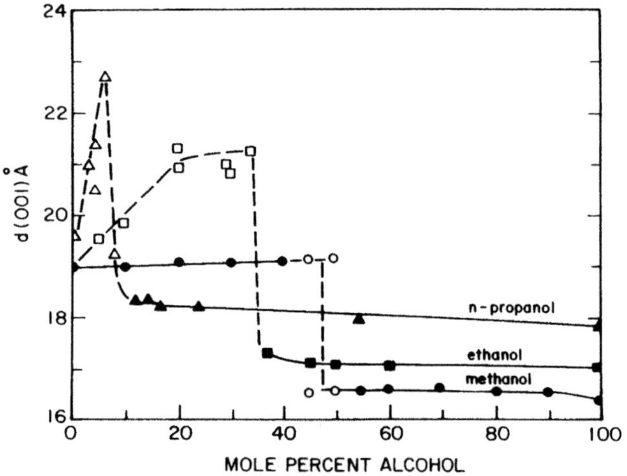

Fig. 11. Interlayer d(001) basal spacing of Ca montmorillonite in equilibrium with different water-alcohol mixtures. *Solid symbols* Average spacing values with deviations from the mean of 0.05 Å or less; *open symbols* average spacing values with deviations from the mean of up to 0.2 Å or more. By converting the mole fraction of the mixture to δ of the mixture, a single broad peak for this clay is obtained at δ between 17 to 25 $(MPa)^{1/2}$ (not shown). (Adapted from Brindley et al. 1969)

A given clay saturated with a cation of equal charge but bigger radius (Cs vs Na) is expected to swell in solvents with lower solubility parameters. Indeed, Cs-montmorillonite displayed maximum swelling in a solvent of lower δ than did Na-montmorillonite (Olejnik et al. 1974; Fig. 10).

Within a clay series, swelling in a given solvent should be highest for the clay with a δ value closest to that of the solvent. For example, in the 2:1 layer silicate mineral series, smectites (layer charge per formula unit 0.25–0.6) display considerable swelling in water, while both the pyrophillite-talc group (layer charge ≈ 0) and mica group (layer charge ≈ 1) do not swell significantly in water. Vermiculites (layer charge 0.6–0.9) exhibit limited swelling in water.

Figure 12A displays the volumetric swelling spectrum of a mixed-layer clay in eleven different solvents with a swelling maximum at an intermediate $\delta \approx 33$ (Graber and Mingelgrin 1994). Figure 12B depicts the same swelling results plotted against fluid dielectric constant. Swelling was very weakly dependent on ε up to $\varepsilon = 110$, followed by a sharp increase at $\varepsilon = 182$ (N-methyl formamide). This relationship between swelling and ε was not as predicted by the DDL theory.

Swelling of Na-montmorillonite increased monotonically with δ, fitting a sigmoid curve and exhibiting greatest swelling in water (Fig. 13A; Graber and Mingelgrin 1994), similar to the results of Olejnik et al. (1974; Fig. 10). The solubility parameters of these clays are apparently larger than or equal to that of water (the solvent with the highest δ of any solvent) and their swelling spectra are necessarily asymmetric. Swelling of montmorillonite plotted versus fluid ε displays a steady increase up to $\varepsilon = 78$ (water), and thereafter a decrease with increasing ε (Fig. 13B; Graber and Mingelgrin 1994) in an obvious departure from the DDL theory.

Cu montmorillonite which underwent spray-drying and thus consisted of highly porous aggregates (Tsvetkov et al. 1993), exhibited an intrinsic permeability maximum in isopropanol and lower permeabilities in mixtures of isopropanol and solvents having both higher and lower solubility parameters (Fig. 14; Graber and

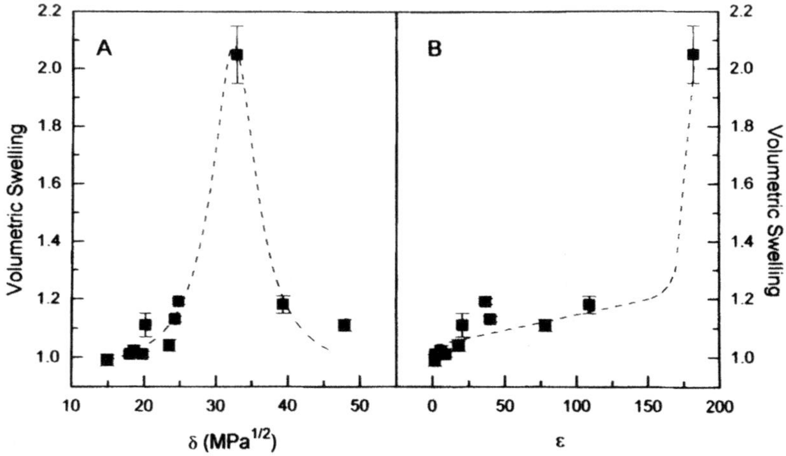

Fig. 12 A,B. Volumetric swelling of a mixed-layer illite-smectite clay in different organic solvents and water as a function of, **A** solvent solubility parameter (δ); **B** solvent dielectric constant (ε). (Graber and Mingelgrin 1994)

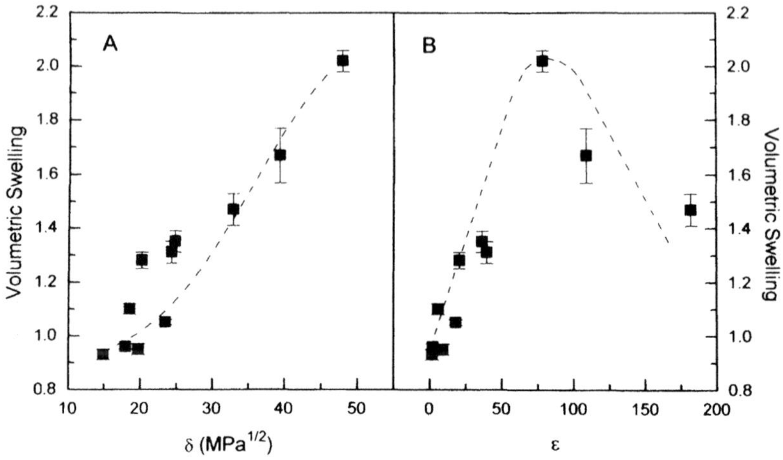

Fig. 13 A.B. Volumetric swelling of a Na montmorillonite in different solvents as a function of, **A** solvent solubility parameter (δ); **B** solvent dielectric constant (ε). (Graber and Mingelgrin 1994)

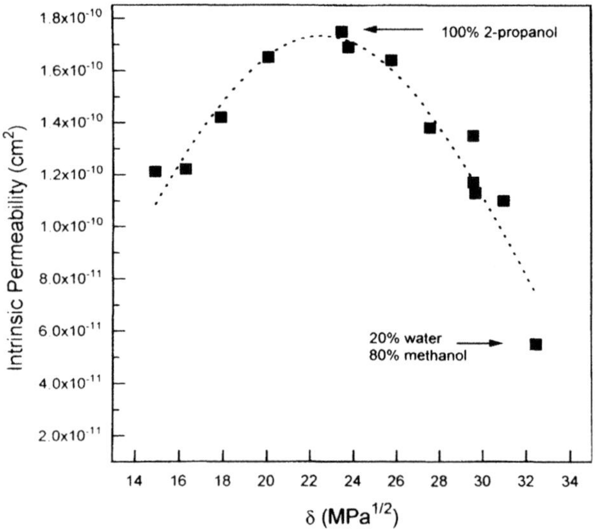

Fig. 14. Intrinsic permeability of spray-dried Cu montmorillonite beads to mixtures of hexane-isopropanol, isopropanol-methanol, and methanol-water as a function of solvent solubility parameter (δ). (Graber and Mingelgrin 1994)

Mingelgrin 1994). A sharp decrease in k was then observed in methanol-water mixtures where water content exceeded 10%. This was interpreted to be the outcome of swelling at edge-to-face bonds in isopropanol with a resultant increase in permeability, followed by swelling at face-to-face bonds in methanol-water mixtures with a resultant decrease in permeability. Graber and Mingelgrin (1994)

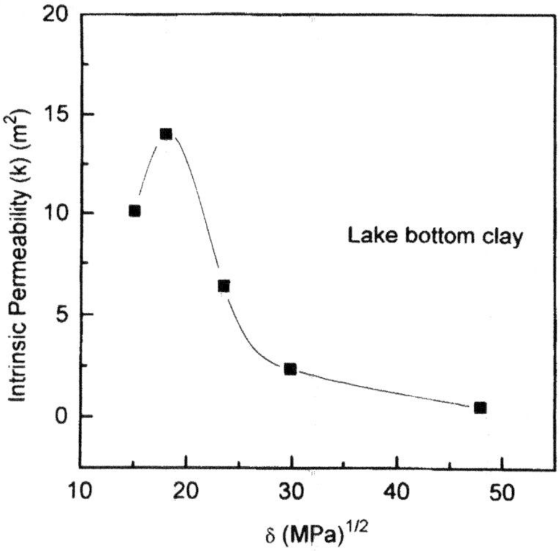

Fig. 15. Intrinsic permeability of lake bottom clay to five solvents as a function of solvent solubility parameter (δ). Intrinsic permeability data from Schramm et al. (1986)

showed that in a fixed volume, permeability can either increase or decrease upon swelling in a medium comprised of two sets of pores. The direction of permeability change will depend upon the geometry and characteristic dimensions of the two pore types. The studied Cu-montmorillonite consisted of porous, quasispherical beads consisting of tactoids (an array of stacked plates) arranged in edge-to-face configurations. Thus two distinct swelling peaks, one at edge-to-face bonds and the second at face-to-face bonds, were anticipated for this system. Multiple swelling maxima were also observed in other clay systems (Berkheiser and Mortland 1975). Similarly, a k maximum in a solvent of relatively low δ (xylene) and a k minimum in water were obtained when the k data of Schramm et al. (1986) were plotted against δ of the tested solvents (Fig. 15). This may also be interpreted to reflect differential swelling at edge-to-face bonds in xylene, versus swelling at face-to-face bonds in water.

6 Conclusions

Models of clay swelling based upon the effect of the pore fluid dielectric constant on the equilibrium distance between clay plates (the DDL approach) do not successfully account for observed expansion and shrinkage in concentrated clay-solvent systems. Swelling of clays in nonionizable solvents can be successfully described in terms of the extended regular solution theory by analogy to the swelling of organic polymers. Swelling maxima are often observed in clay systems immersed in a solvent of intermediate δ value in agreement with that theory.

According to the extended regular solution theory, maximum swelling (solubility) of a polymer should occur when fluid-fluid and solid-solid interaction energy densities are similar to each other, namely, in a solvent whose δ value equals that of the polymer. Accordingly, clays are expected to display swelling maxima in solvents with δ values similar to their own.

Swelling spectra of clay may display multiple peaks, and swelling may result in either an increase or decrease in porous media permeability, depending on the geometry and size of the pores in the particular porous medium and on the orientation of the primary clay particles relative to each other. The often contradictory results obtained in permeability measurements may indeed result from the complex interplay between the opposing effects on permeability of swelling at different types of particle-particle bonds.

Regular solution theory is considerably more successful at explaining experimental observations and predicting swelling in clay systems than is the DDL theory. Regular solution theory is particularly effective in treating nonionizable solvents and concentrated clay-liquid mixtures, as are most environmentally significant solvent-clay systems.

The swelling and shrinkage of clays in contact with organic fluids and concomitant permeability changes can play an important role in organic fluid transport in the subsurface. Demond et al. (1996) recently demonstrated that estimated transport parameters for organic liquids can differ significantly from measured transport parameters as a result of poor estimates for relative permeability. Inadequate permeability estimates may be a direct result of clay volume changes when contact with organic fluids occurs. As such, a better understanding of the mechanism controlling clay volume changes can ultimately aid efforts to describe and simulate organic liquid flow in porous media.

References

Acar YB, Olivieri I (1989) Pore fluid effects on the fabric and hydraulic conductivity of laboratory-compacted clay. Geotechnical engineering. Transport Res Rec 1219:144–159

Acar YB, Seals RK (1984) Clay barrier technology for shallow land waste disposal facilities. Haz Waste 1:167–181

Amoozegar A, Warrick AW, Fuller WH (1986) Movement of selected organic liquids into dry soils Haz Mat Haz Waste 3:29–41

Anderson DC, Brown KW, Thomas JC (1985) Conductivity of compacted clay soils to water and organic liquids. Waste man Res 3:339–349

Aylmore LAG, Quirk JP (1962) The structural status of clay systems. In: Ingerson E (ed) Clays and clay minerals. Earth Science Series, Monograph 11. Pergamon Press, New York, pp 104–130

Bagley EB, Scigliano JM (1976) Polymer solutions. In: Dack MRJ (ed) Solutions and solubilities. Wiley Interscience, New York pp 437–485

Barshad I (1952) Factors affecting the interlayer expansion of vermiculite and montmorillonite with organic substances. Soil Sci Soc Am Proc 16:176–182

Barton AFM (1983) Handbook of solubility parameters and other cohesion parameters. CRC Press, Boca Raton

Bear J (1972) Dynamics of fluids in porous media. Elsevier New York

Berkheiser V, Mortland MM (1975) Variability in exchange ion position in smectite: dependence on interlayer solvent. Clays Clay Miner 23:404–410

Bohn HL, McNeal BL, O'Connor GA (1979) Soil chemistry. John Wiley New York

Bradley WF (1945) Molecular associations between montmorillonite and some poly-functional organic liquids. J Am Chem Soc 67:975–981
Brindley GW (1980) Intracrystalline swelling of montmorillonites in water-dimethylsulfoxide systems. Clays Clay Miner 28:369–372
Brindley GW, Wiewiora K, Wiewiora A (1969) Intracrystalline swelling of montmorillonite in some water-organic mixtures. Clay organic studies, XVII. Am Mineralogist 54:1635–1644
Brown KW, Thomas JC (1984) conductivity of three commercially available clays to petroleum products and organic solvents. Haz Waste 1:545–553
Brown KW, Thomas JC (1987) A mechnaism by which organic liquids increase the hydraulic conductivity of compacted clay materials. Soil Sci Soc Am J 51:1451–1459
Brown KW, Thomas JC, Green JW (1986) Field cell verification of the effects of concentrated organic solvents on the conductivity of compacted soils. Haz Waste Haz Mat 3:1–19
Chen S, Low PF, Cushman JH, Roth CB (1987) Organic compound effects on swelling and flocculation of Upton montmorillonite. Soil Sci Soc Am J 51:1444–1450
Delville A, Laszlo P (1990) The origin of the swelling of clays by water. Langmuir 6:1289–1294
Demond AH, Rathfelder K, Abriola LM (1996) Simulation of organic liquid flow in porous media using estimated and measured transport properties. J Contam Hydol 22:223–239
Emerson WW (1963) The swelling of Na montmorillonite due to water adsorption. Aust J soil Res 1:129–143
Gerstl Z, Galin Ts, Yaron B (1994) Mass flow of a volatile organic liquid mixture in soils. J Environ Qual 23:487–493
Graber ER, Mingelgrin U (1994) Clay swelling and regular solution theory. Environ Sci Technol 28:2360–2365
Green WJ, Lee GF, Jones A (1981) Clay-soils permeability and hazardous waste storage. J Water Pollut Control Fed 53:1347–1354
Hettiaratchi JPA, Hrudey SE (1987) Influence of contaminant organic-water mixtures on shrinkage of impermeable clay soils with regard to hazardous waste landfill liners. Haz Mat Haz Waste 4:377–388
Hildebrand JH, Prausnitz JM, Scott RL (1970) Regular and related solutions. Van Nostrand Reinhold, New York
Israelachvili J, Pashley R (1982) The hydrophobic interaction is long range, decaying exponentially with distance. Nature 300:341–342
Jarsjo J, Destouni G, Yaron B (1997) On the relation between viscosity and hydraulic conductivity values for volatile organic liquid mixtures in soils. J Contam Hydrol 25:113–127
Low PF (1980) The swelling of clay. II. Montmorillonites. Soil Sci Soc Am Proc 44:667–676
Low PF (1981) The swelling of clay. III. Dissociation of exchangeable cations. Soil Sci Soc Am Proc 45:1074–1078
Low PF, Margheim JF (1979) The swelling of clay. I. Basic concepts and empirical equations. Soil Sci Soc Am Proc 43:473–481
MacEwan DMC (1948) complexes of clays with organic compounds. I. Complex formation between montmorillonite and halloysite and certain organic liquids. Trans Faraday Soc 44:349–367
Mark HF, Tobolsky AV (1950) Physical chemistry of high polymeric systems, Chap 8. Interscience Publ New York
Moore CA, Mitchell JK (1974) Electromagnetic forces and soil strength. Geotechnique 24:627–640
Murray RS, Quirk JP (1982) The physical swelling of clays in solvents. Soil Sci Soc Am J 46:865–868
Norrish K (1954) The swelling of montmorillonite. Trans Faraday Soc Lond 18:120–134
Norrish K, Rausell-Colom JA (1961) Low angle x-ray diffraction studies of the swelling of montmorillonite and vermiculite. Clays Clay Miner 10:123–144
Nutting PG (1930) Physical analysis of oil sands. Bull Am Assoc Petroe Geol 14:1337–1349
Olejnik S, Posner AM, Quirk JP (1974) Swelling of montmorillonite in polar organic liquids. Clays Clay Miner 22:361–365
Rowell DL (1965) Influence of positive charge on the inter- and intra-crystalline swelling of oriented aggregates of Na montmorillonite in NaCl solutions. Soil Sci 100:340–347
Schramm M, Warrick AW, Fuller WH (1986) Permeability of soils to four organic liquids and water. Haz Mat Haz Waste 3:21–27

Tsvetkov F, Heller-Kallai L, Mingelgrin U (1993) Potassium halide-treated montmorillonite (KTM) as a solid phase in liquid chromatography. Clays Clay Miner 41:527–536

van Olphen H (1962) Compaction of clay sediments in the range of molecular particle distances. In: Ingerson E (ed) Clays and clay minerals. Earth Science Series, Monograph 11. Pergamon Press, New York pp 178–187

van Olphen H (1963) An introduction to clay colloid chemistry. Interscience Publ New York

Viani BE, Low PF, Roth CB (1983) Direct measurement of the relation between interlayer force and interlayer distance in the swelling of montmorillonite. J Colloid Interface Sci 96:229–244

Viani BE, Roth CB, Low PF (1985) Direct measurement of the relation between swelling pressure and interlayer distance in Li-vermicultie. Clays Clay Miner 33:244–250

5 Sorption-Desorption and Complexation Processes of the Adsorbed and Aqueous Fraction of NAPL: Importance for Complete Cleanup of Aquifers

M. Rebhun[1], C. Rav-Acha[2], I. Sabbah[1], J. Rwetabula[3], and F. DeSmedt[3]

Contents

1 Introduction

In sites contaminated by petroleum products, the contaminant may be present in the following forms: (1) mobile NAPL; (2) immobile NAPL; (3) aqueous phase (dissolved) controlled by aqueous solubility of the NAPL; (4) adsorbed on aquifer solids or soil.

A schematic description of contamination by NAPL and the interactions of the dissolved parts of NAPL (the aqueous phase) is shown in Fig. 1.

After the removal of the nonaqueous liquid phase – a difficult task in itself – the residual contaminant soluble in the aqueous phase will be in equilibrium with the adsorbed part in the solid phase. In groundwater cleanup, the main goal is to return the aquifer to drinking-water quality, which requires concentrations in the range of 10 to 100 mg/l for many organic micropollutants, including petroleum products. The concentration of pollutants in groundwater at such sites is dependent on the their release from the solid phase, determined by adsorption equilibrium and kinetics. The mass of contaminant sorbed in solid form may be greater than that in soluble form. The solid phase acts as a contaminant-releasing reservoir and many cause the water to fail in meeting required standards over a prolonged period.

[1]Technion-Israel Institute of Technology, Haifa 32000, Israel
[2]Ministry of Health, Jerusalem, Israel
[3]Laboratory of Hydrology, Vrije Universiteit, Brussels, Belgium

H. Rubin/N. Narkis/ J. Carberry
Soil and Aquifer Pollution

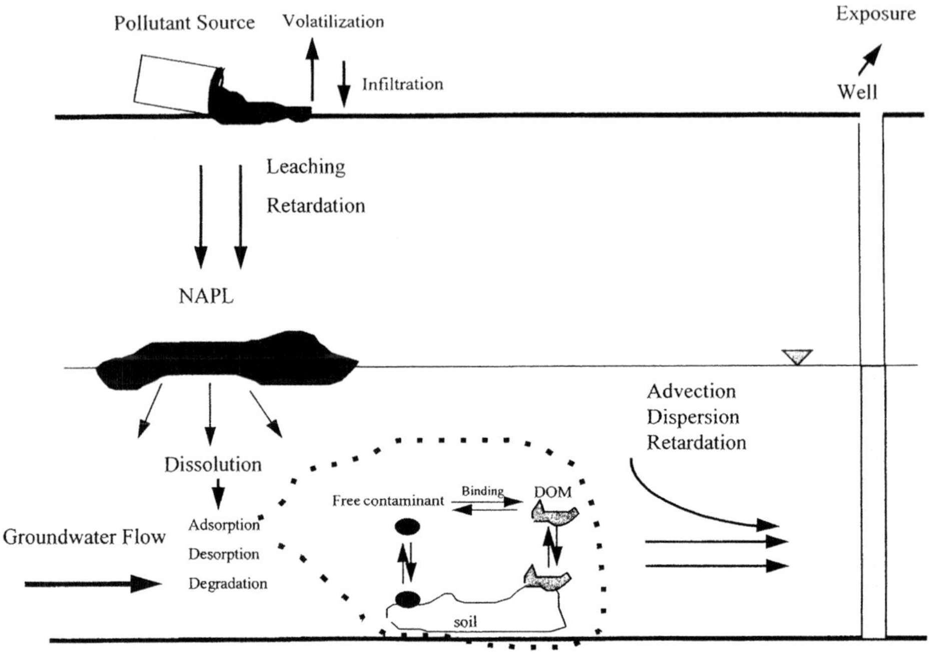

Fig. 1. Schematic of contamination with NAPL and the interactions and transport of the dissolved part (aqueous phase)

The sorption process is responsible for the retardation of pollutants, e.g., reduction in their mobility – usually a positive factor in transport, delaying the arrival of pollutants to extraction wells. In remediation operations, retardation by sorption slows the release-remediation process, thus being a negative, undesirable factor. The stronger the adsorption and, correspondingly, the higher the retardation, the more time is needed for flushing at a constant pumping rate, or, in general terms, the more water is required to accomplish remediation.

In the remediation operations, either by "pump and treat" or by in-situ flushing, the role that sorption and desorption play for the part of contaminant mass that is sorbed to aquifer media is often ignored.

2 Transport of Contaminants

The transport of contaminants is controlled by the following processes; (1) advection; (2) dispersion and reaction, which could be sinks (adsorption, precipitation, decomposition); or (3) sources [generation-desorption, dissolution, bio (oxidation)]. The following is a mathematical expression of transport presented by Bear, Freeze, and Cherry:

$$-V\frac{\partial c}{\partial x} + D\frac{\partial^2 c}{\partial x^2} - R = \frac{\partial c}{\partial t}, \qquad (1)$$

where R = rate of processes and $R = R_{ads.} + R_{dec.} - R_{gen.}$

3 Adsorption

For most soil-water systems linear adsorption can be assumed:

$$q = K_p \cdot c, \qquad (2)$$

where q = concentration of sorbed solute per mass of soil (mass/mass), c = concentration in water at equilibrium (mass/volume), and K_p = adsorption, (sorption), partition constant (volume/mass sorbent).

Adsorption results in the retardation of the sorbing contaminant relative to the movement of water or a conservative (nonsorbing) solute. The following is a simplified mathematical expression of transport with adsorption, assuming rapid and linear adsorption and neglecting dispersion:

$$R = 1 + \frac{\rho}{\varepsilon} K_p = v_r^{1-} \qquad (3)$$

where R is defined as the retardation factor, R is relative residence time and repesents pore volumes to breakthrough of the contaminant, and v_r is the velocity

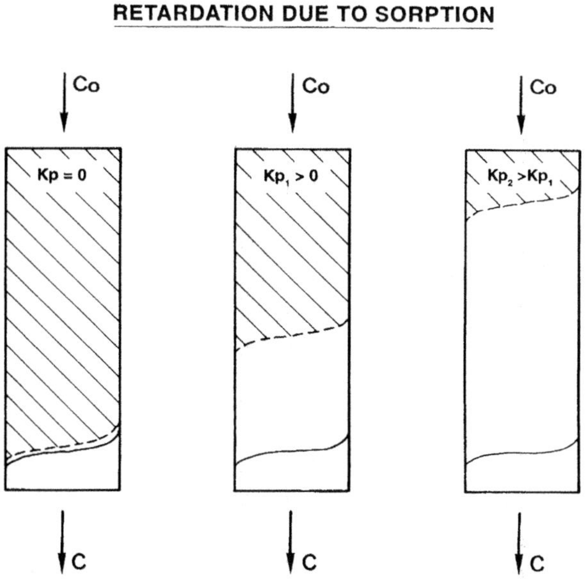

Fig. 2. Retardation due to sorption-concentration profiles of sorbing and nonsorbing solutes

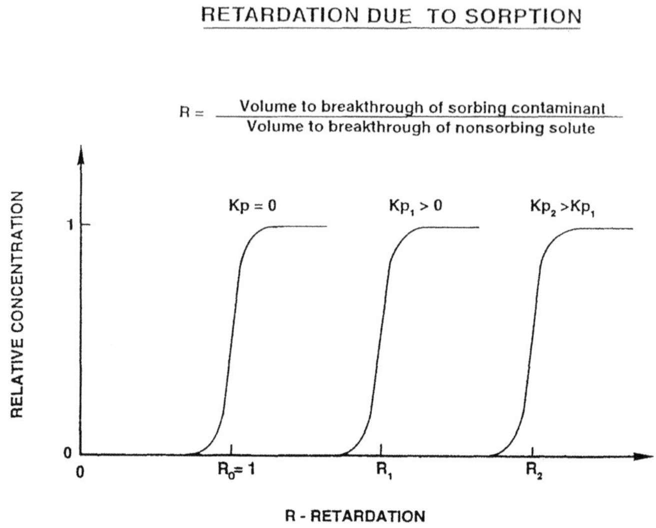

Fig. 3. Breakthrough of nonsorbing and sorbing solutes

of sorbing solute/velocity of water. Figures 2 and 3 show the effect of sorption on retardation.

Adsorption on Mineral-Humic Solids. Soil organic matter is often considered to be the main sorbent. Sorption is viewed as a partitioning of the organic pollutant between the organic (humic) in the solid phase and in the aqueous phase:

$$K_p = \frac{q}{c} \tag{4}$$

$$K_p = f_{oc} \cdot K_{oc}, \tag{5}$$

where q = concentration of sorbed solute per mass of soil, c = concentration of solute in the aqueous phase, f_{oc} = fraction of solid or organic (humic) matter in the soil (in the solid phase) and K_{oc} = partition constant per mass of soil organic (humic), solid phase organic matter.

Sorption of micropollutants onto particles and solids in the aqueous environment is a major process determining the fate and transport of these constituents. The organic fraction, particularly humic substances, of the solid phase (adsorbent) has been considered by many as the major, and often as the sole, adsorbing factor, neglecting the sorption on mineral surfaces.

In many aquatic systems, such as groundwater, the aquifer soil may be low in humics (organics) but may contain a considerable fraction of sorbing clay minerals. This may also be true concerning the solid particulates in some surface waters. Thus, the adsorbing role of the mineral surfaces should also be considered. In such cases, the following equation should be used to account for sorption on the solid organic phase and adsorption on mineral surfaces:

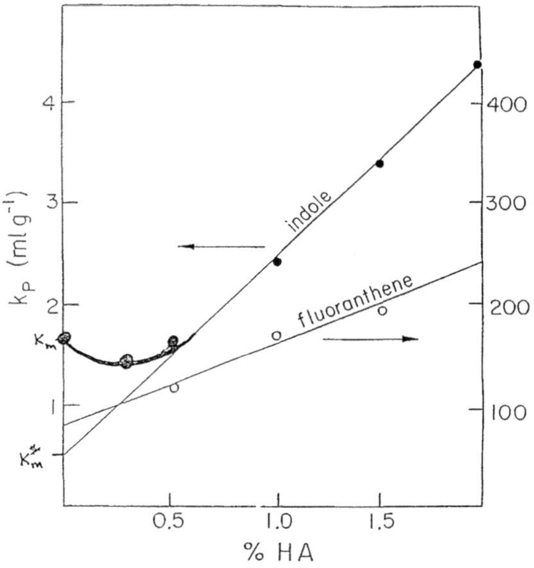

Fig. 4. Adsorption (partition) constants K_p vs humic acid content of soil

$$K_p = K_m f_m + K_{oc} f_{oc}, \tag{6}$$

where K_p = overall sorption constant, K_m = adsorption constant for clay-mineral, and f_m = clay-mineral fraction in soil.

In Fig. 4 the role of both mineral and organic solid fractions are shown (Rebhun et al., 1992) For simplicity of presentation, in present chapter the solid organic fraction (OC) will be considered as the only sorbent.

4 Partition-Sorption Relationships

From hydrophobicity and lipophilicity considerations, a linear free-energy relationship can be developed:

$$\log K_p = a + b \log K_{rs} \text{ and} \tag{7}$$

$$b = \Delta G_{ws} / \Delta G_{wr}, \tag{8}$$

where K_{rs} = partition constant water/reference solvent; a,b = constants; and ΔG_{ws} and ΔG_{wr} = free energy change in passing to a given solvent (ws) and to a reference solvent (wr), respectively.

Octanol is a common reference solvent, thus:

$$\log K_p = a + b \cdot \log K_{ow}, \tag{9}$$

where K_{ow} = octanol/water partition constant and K_p in our case is indeed – K_{oc} – partition constant onto the organic fraction of the soil (solid).

For a given contaminant characterized by its K_{ow} and a soil with an organic fraction f_{oc}, the adsorption (partitioning) constant is:

$$K_p = f_{oc}\,(10^a K_{ow}^b). \tag{10}$$

For remedial desorption we will reduce K_p.

5 Desorption-Remediation

As stated previously, retardation is a positive factor in transport, significantly delaying the arrival of pollutants to extraction wells. In remediation operations, retardation by sorption slows the release-remediation process, thus being an undesirable factor. The greater the retardation (the stronger the sorption), the more time is needed for flushing out at a constant pumping rate or the more water is needed to accompplish the operation.

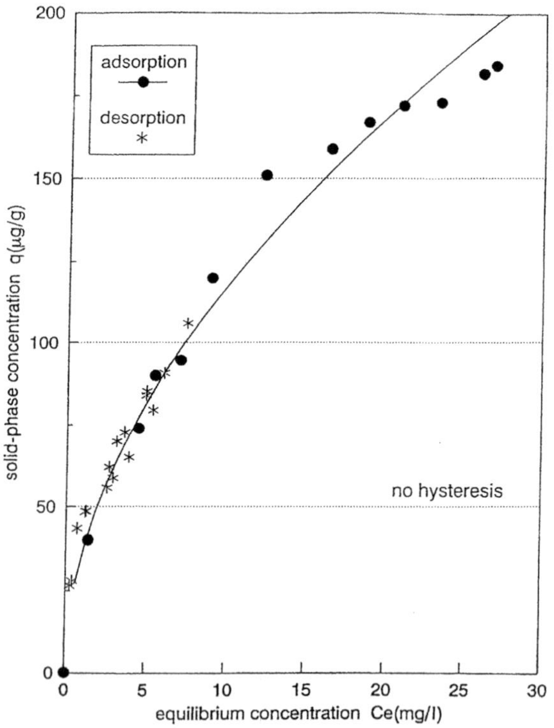

Fig. 5. Adsorption-desorption isotherms on pure clay

5.1 Reversibility and Hysteresis

In some cases, sorption is reversible but still requires large volumes of water for desorption-remediation governed by equilibrium considerations. In other cases (other soils or aquifer media), hysteresis has been observed. Both cases are shown in Figs. 5 and 6 and have been discussed in detail by Sabbah and Rebhun (1996).

5.2 Enhancement of Desorption-Remediation by Aqueous Phase (Dissolved) Humic Substances

Humic substances hydrophobically bind organic microcontaminants, forming aqueous phase bound complexes. If the bound form is not adsorbable (or only slightly adsorbable), then organic contaminant binding to dissolved humics will reduce the adsorption constant from that of the unbound form, enhance decontamination and improve soil flushing. The role of humic substances on adsorption and desorption in aquatic systems is shown in Fig. 7:

Major modes of interaction between organic contaminants and solid-and aqueous-phase humic substances have been described by Rav-Acha and Rebhun (1992; Fig. 8), in which they developed and discussed the equilibrium and quantitative mass balance relationship of binding complexation. The effect of binding complexation by dissolved humic substances (DHS) is a reduction in the adsorption constant as shown below.

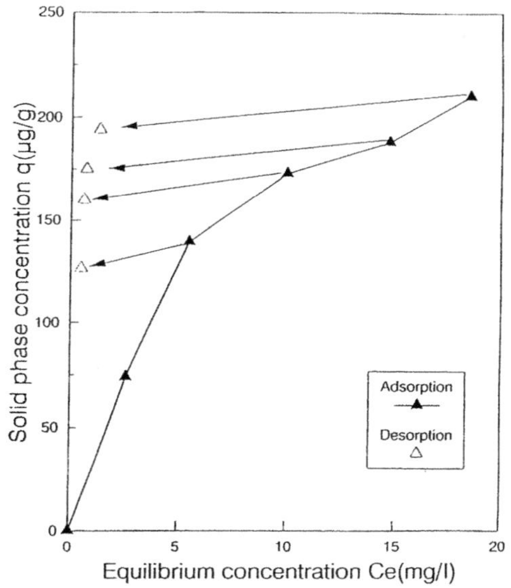

Fig. 6. Desorption hysteresis of TCP on natural soil #1

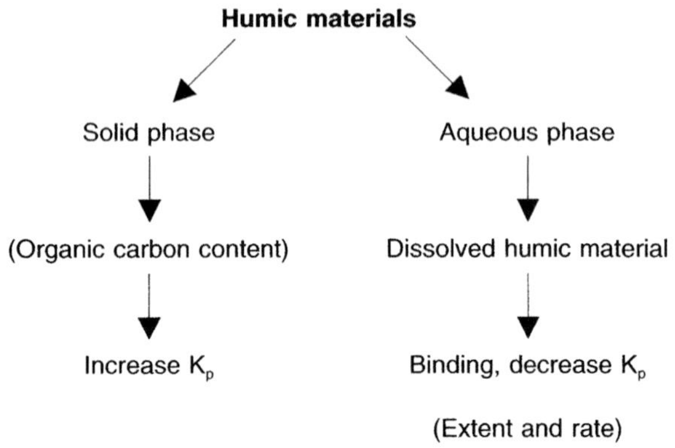

Fig. 7. The role of humic materials in aquatic systems

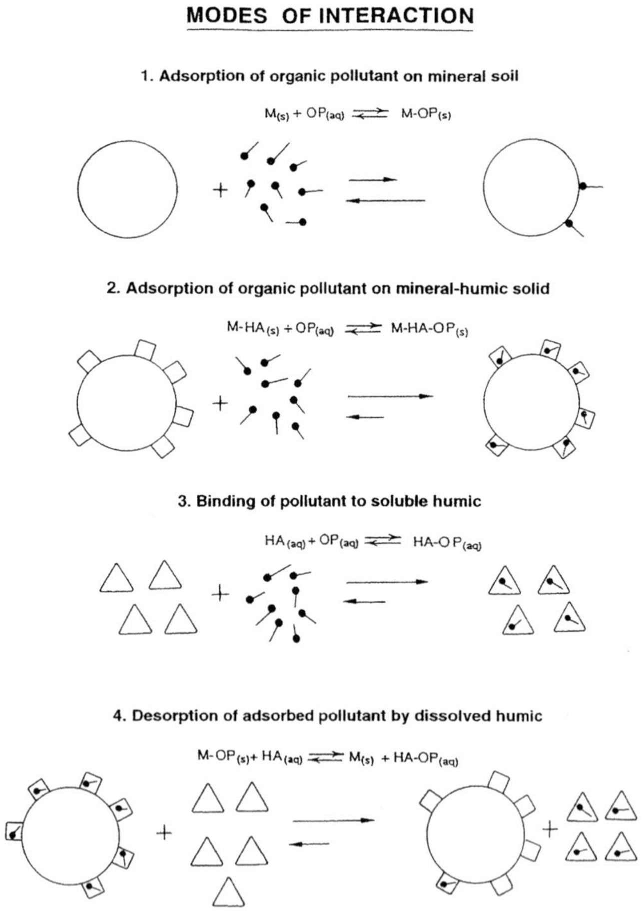

Fig. 8. Modes of interactions between organic solutes (organic pollutants) and humics in solid and aqueous phases

Apparent partitioning constant K_p^* with complexation by DHS:

$$K_p^* = \frac{K_p}{1+\overline{K}[HA]} < K_p, \qquad (11)$$

where K_p^* = the sorption constant including binding complexation by DHS; [HA] = concentration of humic substance; and K_b = the bind-complexation constant to DHS.

Binding-complexation relationships can be derived from hydrophobicity and free-energy relationships, similar to partition-sorption relationships:

$$\log K_b = a' + b' \cdot \log K_{ow}, \qquad (12)$$

where the constants for binding complexation to DHS are a′ = 1.85 and b′ = 0.72. Figure 9 shows such a relationship.

Combining the adsorption and binding-complexation equations and inserting the equation constants and their correlations with K_{ow}, the following set of equations can be solved simultaneously and can predict sorption-desorption and retardation at various scenarios:

$$K_p^* = \frac{K_p}{1+\overline{K}[HA]}, \qquad (13)$$

where $K_p = f_{oc} \cdot K_{oc}$; $K_{oc} = 10^a \cdot k_{ow}^b$; and $K_b = 10^{a'} \cdot K_{ow}^{b'}$

For most organic microcontaminants, K_{ow} is given in tables in the literature; a,b values are also available (Karickhoff et al. 1979; Schwarzenbach and Westall 1981); and a′, b′ are available from this study with some data from the literature. However, more work is needed to establish valid relationships.

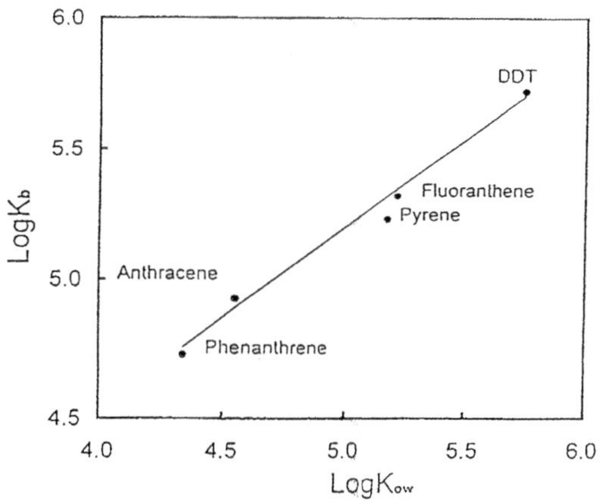

Fig. 9. Correlation between logarithms of binding constants and octanol-water partition constants. (Data from Rav-Acha and Rebhun 1992)

The retardation factor R* as affected by binding complexation of the organic micropollutants by dissolved humic substances (DHS) can be expressed as follows:

$$R^{*}1+\frac{\rho}{\varepsilon}K_{p}^{*}. \tag{14}$$

For estimation of the binding effect of DHS on sorption and retardation, the following ratio may be used:

$$\frac{R^{*}}{R}=\frac{1+\frac{\rho}{\varepsilon}K_{p}^{*}}{1+\frac{\rho}{\varepsilon}K_{p}}. \tag{15}$$

Since in most cases $K_p \gg 1$ and $K_p^* \gg 1$, Eq. (15) can be simplified to:

$$\frac{R^{*}}{R}\cong\frac{K_{p}^{*}}{K_{p}}=\frac{1}{1+K_{b}[HA]}. \tag{16}$$

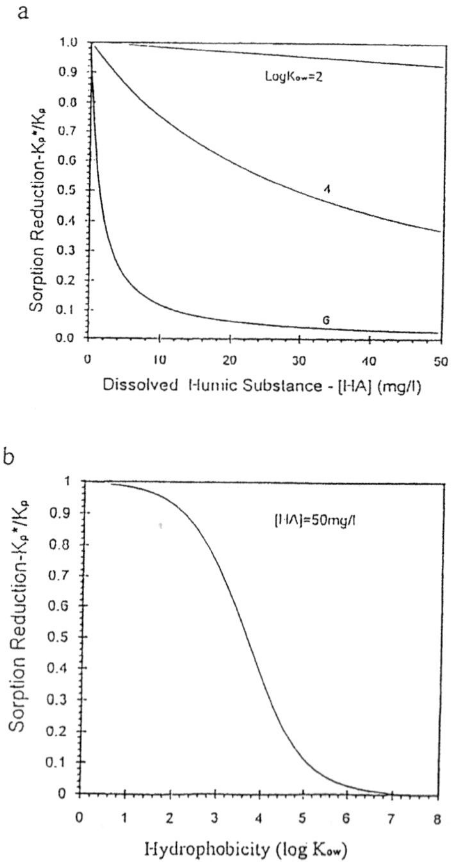

Fig. 10a,b. Sorption reduction by DHS binding of organic solutes. **a** Effect of HDS concentration; **b** effect of solutes hydrophobicity (log K_{ow})

Fig. 11. **a** Breakthrough curves for contaminants of different sorptivity; **b** elution curves for contaminants of different sorptivity

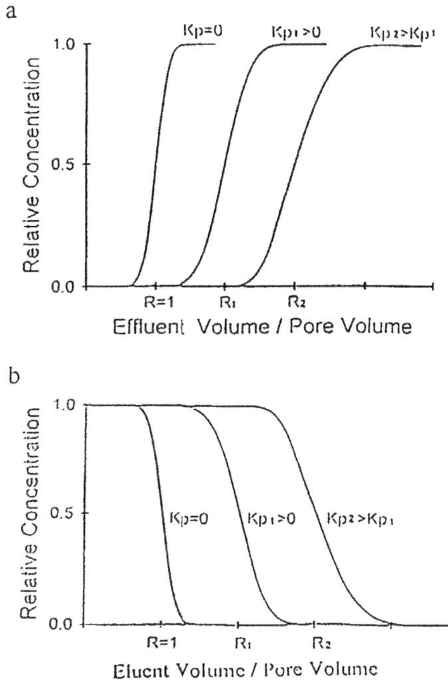

Reduction in adsorption and, consequently, in retardation by binding complexation to DHS is shown in Fig. 10 (Rebhum et al. 1996). The reductions depend on the hydrophobicity of the contaminant and on the DHS concentration. The breakthrough curves in Fig. 11a explain the effect of HDS causing "faster" movement (less retardation) and earlier arrival of contaminants to wells because of enhanced transport – a negative factor.

In remediation operations, the effect of DHS (Fig. 11b), reducing retardation, results in "faster" remediation – a positive factor.

6 Summary

Dissolved humic substances bind many organic solutes through hydrophobic interactions, forming aqueous humic-solute complexes. This phenomenon lowers the concentration of free solute, reducing its adsorption. Thus, a solution of dissolved humic substances can enhance desorption, the flushing-out of adsorbed contaminants, and the remediation of contaminated sites and aquifer materials. Models combining transport, desorption, and, binding processes were used to compute and predict desorption-remediation by dissolved humic substances. Both, the adsorption-desorption constants to solid-phase humic and the binding constant to aqueous phase humics of the contaminant were found in respective correlations to octanol-water partition of the contaminants. Equilibrium situations were considered.

References

Karickhoff W, Brown S, Scott TA (1979) Sorption of hydrophobic pollutants on natural sediments. Water Res 13:214–248

Rav-Acha Ch, Rebhum M (1992) Binding of organic solutes to dissolved humic substances and its effects on adsorption and transport in the aquatic environment. Water Res 26:1645–1654

Rebhun M, Kalabo R, Grossman L, Manka J, Rav-Acha Ch (1992) Sorption of organics on clay and clay-humic complexes simulating aquifer processes. Water Res 26:79–84

Rebhun M, DeSmedt F, Rwetabula J (1996) Dissolved humic substances for remediation of contaminated sites. Binding-desorption model predictions. Water Res 30(9):2027–2038

Sabbah I, Rebhun M (1996) Adsorption-desorption of trichlorophenol in water-soil systems reltated to groundwater contamination and remediation. Environ Water Res (in press)

Schwarzenbach R, Westall J (1981) Transport of nonpolar organic compounds from surface water to groundwater. Environ Sci Technol 15:1360–1367

6 Estimating Mass Transfer of Polyaromatic Hydrocarbons from Coal Tar-Contaminated Soil

L. S. Lee[1], N. D. Priddy[1], and D. C. M. Augustijn[2]

Contents

1 Introduction

Many industrial waste disposal/spill sites are characterized by the presence of multicomponent nonaqueous phase liquids (NAPLs) such as gasoline, diesel fuel, oils, and coal tar. The manufacture of gas from coal and oil from the late 1800s to the early 1950s left several manufactured-gas plant (MGP) sites contaminated with coal-tar wastes. These wastes were commonly left on-site in pits and containers, placed in nearby ponds or lagoons, or taken to off-site areas for land disposal (EPRI 1992; Lee et al. 1992b; Luthy et al. 1994). Constituents of particular interest in coal tar include several polyaromatic hydrocarbons (PAHs) currently on the EPA's priority pollutants list. The threat of the continued release of these organic chemicals to groundwater is resulting in the removal of pure coal tar sources. Given source removal, focus turns to the large amounts of soil that have been contaminated by direct contact with a NAPL or through years of contact with a contaminated aqueous phase.

To aid in making environmentally sound and cost-effective decisions regarding NAPL-contaminated soils, such as the coal tar-contaminated soils found at old MGP sites, reasonable estimates of equilibrium partition coefficients and

[1]Department of Agronomy, Purdue University, 1150 Lilly Hall, West Lafayette, Indiana 47907, USA
[2]Civil Engineering & Management, University of Twente, P. O. Box 217, 7500AE, Enschede, The Netherlands

H. Rubin/N. Narkis/ J. Carberry
Soil and Aquifer Pollution

desorption rates of contaminants are needed. Overall partitioning of contaminants and their release rates from a soil-NAPL matrix to the solution phase will determine the potential for contamination of water sources downfield and the types of remediation that may be successful. Therefore, properly quantifying the distribution of contaminants under equilibrium conditions and assesssing mass-transfer constraints is imperative. This chapter will highlight both dissolution and desorption processes for multicomponent NAPLs and NAPL-contaminated soils with an emphasis on assessing equilibrium and rate-limited desorption characteristics of soils contaminated with coal-tar residuals.

2 Dissolution of Multicomponent NAPLs

Dissolution is defined as the transfer of chemicals from their original phase to a dissolved stage in the contacting solution. This process includes diffusion of the constituents through the NAPL phase, dissolution at the interface, and diffusion away from the NAPL-water interface into the bulk water phase. The rate of dissolution will be determined by the slowest of these three steps. When the NAPL consists of only organic compounds with significant aqueous solubilities (e.g., gasoline), the ganglia will shrink as the dissolution process proceeds. As the ganglia shrink, the interfacial area between NAPL and water becomes smaller, reducing the dissolution rate (Hunt et al. 1988). On the other hand, the presence of a NAPL phase often creates regions of immobile water, but with shrinking ganglia, flow can be established, thereby increasing the effective surface area between the NAPL phase and the mobile water. This may also result in a redistribution of the NAPL phase, which is accompanied by a change in contact area. The contact area between the solution and NAPL phases depends on the spatial distribution of the phases, which is usually difficult to assess, thus further increasing the complexity of modeling dissolution of NAPLs in the environment (Chatzis et al. 1983; Schwille 1988; Wilson et al. 1990; Powers et al. 1992).

For a single-component NAPL, the availability of the compound for dissolution at the NAPL-water interface and diffusion within the organic phase is not limited. Therefore, the dissolution rate is primarily a function of the effective surface area of the NAPL (i.e., NAPL surface in contact with mobile water). In contrast, for multicomponent NAPLs, the constituents first have to diffuse through the organic phase toward the NAPL-water interface under influence of a concentration gradient before dissolution can take place. In this case, the dissolution rate also depends on solute diffusivity through the NAPL and the diffusion path length. For a given surface area, the diffusion path length for the constituents increases with increasing volume of NAPL, consequently, the dissolution rate decreases with decreasing surface to volume ratio of the NAPL phase.

For complex mixtures consisting of a large portion of essentially insoluble compounds, here referred to as "pitch", changes in size and shape of the ganglia can be ignored. When a multicomponent NAPL dissolves, components will be removed at different rates, thus changing the composition and consistency of the

NAPL, which may influence the molecular diffusion of the remaining constituents through the organic phase. Luthy et al. (1993) observed the formation of a semigelatinous film at the coal-tar water interface during aging under quiescent conditions. Nelson et al. (1996) recently reported that this film appeared to be similar in composition to the bulk tar, but enriched with water. The presence of this film will result in an increased interfacial resistance, thus decreasing mass-transfer rates across the NAPL-water interface (Luthy et al. 1993). The complexity of the processes described above has led to the use of simple first-order mass-transfer models in which all spatial and temporal complexity is contained in the first-order mass-transfer rate coefficient.

2.1 Tar-Water Partitioning: Equilibrium Batch Studies

The equilibrium partitioning of a chemical between water and a multicomponent NAPL can be described by the following equation:

$$C_{w,i} = \gamma_{n,i}\, X_{n,i}\, S_{l,i}, \tag{1}$$

where $C_{w,i}$ (mol/l) is the concentration in water; $\gamma_{n,i}$ the activity coefficient in the NAPL phase; $X_{n,i}$ the mole fraction in the NAPL phase; and $S_{l,i}$ (mol/l) the aqueous solubility of the liquid solute i (hypothetical supercooled liquid solubility for crystalline solutes) (Lee et al. 1992a,b). If the composition of the organic mixture is known or can be estimated, the activity coefficients in the NAPL phase can be estimated using various computational schemes, of which the most common is the UNIFAC model (UNIQUAC – Functional-Group Activity Coefficient) proposed by Fredenslund et al. (1975). Several researchers (Cline et al. 1991; Lee et al. 1992a,b; Chen 1993), however, demonstrated that the activity coefficients for many complex waste mixtures can be assumed unity. Under ideal conditions (i.e., $\gamma_{n,i} = 1$), Eq. (1) reduces to Raoult's law:

$$C_{w,i} = X_{n,i}\, S_{l,i}. \tag{2}$$

In assessing the application of Raoult's law to compositionally complex mixtures, it is expedient to use the partition coefficients for the components between the organic and water phases. The use of partition coefficients facilitates examining the ideality of the entire system in a single practical representation. The NAPL-water distribution coefficient (K_{nw}) is defined by the ratio of the chemical concentrations in the NAPL and water phases,

$$K_{nw} = \frac{C_{n,i}}{C_{w,i}}, \tag{3}$$

where $C_{n,i}$ is the concentration (mol/l) of the chemical of interest in the NAPL phase. Substituting the expression derived from Raoult's law [Eq. (2)] for $C_{w,i}$ gives

$$K_{nw} = \frac{C_{n,i}}{X_{n,i} S_{l,i}}. \tag{4}$$

Direct measurement of $X_{n,i}$ for complex wastes is often not feasible because many of the waste components may not be easily quantifiable or not quantifiable at all with current techniques. For mixtures comprised of a large number of constituents, each contributing a small fraction to the total, $X_{n,i}/C_{n,i}$ can be estimated by the molar volume of the NAPL phase (EPRI 1992; Lee et al.1992a). The molar volume (V_n, l/mol) can be approximated by the ratio of the average molecular weight (MW_n, g/mol) and density (ρ_n, g/l) of the NAPL phase (i.e, MW_n/ρ_n). K_{nw} can now be estimated by

$$K_{nw} = \frac{(\rho_n/MW_n)}{S_{l,i}}. \tag{5}$$

Lee et al. (1992b) were able to apply this relationship successfully within a factor of two for the distribution of several PAHs between water and coal tars from eight different MGP sites. In Fig. 1 a sample of the data presented by Lee et al. (1992b) using coal tar ID#4 and the line predicted from Eq. (5), which assumes ideal behavior (i.e., no specific interactions between molecules in the NAPL phase; $\gamma_{n,i} = 1$), is shown. If the latter assumption were invalid, Eq. (5) could be written to include nonideal behavior in the NAPL phase:

$$K_{nw} = \frac{(\rho_n/MW_n)}{\gamma_{n,i} S_{l,i}}. \tag{6}$$

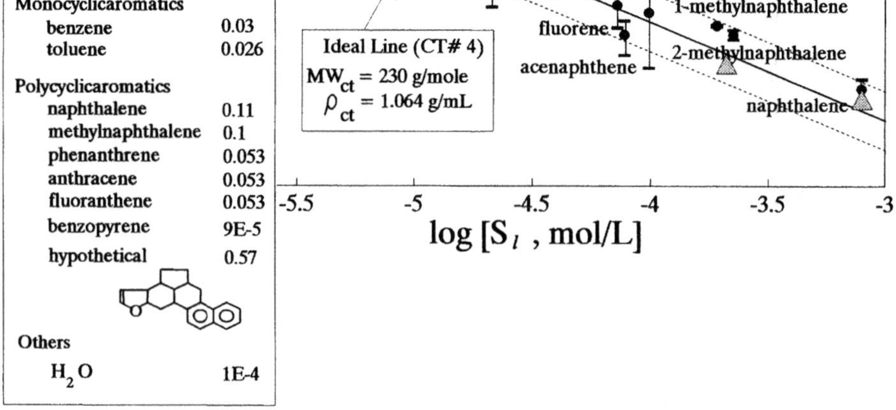

Fig. 1. Measured tar-water partition coefficients (K_{nw}) for coal tar ID #4 compared with predictions based on Raoult's law and using the UNIFAC model to estimate the activity coefficients in the NAPL phase

Also shown in Fig. 1 are log K_{nw} values calculated using the UNIFAC model to estimate $\gamma_{n,i}$. Since only about 40% of the composition of this coal tar was known, the remaining unknown fraction was represented by a single hypothetical compound (as noted in Fig. 1) derived from the information known for this coal tar (i.e., elemental composition, average molecular weight, moisture content, and mass fractions of the PAHs quantified). The UNIFAC simulations suggest that nonidealities are indeed small and that Raoult's law approximation was justified. The coal tars investigated by Lee et al. (1992b) were primarily nonpolar in nature; however, for coal tars that contain significant quantities of polar constituents, deviations from ideal behavior may be noted (i.e., $\gamma_{n,i} > 1$).

2.2 Dissolution Under Flow Conditions

A majority of the information available on the dynamic dissolution of NAPLs is from single-component studies; however, some researchers have studied the more complex problem of dissolution of multicomponent NAPLs under flow conditions (Abriola and Pinder 1985; Corapcioglu and Baehr 1987; Borden and Kao 1989; Kaluarachchi and Parker 1990; Mackay et al. 1991; Zaladis et al. 1991; Borden and Piwoni 1992; Geller and Hunt 1993; Priddle and MacQuarrie 1994). In many of these studies, multicomponent dissolution was described by a constant partition coefficient between the NAPL and water, as described in the previous section for closed batch systems. However, under flow conditions, the continued release of the solubilized components affects the composition of the NAPL, and thus of the partition coefficient [see Eq. (4)].

Augustijn (1993) studied dissolution of PAHs from coal tar-coated sand using a miscible displacement technique. Elution profiles of several PAHs were modeled using an advective-dispersive transport equation modified to account for changing mole fractions in the NAPL. The coal tar used in this study was one of the eight coal tars used by Lee et al. (1992b) identified as ID#4. A few drops of liquid coal tar ID#4 (equal to 0.176 g) were added to 120 g of the 0.3–0.5 mm fraction of builder's sand. The tar and sand were thoroughly stirred until the coal tar was homogeneously distributed on the sand grains. Imaging with an optical microscope revealed that the coal tar was smeared into thin films surrounding the sand grains with some accumulation in the crevices on the mineral surfaces. After stirring the tar and sand, the mixture was immediately packed into a glass column from Kontes (2.5 cm i.d. × 15 cm) and slowly saturated with a degassed 0.01 N $CaCl_2$ solution. The attraction forces between the tar and quartz were strong enough so that the tar was not displaced by water. When the first effluent appeared at the outlet of the column, the flow was stopped. After the column was initially equilibrated for a day, flow was resumed at a pore-water velocity of 15 cm/h. Effluent samples were collected periodically and analyzed using a gas chromatograph (GC) with a ion-trap detector.

Elution profiles of PAHs from the coal tar-coated sand were modeled using an advective-dispersive transport equation for one-dimensional steady-state flow through a saturated homogeneous medium that was modified to account for changing mole fractions in the NAPL:

$$\theta \frac{\partial C_{w,i}}{\partial t} + \frac{MW_i}{S_{l,i}} \frac{\partial (N_o C_{w,i})}{\partial t} = \theta D_h \frac{\partial^2 C_{w,i}}{\partial x^2} - q \frac{\partial C_{w,i}}{\partial x} \tag{7}$$

$$C_{w,i} = \frac{m_i}{N_o} \frac{S_{l,i}}{MW_i} \tag{8}$$

$$N_o = \frac{m_p}{MW_p} + \sum \frac{m_i}{MW_i}, \tag{9}$$

where D_h (cm^2/h) is the hydrodynamic dispersion coefficient, θ (cm^3/cm^3) is the volumetric water content, t (h) is time, x (cm) is distance, q (cm/h) is the Darcy flux, N_o (mol/cm^3) is the total number of moles in the organic phase, m_i and m_p are the masses of solute i and the pitch in the soil (g/cm^3), respectively, and $C_{w,i}$ and $S_{l,i}$ are now expressed in units of g/cm^3. Note that Eq. (8) was derived from Raoult's law [Eq. (2)].

For convenience, the system was assumed to be initially at equilibrium, thus Eq. (8) describes the initial condition and all contaminants were assumed to reside in the NAPL phase coatings. For the inlet boundary ($x = 0$), it is assumed that at $t = 0$ a solution free of contaminants is introduced. This means that the flux-averaged concentration in the column inlet is equal to zero. The outlet boundary condition ($x = L$) for a finite medium is $\partial C_{n,i}/\partial x = 0$ at $t > 0$. The Crank-Nicolson central finite difference method was used to solve the transport equation with an iteration procedure to calculate the number of moles remaining in the NAPL phase.

For simplification, the coal tar was assumed to consist of only ten soluble compounds, whereas the remainder of the tar (80% on mass basis) was considered pitch with an average molecular weight of 260 g/mol. The experimental data and independent model predictions for six different constituents are shown in Fig. 2. Initially the concentration for each component is constant until at least one of the constituents is depleted; the concentration of the remaining components is then raised to a new constant level in response to the changing mole fractions. Considering the high level of uncertainty in the input data and model assumptions, the simulations and experimental data were in good agreement.

The initial effluent concentration for each compound was also in good agreement with the data collected by Lee et al. (1992b) in a batch equilibration study with the same coal tar. The assumption of the equilibrium dissolution of PAHs was reasonable because the contact area between the tar and solution phase was large; the coal tar coated as a thin film the mineral surfaces. Given the density of 1.06 g/cm^3 for the tar, and assuming that the sand grains can be approximated by spheres with an average diameter of 0.4 mm and a density of 2.65 g/cm^3, the average thickness of the tar film was estimated to be approximately 250 nm. A film thickness several hundred times more than the diameter of the soluble constituents caused apparently no significant mass-transfer constraints.

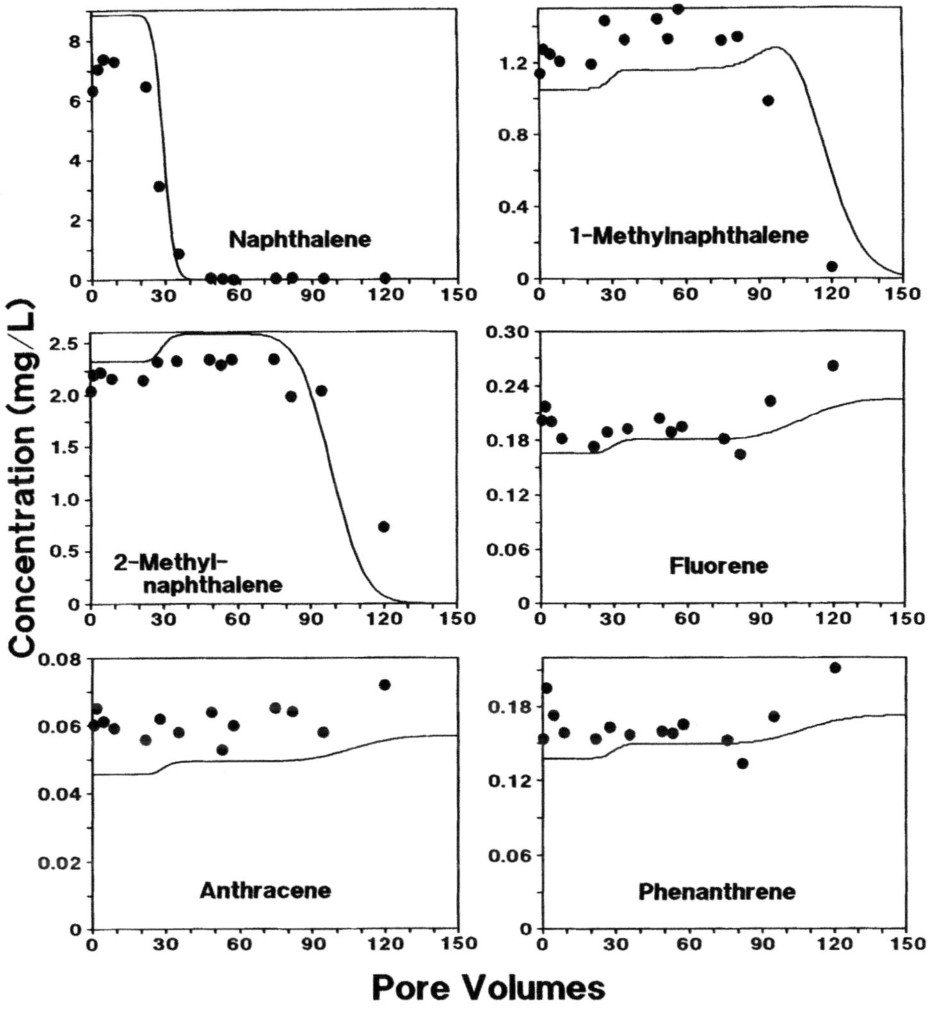

Fig. 2. Elution profiles for six PAHs from a column packed with sand coated with coal tar ID#4

Priddle and MacQuarrie (1994) investigated the dissolution of a few PAHs from creosote using both continuous flow and flow-interrupt techniques. Interpretation of their data from the flow-interrupt experiments suggested that equilibrium was not attained for at least approximately 60 h. To successfully model the rise in concentration with contact time, they applied a one-site model and obtained first-order mass-transfer rates of approximately $0.08\,h^{-1}$ for the two- and three-ring PAHs monitored. In Priddle and MacQuarrie's continual flushing experiments at pore-water velocities of 8 to 15 cm/h, effluent PAH concentrations were proportional to their aqueous solubilities; however, concentrations were much lower than predicted from Raoult's law. This was attributed to mass-transfer constraints and dilution by water which did not come in contact with the NAPL phase. The fact that Augustijn (1993) successfully applied equilibrium assumptions, whereas Priddle and MacQuarrie (1994) could not, is most likely due to the differences in the distribution of NAPL within the columns. In the study conducted by Augustijn (1993), the coal tar and sand was well mixed prior to packing, resulting in thin tar coatings. Priddle and MacQuarrie (1994) used a generator column technique where the glass beads were prepacked in the column before creosote was injected at a 10% residual saturation. This technique resulted in the presence of NAPL globules which have less effective surface area than coatings causing slow mass transfer.

3 Desorption from NAPL-Contaminated Soils

Both dissolution and desorption processes are governed by partitioning, based on the affinity hydrophobic molecules have for hydrophobic domains, and can be limited by mass-transfer constraints. However, the primary factors that limit mass transfer between two phases and the magnitude of those constraints may be different. In soils contaminated with nonaqueous phase liquids (NAPLs) such as coal tar, interactions between the NAPL and the soil matrix may result in a matrix that does not retain the independent characteristics of either phase. The NAPL phase may not behave as a free phase and the original sorption/desorption characteristics of the soil may be altered as well. The NAPL may be adsorbed by the soil matrix followed by various aging processes (e.g., oxidation and polymerization), so that the NAPL residues cannot be distinguished from natural soil organic matter. The adsorbed NAPL will serve as an additional hydrophobic domain that may have a higher affinity for sorption on a carbon-normalized basis than natural soil organic matter. Also, entrapment of NAPLs in micropores after long-term contact will greatly enhance mass-transfer constraints. Therefore, the NAPL-soil matrix may exhibit release kinetics and partitioning behavior different from either the pure NAPL or a natural soil matrix.

For desorption from natural soils, the rate-limiting mechanism is thought to be intrasorbent diffusion to the solid-water interface. For intrasorbent diffusion, mass-transfer rates are primarily controlled by the diffusivity of the diffusing species, the resistance to diffusion associated with the sorbent matrix, and the diffusion path length. The resistance to diffusion is related to the type of organic matrix, while diffusion path length is related to the distribution and structure of the organic material present, e.g., coatings versus patches (Brusseau and Rao 1989). Desorption from a NAPL-soil matrix will be similar. Also, for aged residual NAPLs, the mass of the soluble constituents in the organic matrix may be small relative to the total organic mass present, so that the assumption of a constant partition coefficient may be valid as assumed for nonpolar organic chemicals in soils. In scenarios such as these, sorption/desorption models commonly used for natural soils may be appropriate for describing release of contaminants from a NAPL-contaminated soil.

The bicontinuum sorption model (also known as the two-site model) is one of the most commonly applied models to soil/water systems. Although in reality there is a continuum of sites, this model assumes that sorption is adequately described by two types of domains or regions available for sorption and desorption: an instantaneous equilibrium type (S_1) and a rate-limited domain (S_2). The S_1 domain is considered to be readily accessible for sorption from the solution phase (e.g., regions located on the surface of soil particles or soil organic matter), while the S_2 domain is often conceptualized as being located within the soil organic matter or within the soil particles. Sorption by the S_2 domain is assumed to be described by first-order reversible kinetics.

We investigated the desorption of several PAHs from a coal-tar contaminated soil (ID# UT710) with 17% total organic carbon and approximately 700 mg of known PAHs per kg of soil, using both batch-rate equilibration and flow-interrupt miscible displacement techniques. The bicontinuum sorption model was successfully applied to the batch-rate data to estimate release rates; however, similar success was not met with in simulating the elution curves. A summary of these experiments is presented in the sections that follow.

3.1 Batch-Rate Studies

Aqueous-phase concentrations in the batch-rate studies and the miscible-displacement studies were anticipated to be near or below our analytical limits of detection; therefore, cosolvents were used to facilitate estimation of aqueous-phase equilibrium coefficients and desorption rates. Several studies on solubility, sorption, and transport of organic chemicals have shown that aqueous phase parameters can be successfully estimated from measurements made in cosolvent/water solutions using a log-linear cosolvency model (Rao et al. 1991 and references cited therein). This model assumes that the solubility of a compound, relative to its aqueous solubility, increases in a log-linear manner with increasing volume fraction of cosolvent (f_c). Based on the inverse relationship between solubility and sorption, sorption of a solute is assumed to decrease in a log-linear manner with increasing f_c (Rao et al. 1985; Fu and Luthy 1986). Methanol was chosen as the cosolvent in our studies because deviations from the log-linear model have been shown to be minimal for methanol/water solutions.

Addition of a cosolvent to contaminated soils from an aqueous system will result in dissolution/desorption of a readily available fraction of PAHs. Therefore, in the batch-rate study, moist soils (approximately 5-g samples) were initially equilibrated for three days with 35 ml of the different methanol/water solutions with a 0.005 M $CaCl_2$ matrix, centrifuged at 300 RCF for 15 min, and the supernatant was removed and analyzed. Clean cosolvent-water solution was then returned to the tubes to a volume of 35 ml, and the tubes were placed back on the rotator. By dilution of the solution left in the soil, an initial solution concentration was estimated, and this estimation was considered the concentration at time zero. At each sampling time, the tubes were centrifuged at 300RCF for 15 min and a 5-ml aliquot was taken for analysis. Once an aliquot was taken, the tubes were capped, shaken up, and placed back on the rotator. Aliquots were filtered through a 0.22-μm Teflon filter fitted in a stainless steel holder and extracted with hexane for at least 4 h. Hexane extracts were concentrated under dry nitrogen gas to an approximate volume of 0.15 ml and analyzed using GC with a flame ionization detector (FID).

3.1.1 Desorption Rates

Desorption rates were estimated from the batch-rate data for several PAHs at each f_c, using a bicontinuum sorption model given by Karickhoff (1980) for batch systems. The model was modified for desorption and a term was added for the initial concentration (C_o) in the solution at t = 0. A representative set of data

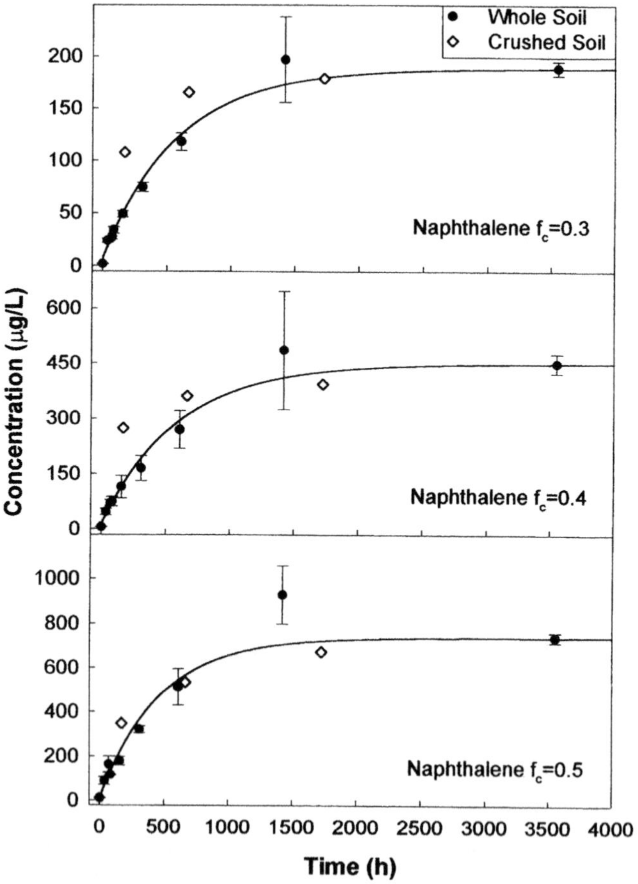

Fig. 3. Desorption batch data for naphthalene at different methanol fractions for whole and crushed UT710 soil. The bicontinuum model fits to the whole soil

collected over time along with a model fit is shown for naphthalene in Fig. 3. For each PAH, the equilibrium distribution coefficients (K_d, l/kg) and the desorption rates (k_2, h^{-1}) estimated from methanol/water solutions were extrapolated to aqueous systems using the log-linear cosolvency model as shown for naphthalene in Fig. 4. For PAHs from naphthalene to pyrene, the extrapolated K_d values ranged from 10^3 to 10^5 l/kg, and k_2 values ranged from $10^{-4.7}$ to $10^{-6.8}$ h^{-1}. Desorption rates from UT710 were almost three orders of magnitude lower than what had been previously reported for release of PAHs from laboratory-contaminated soil (Brusseau and Rao 1989). This suggests that when coal tar is incorporated into a soil matrix, mass-transfer constraints are increased above that which would be expected for either a pure NAPL phase or a natural soil (i.e., an uncontaminated soil).

From the model fits to whole soil data, the fraction of the domain where desorption is considered to be instantaneous was estimated to be from near zero to approximately 3% for the PAHs investigated in all methanol/water solutions employed. This suggests that the fraction of readily available PAHs from coal tar contaminated soils is much smaller than what has been reported for natural soils,

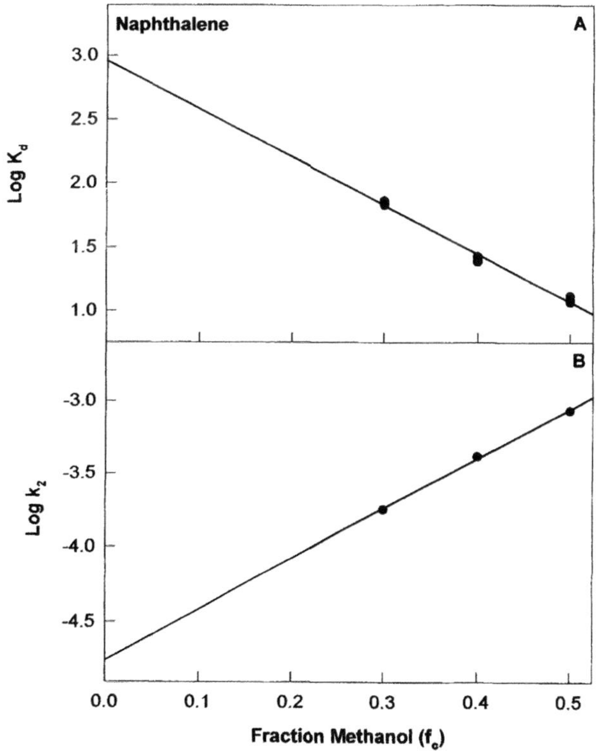

Fig. 4. Log-linear cosolvency model fits to naphthalene cosolvent data to estimate (A) aqueous mass-transfer rates and (B) distribution coefficients

which will have a direct impact on the contaminant concentrations leaving these soils. Also shown in Fig. 3 are data measured for a crushed sample of the same tar-contaminated soil. Crushing the sample exposes internal domains, which decreases the diffusion path length, thereby reducing the time needed to attain equilibrium. Reporting model fits from the crush data is tenuous, given that concentrations were only measured three times; however, F from the crushed soil data was estimated to be approximately 20–25%. This strongly suggests that desorption of PAHs from a coal tar-contaminated matrix is primarily due to diffusion constraints rather than chemical kinetics, which has been hypothesized as one potential explanation for slow release kinetics (Pignatello and Xing 1996).

3.1.2 Equilibrium Distribution Coefficients

The aqueous-phase K_d values estimated from the extrapolation of the methanol/water data measured on day 146 in the batch-equilibration study were used to estimate organic carbon (OC) normalized equilibrium partition coefficients (K_{oc}, l/kg OC) for several PAHs. The log K_{oc} values estimated for UT710 were on an average 0.75 log units higher than what has been reported in the literature for

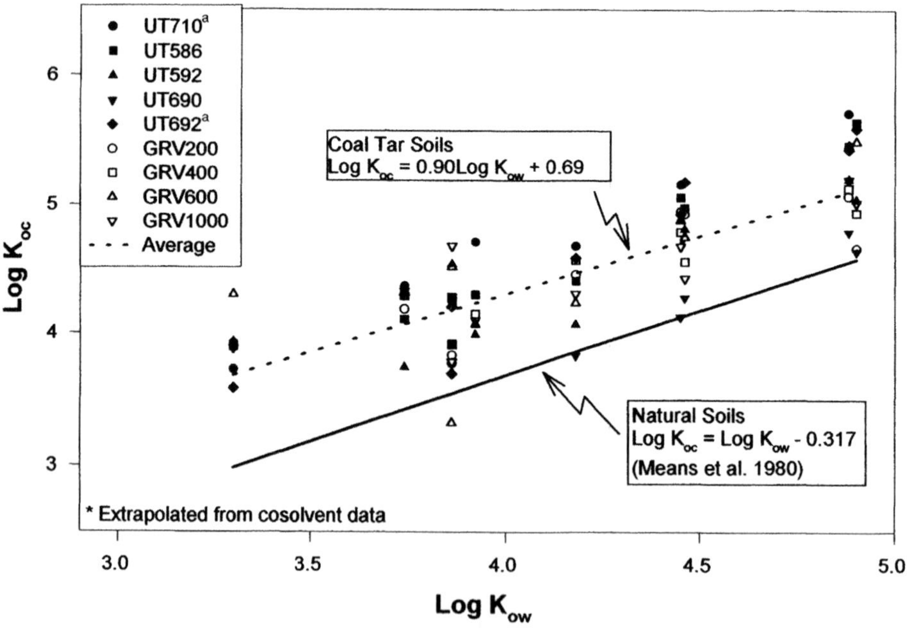

Fig. 5. Carbon normalized distribution coefficients (K_{oc}) versus octanol-water partition coefficients (K_{ow}) for PAHs from several coal tar-contaminated soils compared to the relationship reported for natural soils

natural soils. This suggests that coal tar-derived organic material is more sorptive than organic carbon derived from natural processes. Sun and Boyd (1991) also observed this phenomenon for soils that contained PCB residues.

To further investigate these trends, apparent K_{oc} values were calculated for several PAHs from preliminary data compiled from some of our own ongoing projects related to the cleanup of old MGP sites. The compiled data set represents soils ranging in total known PAH contamination levels of 100–3500 mg/kg and total organic carbon (TOC) contents of 0.5 to 17%. The estimated log K_{oc} values are plotted in Fig. 5 with respect to their log K_{ow} values along with the resulting regression line (long $K_{oc} = 0.9 \log K_{ow} + 0.69$) (dashed). Also shown in Fig. 5 is the relationship reported by Means et al. (1980) for the sorption of PAHs by natural soils ($\log K_{oc} = \log K_{ow} - 0.31$; solid line). Although there is considerable scatter in the data, the average sorptivity of the NAPL residuals is greater than that observed for organic carbon derived from natural processes, as was observed with the data from UT710. After long-term weathering of coal tar-contaminated soil, as indicated by low levels of PAH contamination, the behavior of the NAPL-soil matrix shown in Fig. 5 may approach that observed for natural soils. Further investigation of this hypothesis may prove to be enlightening.

3.2 Column Studies

A flow-interrupt miscible displacement technique similar to that described by Brusseau et al. (1989) was also used with UT710 to estimate desorption rates and compare values obtained from different methods. This technique has a greater sensitivity to nonequilibrium than the standard column techniques, thus a greater potential for improved estimation of model parameters. Moist coal-tar contaminated soil (ID#710) was packed into an Alltech stainless steel Opti-Flow column (1 cm i.d. × 5.5 cm) with 0.5 μm stainless steel frits at the column ends and then saturated with aqueous 0.005 M $CaCl_2$. The column was then switched to a methanol/water solution at a pore-water velocity of 17 cm/h until steady-state flow concentrations were achieved. Flow was then stopped for varying periods of time with steady-state flow reestablished in between each flow interrupt. Effluent samples were collected under hexane and were extracted and analyzed as previously described for the batch-rate study.

Estimation of desorption rates from the elution curves was attempted using a simulation model written by Jessup (pers. comm.). The simulation model is based on the advective-dispersive equation for transport, with the bicontinuum sorption model incorporated for describing sorption/desorption processes (see Brusseau et al. 1991 for details). The following parameters were needed in the simulation model:

$$P = vL/D_h \tag{10}$$

$$R = 1 + \rho_s K_d / \theta \tag{11}$$

$$\beta = \left(1 + F\frac{\rho_s}{\theta}K_d\right)\Big/R \tag{12}$$

$$\omega = k_2(1-F)\frac{\rho_s}{\theta}K_d, \tag{13}$$

where v (cm/h) is the average pore water velocity, P is the Peclet number which represents the dispersive-flux contribution to transport, L (cm) is column length, R is the retardation factor, β is the fraction of instantaneous retardation, ρ_s (g/cm^3) is the soil bulk density, F is the fraction of the sorbent for which sorption is considered instantaneous, and the mass-transfer constraints are expressed in ω (h^{-1}) (Jessup, pers. comm.; Brussseau et al. 1991). R was calculated using K_d values obtained from the batch-rate study, and tritium, a nonreactive tracer, was used to estimate P. The remaining parameters β and ω were optimized for a best-fit simulation.

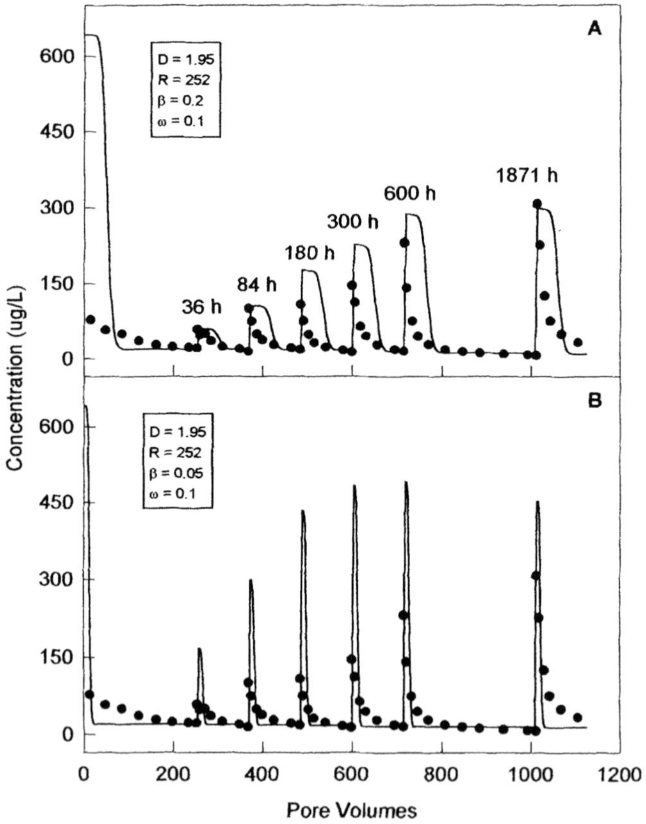

Fig. 6. Elution profiles of phenanthrene from UT710 in 40% methanol and the resulting bicontinuum model fits assuming (A) $\omega = 0.1$ and $\beta = 0.2$; (B) $\beta = 0.05$

A representative elution curve with multiple flow interruptions for the desorption of phenanthrene in 40% methanol is shown in Fig. 6. After each flow interrupt, a rise in concentration is observed indicative of nonequilibirum conditions. In Fig. 6, the duration of the flow interrupt is reported above each peak. Optimization of β and ω could not successfully simulate both the peak heights after flow interrupts and the subsequent elution profile. Variations in β affect both the height and width of the peak height after a flow-interrupt event, while changes in ω affect the peak height and the steady-state flow concentrations. Of the two fitting parameters, only ω affects the simulation of the steady-state flow concentrations in between flow-interrupts; therefore, ω was optimized to the observed steady-state flow concentrations. Using this ω value, attempts were made to optimize β; however, we were unable to fit the tailing of the flow-interrupt peaks. A similar problem was encountered by Augustijn (1993) when attempting to use the bicontinuum sorption model for fitting the breakthrough of benzene and toluene pulses through a column packed with a coal tar soil similar to UT710. The unsuccessful attempts to simulate both peak heights and peak widths could be related to the observed decrease in mass-transfer rates with decreasing porewater velocity (Miller et al. 1990; Brusseau 1992; Powers et al. 1992). This suggests that ω during flow interrupt should be much smaller than what is used to simulate steady-state flow concentrations. An ω value of approximately 0.001 h^{-1} was estimated, using the k_2 value determined from the batch-rate experiment with phenanthrene at $f_c = 0.4$. The smaller ω estimated from the batch-rate data supports the need to assume a smaller ω during times of no flow, which will impact the predicted rise in concentration. Modifying the model to allow for changes in ω with pore-water velocity may improve the utility of the current bicontinuum sorption model.

In the past, applying a local equilibrium assumption (LEA) was considered a conservative approach for assessing the potential for contamination from a waste site. However, when designing a remediation scheme, LEA is not conservative, and the consideration of mass transfer constraints is imperative. In designing an optimal remediation scheme or in assessing the need for remediation, a decision needs to be made regarding the most appropriate parameters. Information (i.e., K_{oc} and k_2) can be derived from the site-specific data, sorption data from natural soils, and/or from other coal tar contaminated site data. Simulations for estimating removal of naphthalene from UT710 in aqueous systems were performed for two scenarios assuming a pore-water velocity of 1 cm/h and no degradation (Fig. 7). In case 1 (dashed line), K_d and k_2 were estimated from the K_{oc}-K_{ow} relationship reported by Means et al. (1980) and the k_2-K_d relationship reported by Brusseau and Rao (1989) for natural soils. In case 2 (solid line), K_d and k_2 were taken from the UT710 data extrapolated to aqueous systems. These results clearly demonstrate the error that may be made if assessment of an MGP site is made based on data from only natural soils.

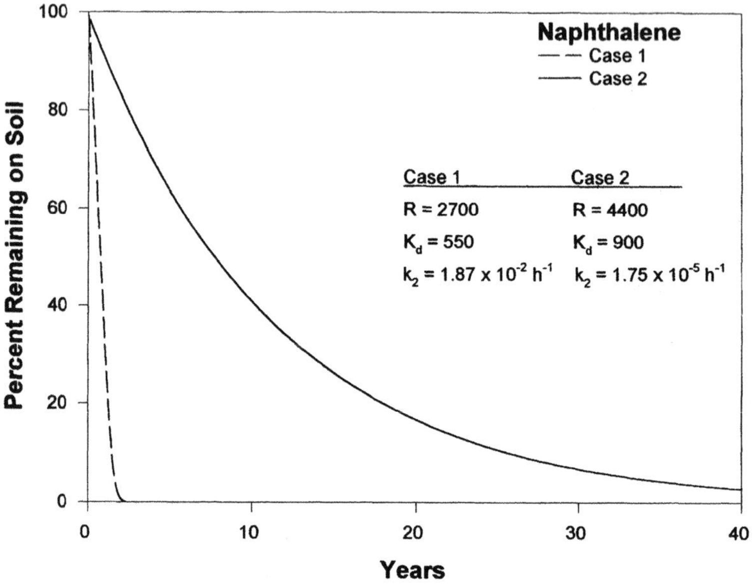

Fig. 7. Simulation of the mass removed of naphthalene assuming no degradation in an aqueous system at a pore-water velocity of 1 cm/h based on data from natural soils (*case 1*) and UT710 (*case 2*)

4 Summary

Extremely slow rates of desorption, high distribution coefficients, and small readily available fractions of PAHs exhibited for coal tar contaminated soils suggest that actual contaminant concentrations being released into the groundwater are small. Furthermore, the possibilities for pump-and-treat remediation of aged residual coal tar contaminated soil may be limited even with the use of cosolvents and surfactants. Enhanced bioremediation may also be limited when the rates of release to the aqueous phase are so slow ($k_{release} \ll k_{degradation}$). On the other hand, this suggests that the acceptable levels of soil contamination to be left on site may be much greater than would be expected, based on data from natural soils, and that natural attenuation may suffice, especially if PAH concentrations are sufficient to stimulate and maintain innate biodegradation. Given the 2000 to 3000 sites in the United States alone that are contaminated with coal tar (Luthy et al. 1994) and the limited data available from tar-contaminated soils, more research on tar-contaminated soils is needed for improving site evaluations at MGP sites. In particular, more research is needed to quantify distribution and desorption rate coefficients and to elucidate sequestration mechanisms.

Acknowledgments. This research was partially supported by Electric Power Research Institute Contracts RP-2879-7 and W09076-03.

References

Abriola LM, Pinder GF (1985) A multiphase approach to the modeling of porous media contaminated by organic compounds, 1. Equation development. Water Resour Res 21:11–18

Augustijn DCM (1993) Chemodynamics of complex waste mixtures: applications to contamination of soils and aquifer media. PhD Diss, University of Florida, Gainesville

Borden RC, Kao C (1989) Water flushing of trapped residual hydrocarbon: mathematical model development and laboratory validaton. In: Proc Petroleum hydrocarbons and organic chemicals in groundwater, Houston, Texas 15–17 Nov 1989, Natl Water Well Assoc, Dublin, Ohio, pp 473–486

Borden RC, Piwoni MD (1992) Hydrocarbon dissolution and transport: a comparison of equilibrium and kinetic models. J Contam Hydrol 10:309–323

Brusseau ML (1992) Nonequilibrium transport of organic chemicals: the impact of pore-water velocity. J Contam Hydrol 9:353–368

Brusseau ML, Rao PSC (1989) The influence of sorbate-organic matter interactions on sorption nonequilibruim. Chemosphere 18:1591–1706

Brusseau ML, Jessup RE, Rao PSC (1991) Nonequilibruim sorption of organic chemicals: elucidation of rate-limiting processes. Environ Sci Technol 25:134–142

Brusseau ML, Rao PSC, Jessup RE, Davidson JM (1989) Flow interruption: a method for investigating sorption nonequilibrium. J Contam Hydrol 4:233–240

Chatzis I, Morrow NR, Lim HT (1983) Magnitude and detailed structure of residual oil saturations. Soc Petrol Eng J 22:311–326

Chen CS (1993) Partitioning of motor oil components into water. Master Thesis, University of Florida, Gainesville

Cline PV, Delfino JJ, PSC Rao (1991) Partitioning of aromatic constituents into water from gasoline and other complex solvent mixtures. Environ Sci Technol 25:914–920

Corapcioglu MY, Baehr AL (1987) A compositional multiphase model for groundwater contamination by petroleum products, 1. Theoretical considerations. Water Resour Res 23:191–200

EPRI (1992) Estimating release of polycyclic aromatic hydrocarbons from coal tar at manufactured-gas plant sites. EPRI. TR-101060, Palo Alto, California

Fredenslund A, Jones R, Prausnitz JM (1975) Group-contribution estimation of activity coefficients in nonideal liquid mixtures. AICHE J 21:1086–1099

Fu J, Luthy RG (1986) Aromatic compound solubility in solvent/water mixtures. J Environ Eng 112:328–345

Geller JT, Hunt JR (1993) Mass transfer from nonaqueous phase organic liquids in water saturated media. Water Resour Res 29:833–845

Hunt JR, Sitar N, Udell KS (1988) Nonaqueous phase liquid transport and cleanup, 1. Analysis of mechanisms. Water Resour Res 24:1247–1258

Kaluarachchi JJ, Parker JC (1990) Modeling multicomponent organic chemical transport in three-fluid-phase porous media. J Contam Hydrol 5:349–374

Karickhoff SW (1980) Sorption kinetics of hydrophobic pollutants in natural sediments. In: Baker RA (ed) Contaminants and sediments, vol 2. Analysis, chemistry, biology. Ann Arbor Science Publ, Ann Arbor Michigan, pp 193–205

Lee LS, Hagwall M, Delfino JJ, Rao PSC (1992a) Partitioning of polynuclear aromatic hydrocarbons into water from diesel fuel. Environ Sci Technol 26:2104–2110

Lee LS, Rao PSC, Okuda I (1992b) Equilibrium partitioning of polynuclear aromatic hydrocarbons from coal tar into water. Environ Sci Technol 26:2110–2115

Luthy RG, Ramaswami A, Ghosal S, Merkel W (1993) Interfacial films in coal tar nonaqueous-phase liquid-water systems. Environ Sci Technol 27:2914–2918

Luthy RG, Dzombak DA, Peters CA, Roy SB, Ramaswami A, Nakles DV, Nott BR (1994) Remediating tar-contaminated soils at manufactured gas plant sites. Environ Sci Technol 28:266–276

Mackay D, Shiu WY, Maijanen A, Feenstra S (1991) Dissolution of non-aqueous phase liquids in groundwater. J Contam Hydrol 8:23–42

Means JC, Wood SG, Hassett JJ, Banwart WL (1990) Sorption of polynuclear aromatic hydrocarbons by sediments and soils. Environ Sci Technol 14:1524

Miller CT, Poirier-McNeill MM, Mayer AS (1990) Dissolution of trapped nonaqueous-phase liquids: mass transfer characteristics. Water Resour Res 26:2783–2796

Nelson EC, Goshal S, Edwards JC, Marsh GX, Luthy RG (1996) Chemical characterization of coal tar-water interfacila films. Environ Sci Technol 30:1014–1022

Pignatello JJ, Xing B (1996) Mechanisms of slow sorption of organic chemicals to natural particles. Environ Sci Technol 30:1–11

Powers SE, Abriola LM, Weber WJ Jr (1992) An experimental investigation of nonaqueous phase liquid dissolution in saturated subsurface systems: steady state mass transfer rates. Water Resour Res 28:2691–2705

Priddle MW, MacQuarrie DTB (1994) Dissolution of creosote in groundwater: an experimental and modeling investigation. J Contam Hydrol 15:27–56

Rao PSC, Hornsby AG, Kilcrease DP, Nkedi-Kizza PJ (1985) Sorption and transport of hydrophobic organic chemicals in aqueous and mixed solvent systems: model development and preliminary evaluation. J Environ Qual 14:376–383

Rao PSC, Lee LS, Wood AL (1991) Solubility, sorption, and transport of hydrophobic organic chemicals in complex mixtures. Environ Research Brief, EPA-600/M-91/009. US EPA, Ada, Oklahoma

Schwille F (1988) Dense chlorinated solvents in porous and fractured media. Lewis, Chelsea, Michigan, 146 pp

Sun S, Boyd SA (1991) Sorption of polychlorophenyl (PCB) congeners by residual PCB-oil phases in soils. J Environ Qual 20:557–561

Wilson JL, Conrad SH, Mason WR, Peplinski W, Hagan E (1990) Laboratory investigation of residual organics from spills, leaks, and the disposal of hazardous wastes in groundwater. EPA/600/6-90/004 US EPA, Ada, Oklahoma, 267 pp

Zalidis GC, Annable MD, Wallace RB, Hayden NJ, Voice Tc (1991) A laboratory method for studying the aqueous phase transport of dissolved constituents from residually held NAPL in saturated soil columns. J Contam Hydrol 8:143–156

7 Transport and Reactions of Petroleum Contaminants in the Unsaturated Soil Zone from Leaking Underground Tanks

J. B. Carberry

Contents

1 Introduction

Groundwaters are increasingly threatened by oil leaks, disposal of hazardous wastes, and landfill leachates migrating through the soil mantle (Wilson et al. 1981; MacKay et al. 1985). Several mechanisms control migration of toxics to underground aquifers (Short 1987), i.e., hydrological transport, adsorption/desorption, volatilization, and decomposition, including biodegradation. The fate and transport of petroleum products in the soil environment are complex and dynamic, including all of the above mechanisms (Grenny et al. 1987; Symons et al. 1988).

Movement of the contaminant peak concentration in the soil will be deeper as time progresses and will decrease in magnitude (Letey and Oddson 1972). Of the factors affecting mass transport of organic chemicals, the most important affecting migration is the adsorption isotherm between the organics and soil. Jury (1986)

Judith Bower Carberry, Inc., 359 N Street, SW, Washington, D.C. 20024, USA

H. Rubin/N. Narkis/ J. Carberry
Soil and Aquifer Pollution

confirmed the importance of the adsorption process for organic chemical movement in soil. Factors affecting adsorption are clay composition and exchange capacity, organic matter content, soil water content, electrical structure, and water solubility.

The vapor retardation by soil organic matter is of minor importance to the volatile organics (Dupont 1986). As diffusion proceeds, the concentration gradient and the rate of volatilization diminish with time (Hamaker 1972).

Biodegradtion is the transformation of an organic chemical to another form by microbial activities (Grady 1987). Petroleum biodegradability (Walker et al. 1975) may depend on the content of n-alkane, asphaltene, nitrogen, sulfur, and oxygen-containing compounds in the petroleum product. Two refined products encountered in the contaminated soil samples used for this study were No. 2 and No. 6 fuel oil. Because they are common in recorded oil spills and leaks from underground storage tanks, these particular petroleum products have been studied previously with respect to their biodegradability (Raymond et al. 1976; Walker et al. 1976). Microbial principels and processes of petroleum degradation and mineralization have been reviewed by Leahy and Colwell (1990).

The purposes and objectives of the project were as follows:

1. To demonstrate the in situ biodegradability of petroleum soil contamination if adequate environmental conditions are provided.
2. To determine the distribution of petroleum in heterogeneous soil conditions between the gas, liquid, and solid phases of two contrasting soil types.
3. To model the behavior of petroleum in order to predict the transport, adsorption, biodegradation, and evaporation in soil columns.

2 Methods and Procedures

2.1 Soil Preparation and Analysis

Two contaminated soil samples were obtained from power plants in 5-gal covered plastic buckets and transported to the laboratory. A 1-g soil sample was separated out and refrigerated until the microbial selection procedure described below could be carried out. For both soil samples, the remainder was spread out on foil-lined broiler pans to air dry. Soil chunks were broken up to facilitate drying. After 2 to 3 days, when the soil was nearly dry, the soil was wetted with water until it was saturated, the slurry was agitated to mix it and autoclaved under standard conditions, chunks were further ground up in a mortar and pestle, and the soil was spread out to dry again. This preparation process was repeated until no bacterial growth occurred on TSA plates. Ten g of soil were used for elemental analysis with a soil testing kit manufactured by Soiltest. Results from this test determined the nitrogen, phosphate, potash, and pH content of the soils. A 500-g sample was used to carry out a soil sieve analysis according to procedures established by ASTM (1985). The latter results determined the soil particle size distribution and the accepted relative amounts of pebbles, gravel, sand, and clay.

2.2 Microbial Selection and Culture

A 1-g soil sample was suspended in a sterilized minimal salts medium, referred to as Glucose Salts Medium minus the Glucose (GSM-G). This aqueous medium contained nitrogen, phosphorus, sodium, potassium, and calcium salts dissolved in tap water, which provided micronutrients. The medium contained no carbon source, so that the petroleum contamination in the soil and the petroleum added later served as the only carbon source. Only microbes which could metabolize the petroleum absorbed to the contaminated soil particles in the sample could survive in this environment, and these microbial communities soon became dominant and proliferated. Successive subcultures were carried out for each soil sample, using fresh medium and directly adding the petroleum specific to each site to the culture flask. Gradually, the soil particles and noncompetitive microbial species were diluted out of the selected culture, and the dominant species degrading the specific petroleum could easily be grown to high concentrations.

2.3 Column Studies

Four Plexiglas columns of 1.75 in. diameter and 34 in. length were constructed, each with five ports at 6-in. intervals down the length of the columns and fitted in the bottom with a fine-mesh screen. Two of these columns were packed with wetted contaminated soil sample which was composed of very fine particles, and two columns were filled with the wetted coarse soil sample. Each pair of columns was operated under biotic or abiotic conditions. The biotic columns initially received measured volumes of stock microbes specifically selected for each petroleum contaminant as described above and then concentrated to a standard titer by centrifugation and resuspension in GSM-G. The biotic columns initially received petroleum and water at the same time as the abiotic columns received petroleum and water. Specifically, mineral oil was added to the tops of columns filled with fine clay soil, and No. 2 and No. 6 fuel oil were added to the tops of columns filled with coarse soil. Periodic additions of water were made to all columns in order to simulate the percolation of rain water through the unsaturated soil column. The simulated leak experiments described above progressed as follows: on day 0, the experiment was begun with the sterile soil in place in the columns. The proper additions of petroleum, water and microbes (to the biotic columns only) were made and allowed to percolate through the columns. On day 1, samples from the columns were taken as described below and analyzed on this day and on day 2. On day 3 additions of water and petroleum were made to the columns, and the analyses of the previous sample set was completed. On day 4 the sampling was repeated, and the 2-day analysis procedure was carried out on the second set of samples. On day 6 further additions to the column were made. This 3-day progression continued until the experiment was terminated after 22 days.

2.4 Sampling and Analyses

First, the gas phase measurements were carried out by attaching a tube onto each port and in sequence attaching the tube to a photoionization detector, model TIP II by Photovac Inc., Canada. This instrument measured the concentration of photoionizable hydrocarbons directly in mg/l. Since there were four columns with five ports each, 20 sequential gas phase measurements were made for each day's sampling. Next, each column was sampled sequentially as follows: the column was rotated from the vertical position to an angle of approximately 45° with the horizontal. A test tube was attached to each port and the port valves were opened sequentially to allow percolated liquid and a small amount of soil to flow out of the port. When the sampling period was finished, the port valves were closed and the column was placed back in the vertical position. The next column was sampled in identical fashion, until 20 samples of admixed liquid and soil were obtained from the four columns. The slurry in each test tube was mixed to obtain a representative sample, and a 1-ml aliquot was removed using sterile techniques. The 1-ml aliquot was diluted and plated to enumerate the bacteria present. The remainder of each heterogeneous sample was separated into liquid and solid fractions by filtering through 0.22μm pore size filter paper. The following analyses were carried out on the solid fraction: total suspended solids and total organic solids, both measured gravimetrically. The microbial population determined from plating (described above) was then correlated to the organic solids determined gravimetrically. The liquid fraction was subjected to the following analyses: (1) pH measurement using a Fisher Model 825 pH meter and long combination electrode, which could be inserted directly into the test tube; (2) dissolved oxygen, again using the Fisher pH meter fitted with the Orion DO probe inserted directly into the test tube; and (3) chemical oxygen demand, using the Hach Chemical Co. method.

2.5 Model Development

For each port on each sampling day, one gas phase measurement was made, two solid phase measurements were carried out, three liquid phase analyses were obtained, and one analysis was conducted on the unfiltered liquid with suspended solids. These data for all analyses at all column depths on all sampling days for all four columns were transferred to floppy disks. These data matrices were analyzed in order to determine rate coefficients for the following naturally occurring processes within each column: evaporation, adsorption/desorption, transport, and biodegradation. The results obtained from each abiotic column were compared with the corresponding results from each biotic column in order to obtain the biodegradation rates for each contaminated soil. A generalized model was developed to accommodate all the variations described, and the model was calibrated using the experimental values determined above. The model function is shown in Eq. 1 as follows:

The net change of petroleum in a control volume	=	The rate of change in petroleum due to transport	+	The rate of change in petroleum due to adsorption	+	The rate of change in petroleum due to evaporation	+	The rate of change in petroleum due to biological activity

$$\left(\frac{\partial C}{\partial t}\right)\partial z = K_T C_L V_\varphi dz + K_S C_S A_S \rho dz + K_G C_V dz + \mu_B C_L dz, \tag{1}$$

where K_T = transport rate constant in liquid phase (day^{-1}); C_L = concentration in liquid phase (mg/l); V_φ = pore space (cm^3/cm^3 soil); z = depth (cm); K_S = mass adsorption/desorption rate const (day^{-1}); C_S = adsorbed initial concentration (mg_o/gm); A_S = surface area (cm^2/gm); ρ = bulk density; K_G = vaporization rate constant (day^{-1}); C_V = concentration in gas phase (ppm); and μ_B = biodegradation rate (day^{-1}).

This expression summarizes all the physical and biological processes taking place within the unsaturated zone of the simulated soil column in one dimension. The resulting concentration of petroleum obtained from numerical integration of the above differential equation would predict petroleum concentration introduced from a petroleum leak to the vadose zone at depth z.

3 Results and Significance

3.1 Soil Analysis

Results of the soil sieve and chemical analyses are reported in Table 1. According to the ASTM criteria for soil composition definition, the sieve analyses indicated

Table 1. Results of soil sieve and chemical analyses

Analyses		Fine Soil	
Sieve	% >2 mm	9.6	43.5
	2 mm < % < 0.2 mm	67.4	47.9
	% < 0.2 mm	22.3	8.0
Physical	Moisture, %	18.2	8.1
	Spec. gravity	2.8	2.6
	Porosity	0.5	0.5
	Spec. surface area	83.5 cm^2/g	43.1 cm^2/g
	Surf. area/control volume	16.3 mm^2	8.0 mm^2
Chemical	pH	5.5	6.5
	Nitrogen	M	M
	Phosphate	H	H

that soil 1 was primarily a fine clay. This soil, as shown in additional tests, retained moisture well, and had a large surface area. These qualities would provide favorable conditions for indigenous bacteria to attach to the soil particles and degrade any adsorbed or percolating petroleum contaminant. Chemical tests demonstrated that sufficient nitrogen and phosphate were present in order for these bacteria to carry out the petroleum biodegradation. The pH, however, was borderline for favorable biodegradation, since aerobic metabolism produces acidic metabolites which might further depress the pH below an acceptable level.

Soil 2 contrasted nicely to soil 1 in that it was composed principally of coarse particles which provided little capability for retaining the moisture necessary for biodegradation. In addition, Soil 2 contained approximately half the surface area of soil 1 for the attachment of indigenous bacteria. Soil 2 also possessed sufficient nitrogen and phosphate for the biodegradation process to be implemented, and it had a more favorable pH than soil 1 for bioremediation.

The porosity of soil 1 was only marginally less than soil 2, even though the particle-size distribution was very different. The porosity of approximately 50% indicated that in situ bioremediation would be physically possible.

3.2 Column Studies

For each simulated leak experiment conducted with intermittent introduction of petroleum to the soil columns throughout the length of the experiment, all results were coded as follows:

1. Column A: abiotic column of fine clay soil receiving additions of insulating oil and water only.
2. Column B: biotic column of fine clay soil receiving additions of insulating oil, microbes, and water.
3. Column C: abiotic column of coarse soil receiving additions of fuel oils number 2 and 6, and water only.
4. Column D: biotic column of coarse soil receiving additions of fuel oil numbers 2 and 6, microbes, and water.

In addition, each column port was labeled 1 through 5 from the top to the bottom of each column. For example, for any sampling day, the port at middepth in the abiotic column A for the fine clay soil will have the label A3 on all the plots of gas, liquid, and solid parameters measured.

3.2.1 Gas Phase

In the fine clay soil under abiotic conditions, concentrations of volatile hydrocarbon built up to a concentration of 300 mg/l after a 10-day period until the end of the experiment. Under biotic conditions for this soil, no significant hydrocarbon concentrations were present at any time or location during the entire experiment.

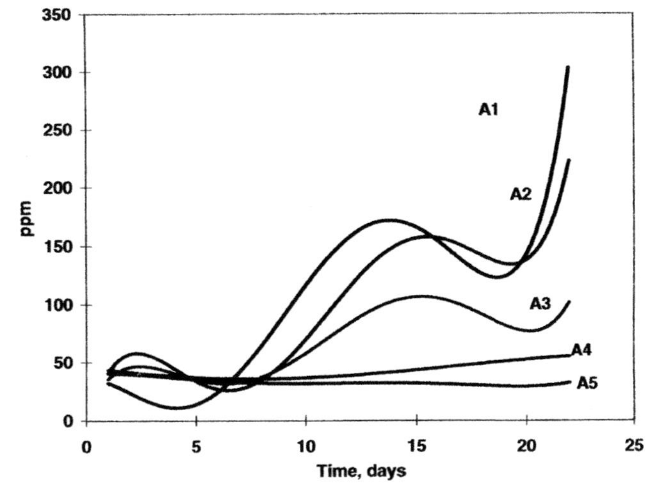

Fig. 1. Concentration of petroleum vapor in abiotic column of fine soil

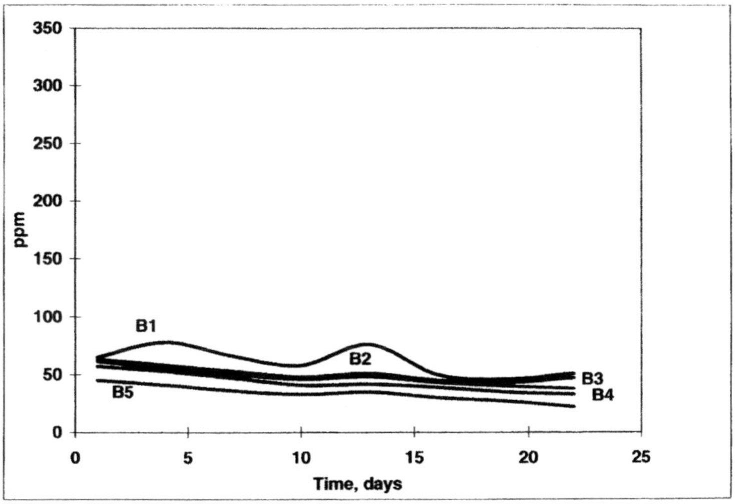

Fig. 2. Concentration of petroleum vapor in biotic column of fine soil

These results are compared in Figs. 1 and 2 and indicate that the selected microbes were able to reduce the concentration of hydrocarbons enough so that no significant equilibrium could be established between the solid and gas phases within the biotic column B. These results indicated, therefore, that the selected microbes could probably metabolize any petroleum vapor, a phenomenon documented by other research (Vercellone-Smith and Herson 1996), but not demonstrated here. Furthermore, the absence of any significant vapor produced in the biotic system indicates that no reductive metabolism occurred under anaerobic conditions. For the coarse soil, there was no difference between the abiotic and biotic columns. In both cases, the concentrations of vapor phase petroleum peaked on day 7 at about 200 mg/l and then were present only in minimal concentrations throughout the rest of the 22-day experiment. These results demonstrated that the fuel oil added to the soil columns contained volatile components which could dissipate through the soil interstitial pore spaces. In addition, since there was little difference between the biotic and abiotic columns, these results demonstrated that the selected microbes in this system could not degrade the vapor phase petroleum components as fast as it could be produced and dissipated.

3.2.2 Solid Phase

For the abiotic column of fine clay soil, the concentration of hydrocarbon adsorbed onto the soil surface decreased until day 13 and then increased until the end of the experiment. Likewise, the biotic column showed a decrease in adsorbed hydrocarbon until day 7, and these concentrations remained low throughout the remainder of the experiment. The differences between the biotic and abiotic columns are shown in Figs. 3 and 4; they demonstrate that the petroleum was adsorbed onto the soil particle surfaces and was easily desorbed during the simulated rainwater percolation processes in both columns, but degraded only in the biotic column. Similar results were observed for the coarse soil, and therefore the bioremediation process is likely to be successful in both fine and coarse soils. Both systems have advantages and disadvantages which impede and enhance the

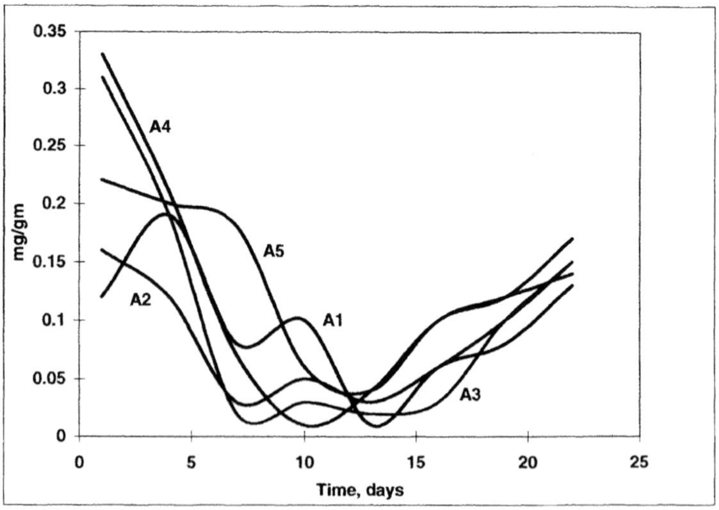

Fig. 3. Concentration of adsorbed petroleum on fine soil in abiotic column

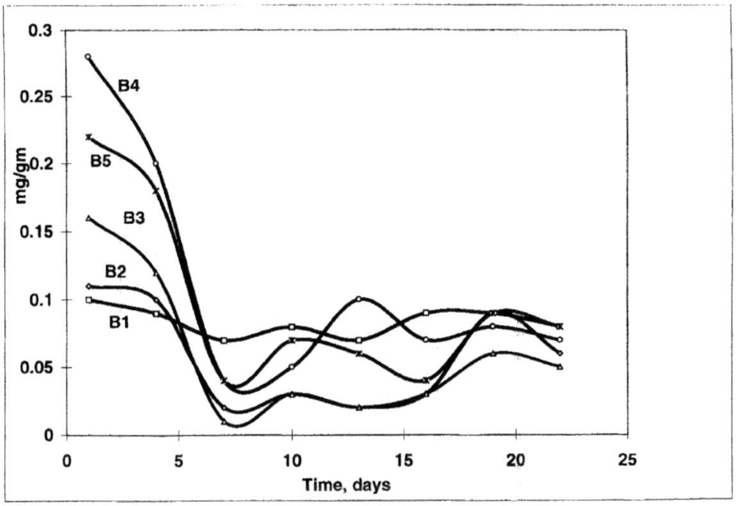

Fig. 4. Concentration of adsorbed petroleum on fine soil in biotic column

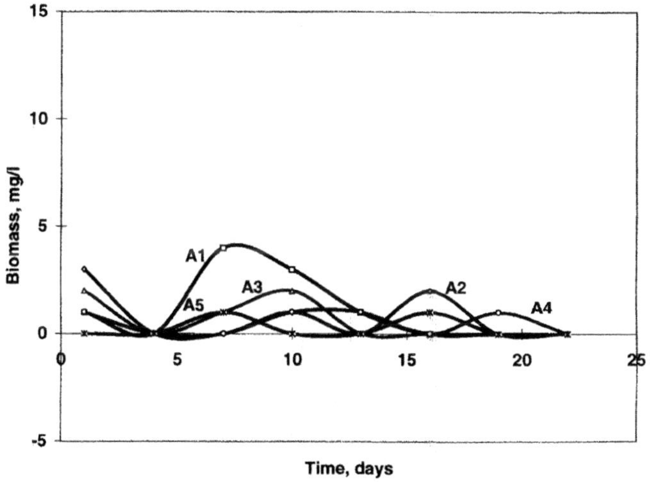

Fig. 5. Concentration of microbes in abiotic column of fine soil

bioremediation process. The large surface area on the fine soil particles provides an opportunity for the biologically active microbes to attach and degrade any adsorbed petroleum, but the decreased porosity impedes the migration of both the microbes and the percolating petroleum. And conversely, the coarse soil provides little secure attachment because the surface area is small and the shear stresses are great in the larger pore spaces. The microbes and substrate can disperse more uniformly throughout the contaminant plume, however, so the concentration will

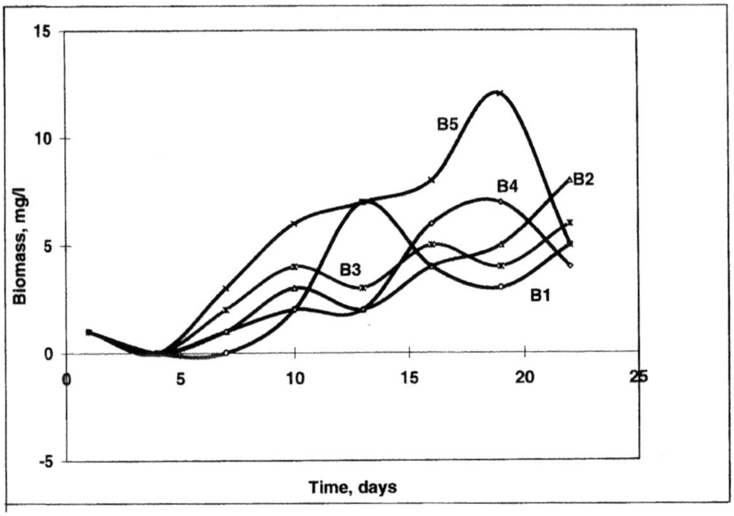

Fig. 6. Concentration of microbes in biotic column of fine soil

be low at any one sampling location. The concentrations of microbes isolated in these soil samples corroborates the previous observations. The concentrations of microbes increased after a period of acclimation in the biotic column of fine soil, in comparison to the abiotic column, as shown in Fig. 5 and 6. Little difference in the microbial concentration of the biotic and abiotic columns could be measured in the coarse soil. These observations suggest that the bioattenuation technique recently implemented to degrade NAPLS at aquifer interfaces could be carried out in porous soils.

3.2.3 Aqueous Phase

The exponential decreases in aqueous petroleum concentrations are presented in Figs. 7 and 8 for the coarse soil. Although it is not apparent from these plots, a linear regression of the logarithmic plots of these data produced significantly greater degradation rates at all depths but one for the biotic case than for the abiotic case. This significant difference in rates between the biotic and abiotic systems can be attributed to the biodegradation process. Furthermore, the regression analyses of all the curves were characterized by R^2 values greater than 0.82, so that both the advection and biodegradation effects could be modeled with a first order rate constant. For the fine clay soil, the decreases in petroleum concentration early in the experiment were approximately the same in the biotic and abiotic systems. In the biotic system, there was no notable microbial concentration initially, so that degradation was due primarily to advective forces and continuous desorption from the soil-particle surfaces. After the seventh day, when microbial concentrations increased, the petroleum concentration decreased faster in the

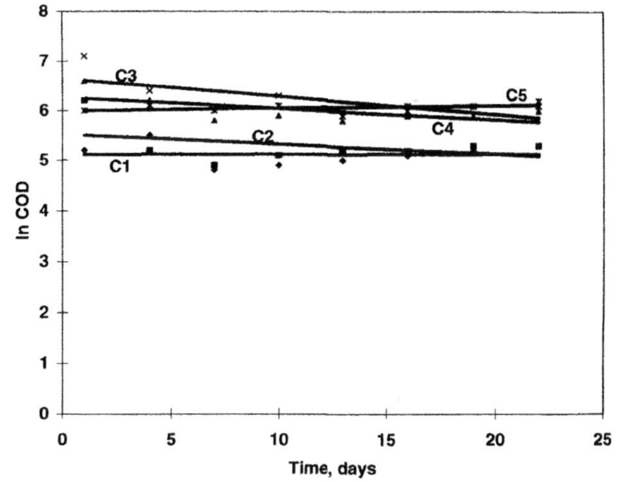

Fig. 7. Concentration of aqueous phase petroleum in abiotic column of coarse soil

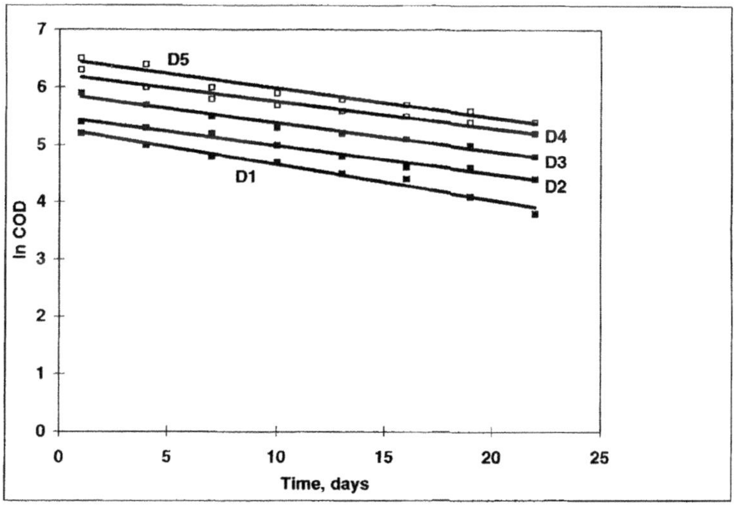

Fig. 8. Concentration of aqueous phase petroleum in biotic column of coarse soil

biotic system compared with the abiotic system. Presumably, if the experiment had been continued longer, the biodegradation effects would have continued to be important and perhaps become controlling.

The pH at all depths in all columns remained between 6.3 and 8.2 during these experiments. The DO fluctuated between 7.0 and 0.5 mg/l at all depths in all columns, indicating that the systems were never anaerobic.

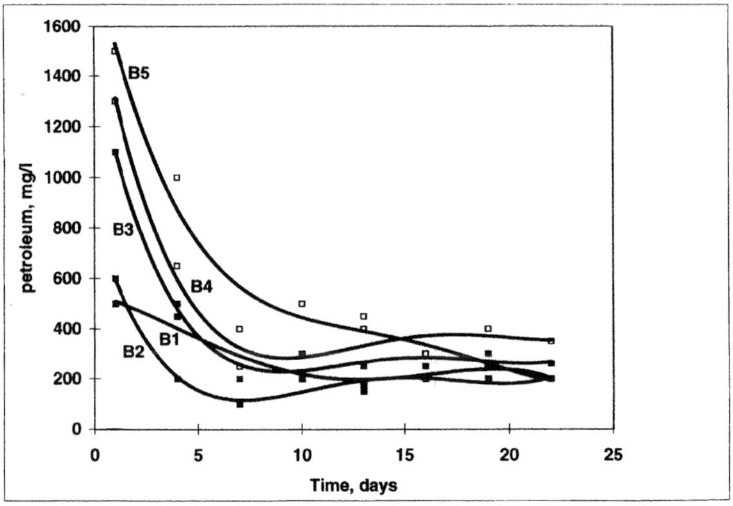

Fig. 9. Predicted values (*lines*) and real values (*points*) of aqueous phase petroleum in biotic column of fine soil

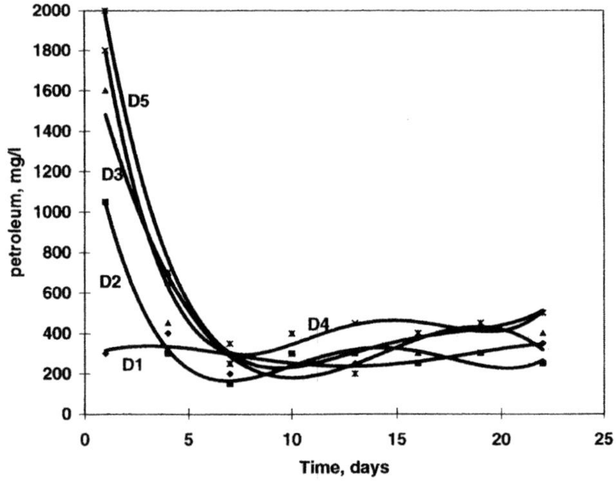

Fig. 10. Predicted values (*lines*) and real values (*points*) of aqueous phase petroleum in biotic column of coarse soil

3.3 Computer Model

Laboratory data were correlated in order to obtain rate constants for the processes described in Eq. (1). Rate constant correlation, measured by R^2 values, in the biotic columns of both the fine and coarse soils were consistently higher than for the abiotic columns. These results indicate that the model could predict bioremediation results rather accurately if selected microbes were injected into a petroleum plume and conditions were conducive for microbial metabolism. Furthermore, the predicted and actual results were better correlated near the end of the experiments, where potential relative errors would be largest. Typical results are presented in Figs. 9 and 10, where lines indicate predicted values and points represent the measured values. In addition, for the biotic column of fine soil, the correlations between predicted and actual values were better at the top of the column than at the bottom, indicating that most of the biological activity occurred at the top of the column. In the biotic column of coarse soil, correlations were approximately the same at all depths. These results indicate that in situ bioremediation projects in the vadose zone of coarse soils risk polluting the underlying aquifer, and precautions must be taken to prevent such contamination of potential drinking water sources.

4 Conclusions

1. The semicontinuous addition of petroleum contaminant to the columns simulated an underground leak and resulted in significant differences between both gas phase and solid phase petroleum concentrations in the biotic and abiotic columns.
2. These results indicated that microbial biodegradation in the solid phase was sufficient to reduce volatilization of the contaminant to the gas phase in fine soil, or to where it could escape to the atmosphere in coarse soil.
3. Few differences resulted in the biotic and abiotic columns for the liquid phase, because the petroleum solubility in the aqueous phase was so low. In the fine soil, biodegradation occurred principally in the upper layers, whereas in the coarse soil, biodegradation occurred throughout the soil column depth.
4. Overall, physical effects and microbial degradation of petroleum could be predicted with some limitations by the one-dimensional model used in this work. A long-term validation study would be required to confirm the short-term results observed here.

References

American Society for Testing Materials (1985) Standard method for particle-size analysis of soils. D422-63. 04.08:117–127

Cohen Y (1986) Organic pollutant transport. Environ Sci Technol 20(6):538–544

Dupont RR (1986) Evaluation of air emission release rate model predictions of hazardous organics from land treatment facilities. Environ Prog 5(3):197–206

Grady CPL Jr (1987) Biodegradation: its measurement and microbiological basis. Biotechnol Bioeng 27:660–674

Grenny WJ, Caupp CL, Sims RC (1987) A mathematical model for the fate of hazardous substances in soil: model description and experimental results. Hazard Waste Hazard Mater 4(3):223–239
Hamaker JW (1972) Diffusion and volatilization. In: Goring CAI, Hamaker JW (eds) Organic chemicals in the soil environment. Marcel Dekker, New York, pp 341–393
Jury WA (1986) Adsorption of organic chemicals onto soil. In: Hem ST, Melanaon SM (eds) Vadose zone modeling of organic pollutants. SM Lewis, Chelsea, Michigan, pp 3–36
Leahy JG, Colwell RR (1990) Microbial degradation of hydrocarbons in the environment. Microbiol Rev 54:305–315
Letey J, Oddson JK (1972) Mass transfer. In: Goring CAI, Hamaker JW (eds) Organic chemicals in the soil environment. Marcel Dekker, New York, pp 399–442
MacKay DM, Roberts PV, Cherny JA (1985) Transport of organic contaminants in groundwater. Environ Sci Technol 19(5):384–392
Raymond RL, Hudson JD, Jamison VW (1976) Oil degradation in soil. Appl Environ Microbiol 31(4):522–535
Short T (1987) Movement of contaminants from oily wastes during land treatment. In: Calabrese EJ, Kostecki PT (eds) Soils contaminated by petroleum. John Wiley, New York, pp 317–330
Symons BD, Sims RC, Grenny WJ (1988) Fate and transport of organics in soil: model predictions and experimental results. J Water Pollut Control Fed 60:1684–1693
Vercellone-Smith P, Herson DS (1996) Toluene elicits a carbon starvation response in *Pseudomonas putida* mt-2 containing TOL-plasmid pWWO. Appl Environ Microbiology 63:1925–1932
Walker JD, Petrakis L, Colwell RR (1975) Microbial petroleum degradation application of computerized mass spectrometry. Can J Microbiol 21:1760–1767
Walker JD, Petrakis L, Colwell RR (1976) Comparison of the biodegradability of crude oil and fuel oils. Am J Microbiol 22:598–605
Wilson JT, Enfield CG, Dunlap WJ, Cosby RL, Foster DA, Baskin LB (1981) Transport and fate of selected organic pollutants in a sandy soil. J Environ Qual 10(4):501–506

8 Aqueous Dissolution of the Constituents of Composite Non-Aqueous Phase Liquid Contaminants

W. J. Weber, Jr.[1], S. Mukherji[1], and C. A. Peters[2]

Contents

1 Introduction

Environmentally important nonaqueous phase liquids (NAPLs) are commonly composite mixtures comprised of similar classes of organic compounds. Such composite NAPLs (CNAPLs) are also usually variable in composition and incompletely characterized in terms of constituents and property values. Coal tars, creosotes, and other oily residuals from the combustion of hydrocarbon fuels, for example, usually include a wide array of polynuclear aromatic hydrocarbons (Mueller et al. 1989; Luthy et al. 1994). Because many of the constituents of such CNAPLs are considered toxic and carcinogenic (Neff 1979), knowledge of their

[1] Environmental and Water Resources Engineering, Department of Civil and Environmental Engineering, The University of Michigan, Ann Arbor, Michigan, USA
[2] Department of Civil Engineering and Operations Research, Princeton University, Princeton, New Jersey USA

H. Rubin/N. Narkis/ J. Carberry
Soil and Aquifer Pollution

aqueous dissolution rates and dissolution equilibria is crucial for predicting environmental impacts and for assessing the effectiveness of water-based remediation technologies.

In this chapter we describe the results of experimental investigations and model simulations designed to characterize the phase equilibria and rates of mass transfer of polynuclear aromatic hydrocarbon (PAH) components from CNAPL mixtures. In the absence of thorough characterization, which is often not feasible for environmental CNAPLs, phase equilibria and mass transfer studies cannot be completely rigorous. We therefore discuss first a novel experimental approach for producing synthetic PAH-CNAPLs that can serve not only as chemically and phenomenologically representative models for complex coal tars and creosotes, but also are additionally and uniquely well characterized. Although most PAHs are solids in their pure form at ambient temperature, we demonstrate how the principle of liquefaction through mixing can be exploited to produce PAH-containing nonaqueous liquids. Investigations of phase equilibria in completely mixed batch reactors (CMBRs) with CNAPL droplets dispersed in water are then discussed and analyzed. The CNAPL phase activity coefficient of each component is experimentally determined. The third major topic considered in the chapter is the simultaneous determination of phase equilibria and intrinsic mass-transfer rates. To facilitate these investigations we designed and fabricated a reactor system that provides a quantifiable and essentially constant interfacial area between internally well mixed but segregated aqueous and CNAPL phases. The CNAPL-phase activity coefficient and the intrinsic mass-transfer coefficient for each component of several CNAPL mixtures are quantified by simulating aqueous-phase solute concentration profiles with a single-impedance mass-transfer model using a statistical parameter search and data-fitting routine.

2 Dissolution Theory

2.1 Dissolution Equilibria

The fugacities of each species of a CNAPL in contact with an aqueous phase are equal in both phases under conditions of thermodynamic equilibrium, so that

$$\alpha_i^n X_i^n = \alpha_i^a X_i^a, \tag{1}$$

where the subscript i denotes quantities pertaining to component i, the superscripts a and n denote quantities pertaining to the aqueous and CNAPL phases, respectively, α is the activity coefficient for component i, and X is its mole fraction.

For a CNAPL composed of sparingly soluble hydrophobic components that form only dilute aqueous solutions, the aqueous-phase activity coefficient for component i can be expressed as

$$\alpha_i^a = \frac{1}{V_M^a C_{s,i}} \left(f_i^s \Big/ f_i^l \right), \tag{2}$$

where V_M is molar volume, C_S is the aqueous solubility of the pure component, and the quantity within the parentheses is the ratio of the solid- and liquid-phase fugacities of the pure component. The ratio of the aqueous solubility to the solid-liquid fugacity ratio is the supercooled liquid solubility. This equation assumes that the aqueous-phase activity coefficient for dissolution of a component of a CNAPL is the same as that for dissolution from its pure liquid state at a given temperature (Peters and Luthy 1993). The aqueous-phase concentration of species i in equilibrium with the CNAPL can thus be expressed as

$$C^a_{e,i} = \frac{X^n_i \alpha^n_i C_{s,i}}{\left(f^s_i / f^l_i \right)} . \tag{3}$$

Equilibrium is often described in terms of a CNAPL-water partition coefficient, $K_{P,i}$, given by the ratio of the concentration of component i in the NAPL phase to its corresponding concentration in the aqueous phase at equilibrium. This quantity can be expressed in terms of Eqs. (2) and (3) as

$$K_{P,i} = \frac{C^n_i}{C^a_i} = \frac{1}{V^n_M \alpha^n_i C_{s,i}} \left(f^s_i / f^l_i \right). \tag{4}$$

Another common assumption for structurally similar CNAPLs is that the nonaqueous-phase activity coefficient is unity. This assumption, referred to as the Raoult's law assumption, implies ideal behavior in the CNAPL phase, i.e., the molecules of a species in a CNAPL behave as if they existed in their pure liquid phase (Schwarzenbach et al. 1993). Differences in the sizes and shapes of molecules and intermolecular forces acting between the different components of the mixture may invalidate this assumption. If the assumption is valid, however, and if the nonaqueous phase is compositionally well defined, the equilibrium aqueous-phase concentration of a species and its CNAPL-water partition coefficient can be determined from Eqs. (3) and (4). For environmental CNAPLs of poorly characterized composition, uncertainties involved in estimates of bulk property values such as molecular weight and molar volume can lead to errors in equilibrium predictions.

2.2 Dissolution Rates: Single-Impedance Mass-Transfer Model

For the purpose of this study, mass transfer is conceptualized as being motivated by a linear driving force through a single-impedance domain comprised of an interfacial film between the bulk CNAPL and aqueous phases, such that the source and sink terms can be expressed as

$$\left(\frac{d\left(C^a_i V^a\right)}{dt} \right) = -\left(\frac{d\left(C^n_i V^n\right)}{dt} \right) = k_{f,i} A^o \left(C^a_{e,i} - C^a_i \right), \tag{5}$$

where C is the concentration corresponding to time, t, V is the volume of a phase, A^o is the interfacial area between phases, and k_f is the film mass-transfer coefficient. Assumptions inherent in this conceptual model are: (1) mass transfer through the film constitutes the rate-limiting step, (2) equilibrium conditions exist where the film contacts the bulk nonaqueous phase, and (3) the concentration gradient is linear across the thickness of the film (Weber and DiGiano 1996).

3 Phase Stability of Multicomponent Liquids

Phase-stability criteria for multicomponent liquids containing components that are solids under standard conditions have been discussed by Peters et al. (1997). The addition of an excess of a solid component to a composite nonaqueous liquid phase will eventually lead to precipitation. CNAPL-phase composition stability criteria for solid components are based on the condition that the aqueous-phase concentration of a component cannot exceed a value corresponding to the equilibrium concentration that would be obtained if the component were present as a pure compound, that is

$$X_i^n \leq \frac{1}{\alpha_i^n}\left(f_i^s / f_i^l\right). \tag{6}$$

The fugacity ratio, indicative of the energy required to convert a solid to a liquid, can be estimated using readily available compound-specific data (Schwarzenbach et al. 1993). If the Raoult's law assumption is valid, Eq. (6) can give an estimate of the maximum mole fraction of a solid component that can be achieved in a compositionally well-defined CNAPL. Because the fugacity ratio for liquid components is unity, such components have no maximum mole fraction limit.

4 Materials and Methods

4.1 Synthesis of PAH-Containing Composite Nonaqueous Phase Liquids

Most individual PAH compounds exist as solids at ambient temperature. Using the stability criterion defined in Eq. (6) for each PAH component, CNAPLs were developed to serve as models for environmental coal tar/creosote mixtures. A stock CNAPL mixture was prepared initially by dissolving predetermined amounts of 1-methylnaphthalene, 2-ethylnaphthalene, acenaphthene, fluorene, phenanthrene, fluoranthene, and pyrene in toluene. The bulk of the toluene was then removed, first by evaporation in a rotary evaporator and then by purging with nitrogen, to yield a residual toluene mole fraction of about 0.05. Three additional synthetic CNAPLs were then produced by adding varying amounts of naphthalene to aliquots of the stock, yielding mixtures having naphthalene mole fractions of 0.05, 0.1, and 0.25.

4.2 Equilibrium Experiments

The CNAPL–water equilibrium experiments were conducted in CMBRs comprised of 25-ml glass centrifuge tubes containing approximately 0.5 g of CNAPL and 25 ml of milli-Q water. The tubes were placed in a tumbler device so that CNAPL droplets were dispersed in the water while the two phases were equilibriated over a 2-week period. The phases were then separated by centrifugation at 7500 rpm for 30 min (Sorvall Centrifuge Model RC-5B, Dupont, Inc.). Samples were collected from each of the two phases, diluted in methanol (using a dilution factor of about 1.5–2 for aqueous samples and 1000 for CNAPL samples) and stored at 5 °C to minimize PAH losses prior to analysis.

4.3 Mass-Transfer Experiments

The reactor depicted in Fig. 1 was designed to maintain a known and virtually constant interfacial area of 4.9 cm^2 between the aqueous and CNAPL phases at an approximately fixed position at the top of the lower reservoir and to provide complete internal mixing (at selected speed settings) in the bulk phases on each side of the interface. At the outset of each mass-transfer experiment, the reactor was partially filled with water and a 20-ml volume of CNAPL was introduced gradually through the water using a syringe, so that an undisturbed interface was formed at the top of the lower reservoir. The reactor was mounted on a magnetic stirrer and filled with water (700–750 ml) to the level of the overflow tube using a side inlet. The aqueous phase was stirred mechanically at 39 rpm and the CNAPL phase was stirred magnetically at a fixed setting. The experiments were conducted by allowing the aqueous-phase concentrations to approach equilibrium over

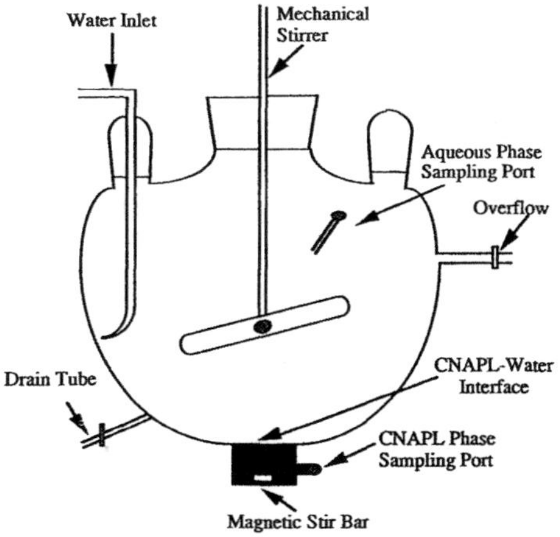

Fig. 1. Schematic of reactor used in mass-transfer experiments

five to seven days while operating the reactor in batch mode. The aqueous phase was sampled regularly. The CNAPL phase was sampled at the beginning and at the end of a study. All samples were diluted in methanol immediately after collection.

4.4 Analytical Methods

Because of the large differences in concentration ranges involved, different analytical techniques were used for analyzing CNAPL- and aqueous-phase samples. Aqueous-phase concentrations were analyzed using a Hewlett Packard 1090 high performance liquid chromatigraph (HPLC), equipped with a diode array detector (DAD), a fluorescence detector (FLD), and a 125 × 3.2 mm reverse phase octadecylsilane column (Envirosep-PP, Phenomenex). A 15-μl volume sample was injected, and the chromatographic conditions were programmed for an acetonitrile-water gradient elution at a constant flow rate of 0.5 ml/min and a constant temperature of 35 °C over a 35-min run. The CNAPL-phase concentrations of each component were determined using a Hewlett Packard 5890A gas chromatograph (GC) equipped with a flame ionization detector and a 5%-phenyl methyl DB5 (J&W Scientific) megabore column (length 30 m, inner diameter 0.53 mm, and film thickness 1.5 μm). Oven-temperature programming was set at 60 °C for 5 min, followed by a linear increase of 10 °C per minute to 250 °C, at which point the temperature was held constant through the end of the 35-min run. The injector and detector temperatures were set at 250 and 300 °C, respectively and an injection volume of 3 μl was used. Mixed PAH external standards were used and duplicate injections were made for all samples and standards.

4.5 Data Analysis

4.5.1 Equilibrium Experiments

Because the CNAPLs were comprised solely of the compounds identified and impurities were negligible, it was possible to use Eq. (4) and the Raoult's law assumption of unity for nonaqueous-phase activity coefficient values to predict the CNAPL-water partition coefficient. The predicted values were then compared to the experimentally determined quantities. Alternately, the CNAPL-phase activity coefficients were estimated by rearranging Eq. (3) and substituting experimentally determined values of the nonaqueous-phase mole fraction and corresponding aqueous-phase concentration at equilibrium. Pure compound solubility and fugacity ratio values at 25 °C obtained from Mackay et al. (1992) were used in all compulations.

4.5.2 Mass-Transfer Experiments

For dissolution of a CNAPL in water in a CMBR containing constant phase volumes and a relatively large nonaqueous- to aqueous-phase volume ratio, the change in the CNAPL-phase composition due to dissolution may be assumed to be neg-ligible, such that the mole fraction of a component and, correspondingly, the aqueous-phase concentration at equilibrium remains constant. With these simplifying assumptions, the aqueous-phase concentration profile due to dissolution is given by

$$C_i^a = C_{e,i}^a - \left(C_{e,i}^a - C_{o,i}^a\right) exp\left(-\frac{A^o k_{f,i}}{V^a}\left(t - t_o\right)\right). \tag{7}$$

Experimentally measured aqueous-phase concentration-time profiles were fitted with Eq. (7). Values for the aqueous-phase concentration in equilibrium with the CNAPL, $C_{e,i}^a$, and the single-impedance film mass-transfer coefficient, $k_{f,i}$, along with their respective confidence intervals, were obtained using a nonlinear parameter estimation routine (SYSTAT Inc., Evanston, Illinois). The interfacial area was taken as $4.9\,cm^2$, the cross-sectional area of the reactor at the interface location, with the assumption that potential contributions of roughness and curvature of the liquid-liquid interface to interfacial area were negligible. The CNAPL-phase activity coefficient was then determined by rearranging Eq. (3) and substituting the experimentally measured CNAPL-phase mole fraction, the fitted value of the equilibrium aqueous-phase concentration, and appropriate compound-specific property values.

5 Results

5.1 Composite Nonaqueous Phase Liquid Formulations

The compositions and properties of the CNAPLs used in this study are summarized in Table 1. Mole fractions are based on CNAPL-phase concentration measurements. Estimated standard errors for these measurements are typically less than 1% for most compounds other than toluene, for which the standard error is about 8%.

5.2 Results of Equilibrium Experiments

The estimated standard error for aqueous-phase concentration measurements was typically less than 5% for most compounds, but as high as 20% for pyrene and fluoranthene. Activity coefficients determined in equilibrium experiments with CNAPL I and CNAPL III are given in Table 2. Comparisons of experimentally obtained CNAPL-water partition coefficients for the various compounds with Raoult's law predictions for CNAPL I are presented in Fig. 2.

Table 1. Composition and bulk properties of synthetic CNAPLs

Components	CNAPL I	CNAPL II	CNAPL III	CNAPL IV
	Mole fractions			
Toluene	0.05	0.05	0.04	0.03
Naphthalene	0.00	0.05	0.10	0.25
1-Methylnaphthalene	0.29	0.28	0.26	0.22
2-Ethylnaphthalene	0.14	0.13	0.13	0.11
Acenaphthene	0.15	0.14	0.14	0.11
Fluorene	0.07	0.07	0.06	0.05
Phenanthrene	0.13	0.12	0.12	0.10
Fluoranthene	0.12	0.11	0.10	0.09
Pyrene	0.05	0.05	0.05	0.04
Mixture properties	Average property values			
Molecular weight (g/mol)	160.2	158.5	157.0	152.2
Density (g/cc)	1.065	1.062	1.060	1.055
Viscosity × 10^2 (g cm^{-1}-s^{-1})	5.05	4.94	4.65	4.00

Table 2. Activity coefficients determined in equilibrium experiments

Compound	CNAPL phase activity coefficients[a]	
	CNAPL I: $X_{nap} = 0.0$	CNAPL III: $X_{nap} = 0.1$
Toluene	0.98	0.85
Naphthalene	–	0.88
1-Methylnaphthalene	1.09	1.07
2-Ethylnaphthalene	1.05	1.03
Acenaphthene	0.86	0.86
Fluorene	0.77	0.76
Phenanthrene	0.69	0.69
Fluoranthene	0.72	0.82
Pyrene	0.48	0.49

[a] Averages of triplicate runs.

5.3 Results of Mass-Transfer Experiments

The best-fit values of the mass-transfer parameters determined in three replicate mass-transfer runs for all compounds in CNAPL I (devoid of naphthalene) are listed in Table 3.

The numbers within the parentheses represent 95% confidence intervals on the parameter estimates. Variations in mole fractions over the course of a mass-transfer run were found to be negligible, thereby validating the assumption inherent in Eq. (7). The absence of values for pyrene and fluoranthene in run 2 relates

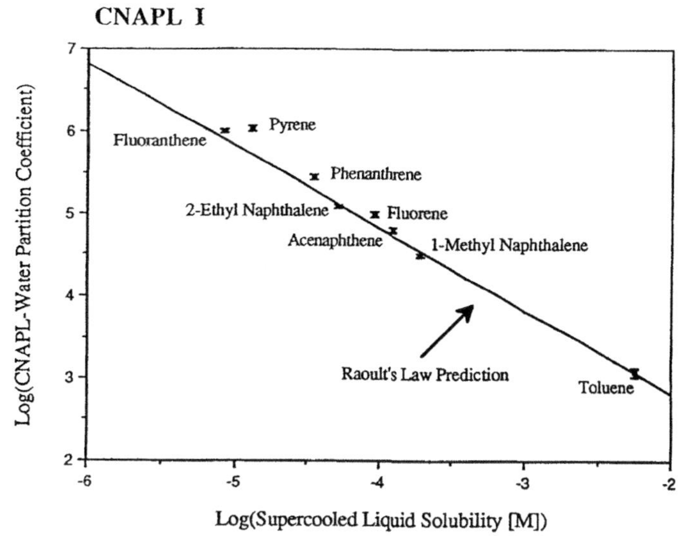

Fig. 2. Partition coefficient versus supercooled liquid solubility: CNAPL I

Table 3. Mass-transfer parameters for CNAPL I

Components	Run	Activity coefficient α_i^n (± 95% CI)	Mass-transfer coefficient $k_{f,i}$ (± 95% CI) (cm/s × 10^3)
Toluene	1	1.27 (1.25–1.30)	1.07 (0.98–1.16)
	2	1.25 (1.22–1.28)	1.35 (1.21–1.48)
	3	1.15 (1.14–1.17)	1.62 (1.54–1.71)
1-Methylnaphthalene	1	1.06 (1.05–1.07)	1.15 (1.11–1.19)
	2	1.10 (1.08–1.12)	1.37 (1.25–1.50)
	3	1.17 (1.15–1.18)	1.42 (1.36–1.49)
2-Ethylnaphthalene	1	0.98 (0.97–1.00)	1.29 (1.20–1.37)
	2	1.02 (1.00–1.05)	2.01 (1.66–2.35)
	3	1.20 (1.18–1.22)	1.36 (1.2901.44)
Acenaphthene	1	0.79 (0.78–0.80)	1.29 (1.24–1.35)
	2	0.84 (0.82–0.85)	1.53 (1.37–1.69)
	3	0.96 (0.94–0.98)	1.28 (1.21–1.36)
Fluorene	1	0.66 (0.65–0.67)	1.21 (1.14–1.27)
	2	0.69 (0.67–0.70)	1.58 (1.38–1.79)
	3	0.77 (0.75–0.80)	1.38 (1.24–1.51)
Phenanthrene	1	0.62 (0.61–0.63)	1.62 (1.50–1.74)
	2	0.64 (0.62–0.65)	2.26 (1.87–2.65)
	3	0.66 (0.65–0.67)	1.48 (1.42–1.55)
Fluoranthene	1	0.65 (0.64–0.67)	2.57 (2.15–2.99)
	2	–	–
	3	0.55 (0.54–0.55)	1.39 (1.33–1.45)
Pyrene	1	0.40 (0.39–0.41)	2.02 (1.72–2.33)
	2	–	–
	3	0.35 (0.34–0.36)	1.54 (1.43–1.65)

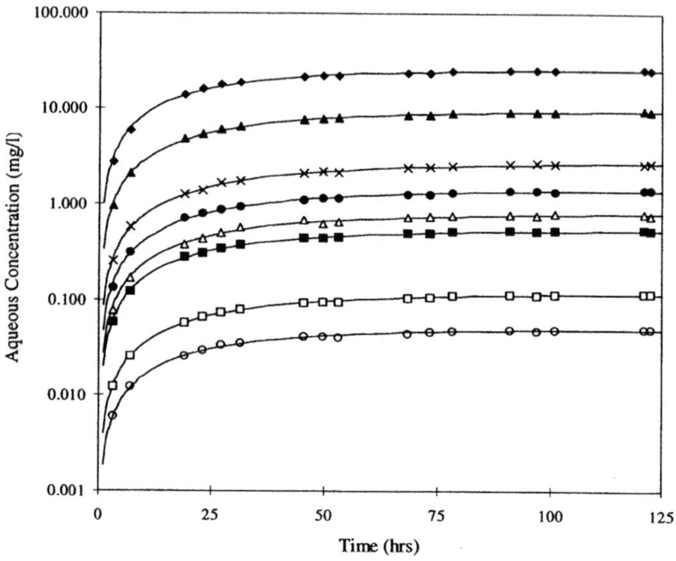

Legends:

◆ Toluene

▲ 1-Methylnaphthalene

● 2-Ethylnaphthalene

x Acenaphthene

Δ Fluorene

■ Phenanthrene

□ Fluoranthene

o Pyrene

Fig. 3. Model fit to aqueous concentration profiles: CNAPL I

to the fact that Eq. (7) could not be used to fit their respective aqueous concentration profiles. With these two exceptions, Eq. (7) provided good fits to experimentally generated aqueous-phase concentration profiles for all compounds. Figure 3 illustrates an example of experimental data and corresponding model fits for Eq. (7) for CNAPL I – run 3. A logarithmic ordinate scale is used in order to depict data and model fits for all components on the same plot. The best-fit mass-transfer parameter values for naphthalene for duplicate runs in CNAPLs II, III, and IV (with naphthalene mole fractions of 0.05, 0.1, and 0.25, respectively), are compiled in Table 4.

Table 4. Mass-transfer parameters for naphthalene

Naphthalene mole fraction $X^{n}_{i,nap}$	Run	Activity coefficient α^{n}_{i} (± 95% CI)	Mass-transfer coefficient $k_{f,i}$ (± 95% CI) (cm/s × 10^3)
CNAPL II: 0.05	1	0.84 (0.79–0.86)	1.24 (1.17–1.31)
	2	0.88 (0.86–0.91)	1.15 (1.06–1.24)
CNAPL III: 0.1	1	0.92 (0.91–0.93)	1.08 (1.04–1.13)
	2	0.92 (0.91–0.93)	1.24 (1.18–1.31)
CNAPL IV: 0.25	1	0.80 (0.83–0.80)	1.22 (1.11–1.33)
	2	0.84 (0.83–0.85)	1.11 (1.04–1.18)

6 Discussion and Conclusions

We have demonstrated that stable CNAPLs can be formed if the stability criterion provided in Eq. (6) is conformed to by each constituent. Assuming that Raoult's law (i.e., nonaqueous-phase activity coefficients of unity) is valid for CNAPLs comprised of structurally similar constituents, theoretically estimated values of fugacity ratios can be used to estimate the maximum mole fraction of a solid component that can dissolve in a CNAPL. For a complex CNAPL, this stability criterion also implies that, as lower molecular weight compounds are lost via such processes as volatilization and biodegradation, the higher molecular weight compounds will increase in relative abundance. The mole fractions of these compounds will eventually exceed their fugacity ratios, thereby leading to CNAPL-phase instability and precipitation of solids. This is likely to have an impact on contaminant concentrations in the aqueous phase.

The relationship between partition coefficients and supercooled liquid solubilities obtained from CMBRs in which CNAPL droplets were dispersed in water is demonstrated in Fig. 2 for the constituents of CNAPL I. While it appears from this figure that the experimentally determined values agree well with Raoult's law predictions, it is to be noted that the plots are on a log-log scale and that the activity coefficient values determined in these same experiments varied by up to a factor of about two from the Raoult's law prediction of unity. The CNAPL-phase activity coefficient deviation from unity is greatest for pyrene, which coincidentally deviates most from the mixture average in terms of molecular size. The activity coefficients of compounds in CNAPL I and CNAPL III are very similar. This can be expected as the addition of 0.1 mol of naphthalene to CNAPL I (to obtain CNAPL III) causes only a slight reduction in the mole fraction of other compounds, as seen in Table 1.

Table 3 presents a comparison of mass-transfer parameters across compounds obtained from the mass-transfer experiments with CNAPL I. There appears to be a compound-specific trend in the deviation of the CNAPL-phase activity coefficient from the Raoult's law prediction of unity. Toluene shows a positive deviation from Raoult's law, with activity coefficient values in the range 1.1–1.3; 1-

methylnapthalene and 2-ethylnaphthalene are in good agreement with Raoult's law; and all other compounds exhibit negative deviations with activity coefficients less than unity. Acenaphthene, which has a molecular weight (MW = 154) close to the mixture average (MW of CNAPL I = 160), shows a slight negative deviation with an activity coefficient in the range 0.79–0.96, whereas pyrene (MW = 202), which deviates most significantly from the mixture average, shows a significant negative deviation with an activity coefficient in the range of 0.35–0.4. These compound-specific variations in activity coefficient values are similar to those observed in the equilibrium experiments with CNAPL droplets dispersed in water. This compound-specific trend in variation of CNAPL-phase activity coefficients from unity may reflect either deviations from Raoult's law or errors in inputs used in obtaining Raoult's law predictions, e.g., pure compound solubility and fugacity ratio estimates. As shown in Table 4, the activity cofficient of naphthalene was found consistently to be in the range 0.8–0.92 across the various CNAPLs, despite the fact that its mole fraction varied from 0.05–0.25. Mass-transfer coefficient values were found to be in the range $(0.9–3.0) \times 10^{-3}$ cm/s (based on an interfacial area estimate of 4.9 cm^2) for all CNAPL constituents. No compound-specific trend was observed, i.e., variations between replicate runs were greater than variations across compounds. This result can be explained by the narrow range of diffusivity values represented by the compounds comprising the CNAPLs ($6.3–9.1 \times 10^{-6}$ cm^2/s). From Table 2 it appears that the single-impedance mass-transfer coefficient for naphthalene remains essentially constant, as the mole fraction of this component is varied in the range 0.05–0.25. There thus seems to be no significant trend with respect to variations in the mass-transfer coefficient of a component with its CNAPL-phase mole fraction as long as the CNAPL-phase characteristics do not change significantly. It is common in the literature to quantify mass transfer in terms of a system-specific lumped mass-transfer coefficient. In most such cases this parameter cannot be converted to an intrinsic film-transfer coefficient due to variable and/or unquantified interfacial areas involved. However, in cases for which interfacial area has been estimated, reported film-transfer coefficient values for PAH mass transfer from pure solid phases (Grimberg et al. 1995) and from nonaqueous liquid phases (Efroymson and Alexander 1994; Ghoshal et al. 1996) are of similar order as those obtained in this study, although mixing conditions and system configurations employed in those studies were notably different.

The primary implications of this study can be summarized as follows: (1) even in CNAPLs comprised of structurally similar components, variatons from Raoult's law predictions by up to a factor of two can be expected in the activity coefficients for components deviating most significantly from the mixture average; (2) the single-impedance mass transfer coefficient determined for one component can serve as a reasonably representative value for other similar components, particularly if their respective free-liquid diffusivities are comparable; and (3) predictions of component dissolution may be limited if a CNAPL is incompletely characterized, thereby leading to uncertainties in bulk-property values and mole-fraction composition, as well as uncertainties as to whether a particular component is structurally similar to the overall CNAPL phase (as defined by all of its components).

Acknowledgments. The authors thank Alois M. Gerlach and Ryan O'Connor for their laboratory assistance in the experimental aspects of this study and Harald W. Eberhart for fabricating the reactor. This research was funded in part by Grant ES04911 from the US National Institute of Environmental Health Sciences.

References

Efroymson RA, Alexander M (1994) Role of partitioning in biodegradation of phenanthrene dissolved in non-aqueous phase liquids. Environ Sci Technol 28:1172–1179

Ghoshal S, Ramaswami A, Luthy RG (1996) Biodegradation of naphthalene from coal tar and heptamethylnonane in mixed batch systems. Environ Sci Technol 30:1282–1291

Grimberg SJ, Nagel J, Aitken MD (1995) Kinetics of phenanthrene dissolution into water in the presence of nonionic surfactants. Environ Sci Technol 29:1480–1487

Luthy RG, Dzombak DA, Peters CA, Roy SJ, Ramaswami A, Nakles DV, Nott BR (1994) Remediating tar-contaminated soils at manufactured gas plant sites. Environ Sci Technol 28:266A–276A

Macay D, Shiu WY, Ma KC (1992) Illustrated handbook of physical and chemical properties and environmental fate for organic chemicals, vol 1–2, Lewis. Boca Raton

Mueller JG, Chapman PJ, Pritchard PH (1989) Creosote-contaminated sites. Environ Sci Technol 23:1197–1201

Neff JM (1979) Polycyclic aromatic hydrocarbons in the aquatic environment: sources, fates and biological effects. Applied Science Publ London, pp 102–115

Peters CA, Luthy RG (1993) Coal tal dissolution in water-miscible solvents: experimental evaluation. Environ Sci Technol 27:2831–2843

Peters CA, Mukherji S, Weber WJ Jr (1996) Phase stability of multicomponent NAPLs containing PAHs and a novel technique for laboratory investigations. (submitted)

Peters CA, Mukherji S, Knightes CD, Weber WJ Jr (1997) Phase stability of multicomponent NAPLs containing PAHs. Environ Sci Technol (submitted)

Schwarzenbach RP, Gschwend PM, Imboden DM (1993) Environmental organic chemistry, chaps 4, 5. John Wiley, New York

Weber WJ, Jr, DiGiano FA (1996) Process dynamics in environmental systems, chap 4. John Wiley, New York

9 Release of Soil Organics by Chlorophenols

N. I. Galil[1] and J. T. Novak[2]

Contents

1 Introduction

The relative resistance of chlorophenols to degradation creates pollution problems when these compounds enter the environment (McGinnis et al. 1989). Due to their hydrophobic characteristics, chlorophenols are expected to be adsorbed from water solutions in both surface and ground water systems. Sorption of pentachlorophenol from aqueous solutions has been widely studied (Schellenbery et al. 1984; Johnson et al. 1985; Patterson and Liebscher 1987 and others). Lee et al. (1990) showed that for a pH less than 7, PCP sorption increases with an increase in the fractions of the PCP neutral species while for a pH greater than 7, it appears that anionic PCP is sorbed as a neutral ion pair. Regardless of the pH-dependent mechanisms, sorption increased with increasing soil organic matter.

In addition to sorption, the distribution of an organic contaminant between the solid and aqueous phases is also influenced by its association with solution colloids and cosolvency due to the presence of other dissolved constituents. Pinal et al. (1990) showed that in an aqueous environment, partially miscible organic solvents with strong polar functional groups are most likely to impart a significant

[1] Department of Civil Engineering, Environmental and Water Resources Engineering, Technion – Israel Institute of Technology, Haifa 32000, Israel
[2] Department of Civil and Environmental Engineering, Virginia Polytechnic Institute and State University, Blacksburg, Virginia 24061 USA

H. Rubin/N. Narkis/ J. Carberry
Soil and Aquifer Pollution

cosolvency effect. An increase in the solubility of hydrophobic organic compounds (HOC) in the presence of cosolvents is coupled with a decreased sorption by soils. Important cosolvents of natural origin are humic and fulvic acids. Choi and Aomine (1972, 1974a,b) reported enhanced solubility of some organic pollutants and pesticides in the presence of humic and fulvic acids. They reported that the extent of solubility enhancement depended on the concentration, source and molecular weight of the dissolved organic matter.

Mobile colloidal particles may be an important carrier of contaminants through the vadose and within the saturated zone of the subsoil. McCarthy and Zachara (1989) suggested that because the composition of colloids in ground water is expected to be chemically similar to that of the surfaces of the immobile aquifer, these particles also could sorb organic contaminants and stabilize them in the mobile phase. In effect, colloids act as a third phase, enhancing the mobility of colloidal bound HOCs.

As part of a study dealing with biodegradability of sorbed organics, it was noted that exposure of a soil to aqueous HOCs appeared to release soil organics into solution. Because this apparent release was of potential import as a source of material for enhanced transport, a study focusing on this aspect of soil and aqueous HOC interaction was formulated. Specific objectives of this study were to determine the influence of pentachlorophenol (PCP) and monochlorophenol (MCP) on the solubility of soil organic matter for several soils.

2 Materials and Methods

Two soils were used in this study. One soil, obtained from a site in Newport News, Virginia, at a depth of 0.5–1.0 m, was characterized by a pH of 5.35, a total organic content of 1.8% and an extractable humic matter of 0.2%. The other soil was a surface soil obtained at Eagle Point, New Jersey. This soil was characterized by a total organic content of 8.25% and an extractable humic content of 0.6%. Its pH was 7.23. The characteristics of the soils used in the experiments are summarized in Table 1.

The total organic content was determined by weight, using loss-on-ignition after heating soil samples to 400 °C for 24 h (Davies 1974). The humic matter was determined by extraction from the soil using NaOH at pH > 12, under a nitrogen atmosphere (Schnitzer and Khan 1978), separating the liquid by centrifugation and then adjusting the pH with HCl to less than 2, for 24 h. Finally, the humics were separated by filtration and weighed

PCP and MCP analyses were performed following the EPA method (US EPA 1986). In this case, PCP and MCP were separately extracted from the water solutions by methylene chloride and injected to a gas chromatograph using a capillary column (DB-5, 30 m × 0.53 mm i.d., J&WP/N 125-5032) and a flame ionization detector.

The water solution used in the experiments was 95% tap water (containing 10 mg/l calcium, 4 mg/l magnesium, 7 mg/l sodium, 9 mg/l chloride, 7 mg/l sulfate,

Table 1. Characteristics of the soils used in the experiments

Parameters	Newport News	Eagle Point
pH	5.35	7.25
Total organics (%) (combusted at 400 °C)	1.78	8.25
Humic acids (%)	0.18	0.56
Physical composition:		
Sand (%)	57.0	18.5
Silt (%)	27.2	51.5
Clay (%)	15.8	30.0

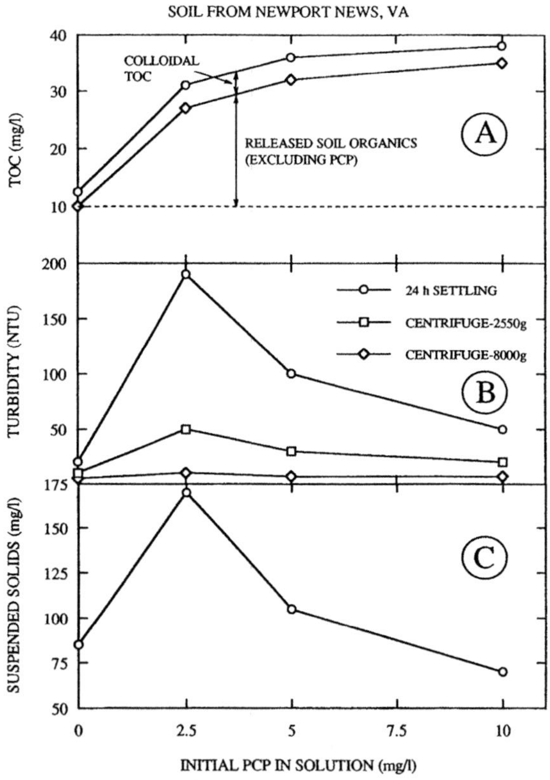

Fig. 1. **A** Solubilization of soil organics. **B** Stabilization of colloids. **C** Suspended solids resulting from contact of aqueous PCP with Newport News, Virginia, soil

and 36 mg/l bicarbonate) and 5% buffer solution (containing 8500 mg/l KH_2HPO_4, 21 700 mg/l K_2HPO_4, 33 400 mg/l $Na_2HPO_47H_2O$, and 1700 mg/l NH_4Cl), resulting in a final solution with an electrical conductance of about 1000 μs/cm and an ionic strength of about 0.017 M.

3 Results

3.1 Experimental

Experiments were conducted using, separately, PCP and MCP, in order to characterize the release of soil organic matter, resulting from the contact with chlorophenols in aqueous solutions. Samples of 100 ml containing different concentrations of either PCP or MCP were contacted with 10 g of soils the same origin (Newport News or Eagle Point) with a contact time of 24 h and mixed using a shaker table.

Separate previous analyses of MCP and PCP solutions enabled the organic carbon of the chlorophenols to be subtracted from the total TOC present at equilibrium so that the concentration of extracted organic matter in solution could be determined.

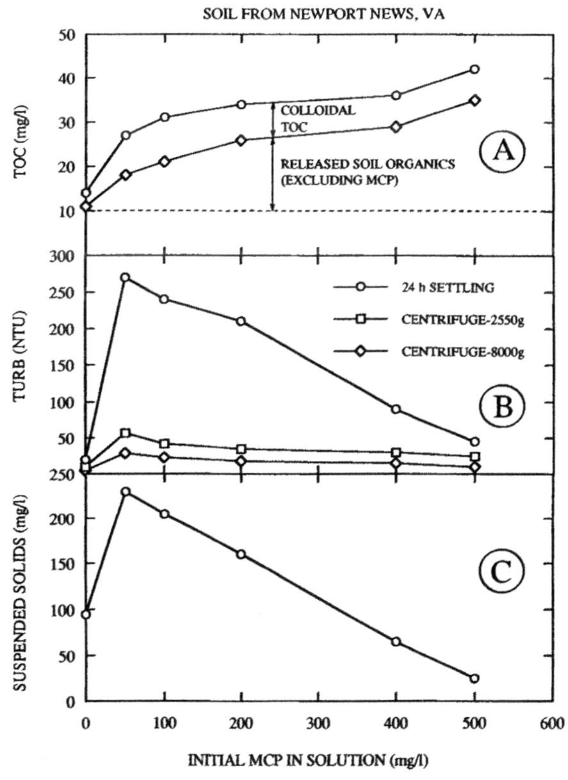

Fig. 2. **A** Solubilization of soil organics. **B** Stabilization of colloids. **C** Suspended solids resulting from contact of aqueous MCP with Newport News, Virginia, soil

3.2 Release of Soil Organics

Parts A in Figs. 1, 2, 3, and 4 describe the amounts of soil organics released to the aqueous solution as a result of contacting the soil samples with different concentrations of MCP or PCP. The contact of aqueous solutions, without additions of chlorophenols, resulted in the release of about 10 mg/l of dissolved TOC from the Newport News soil and about 50 mg/l dissolved TOC from the Eagle Point soil.

The addition of chlorophenols to the aqueous solutions always resulted in increased concentrations of dissolved organic matter, after being contacted with soil samples. PCP in concentrations of 2.5 to 10 mg/l increased the dissolved TOC up to about 35 mg/l with the Newport News soil and up to 125 mg/l with the Eagle Point soil. MCP in concentrations of 50 to 500 mg/l increased the dissolved TOC up to 40 mg/l and above with Newport News and up to 145 mg/l with Eagle Point.

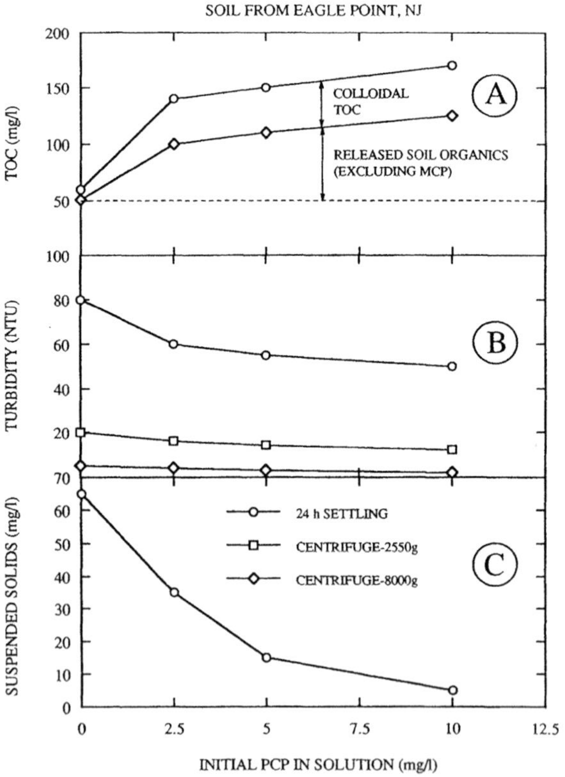

Fig. 3. **A** Solubilization of soil organics. **B** Stabilization of colloids. **C** Suspended solids resulting from contact of aqueous PCP with Eagle Point, New Jersey, soil

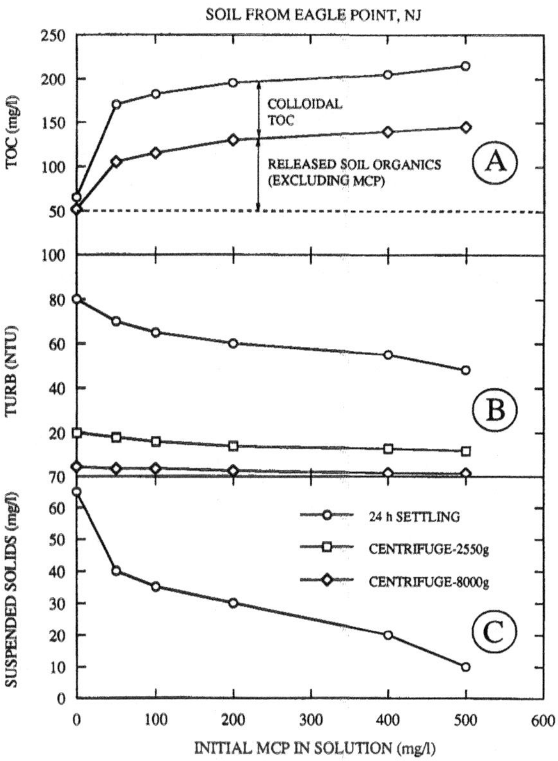

Fig. 4. A Solubilization of soil organics. **B** Stabilization of colloids. **C** Suspended solids resulting from contact of aqueous MCP with Eagle Point, NerJersey, soil

3.3 Release of Soil Colloids

The mixing stage was followed by a 24-h settling period and then supernatant turbidity and suspended solids content were measured. After separation by settling, the supernatant organic content and turbidity were measured. The results (Figs. 1, 2, 3 and 4, parts B) show high turbidity values for the soil originating from Newport News, at initial PCP concentrations of 2.5 mg/l and MCP concentrations of 50 mg/l. Centrifugal separation at 2550 *g* reduced the turbidity, but substantial levels of turbidity still remained and were always highest at 2.5 mg/l PCP and 50 mg/l MCP, in the case of the Newport News soil. At 8000 *g* the turbidity in all the samples was reduced to less than 10 NTU.

Based on microscope observations, the colloids brought into the suspension by PCP or MCP addition and mixing were clay particles. As the solution concentration was increased above 2.5 mg/l PCP and above 50 mg/l MCP, physical separation of the colloids became easier.

This experiment was repeated under similar conditions and the same buffer solution, using the soil from Eagle Point, New Jersey. This soil had a higher organic content than the Newport News soil. In the case of the Eagle Point soil, the organics released by additions of a 2.5 mg/l PCP solution or a 50 mg/l MCP solution resulted in increased organic matter released to the solutions. Colloidal TOC was considered to be the difference between the settled TOC and the TOC following centrifugation at 8000 g. Higher concentrations of either PCP or MCP in solution resulted in additional release of soil organics, including colloidal particles (Figs. 1, 2, 3 and 4, parts A). Contacts between PCP or MCP with the Eagle Point soil did not result in as high turbidity as occurred with the Newport News soil and the colloidal separation by centrifugation was easier (Figs. 3 and 4, parts B).

3.4 Suspended Solids in Aqueous Solutions

Concentrations of suspended solids were measured in the aqueous solutions after the 24-h periods of settling, following the contact experiments with soil samples, The Newport News soil released the highest suspended solids concentration when contacted with 2.5 mg/l PCP or with 50 mg/l MCP; 170 and 230 mg/l, respectively (Figs. 1 and 2, parts C). The Eagle Point soil released the highest suspended solids concentration of about 65 mg/l when no chlorophenols were present in the solutions (Figs. 3 and 4, parts C). As the chlorophenols concentration in the aqueous solutions was increased, the suspended solids separation was improved.

3.5 Influence of pH

A matrix study was conducted in which samples of Newport News and Eagle Point soils were separately contacted with PCP and MCP, at different concentrations and

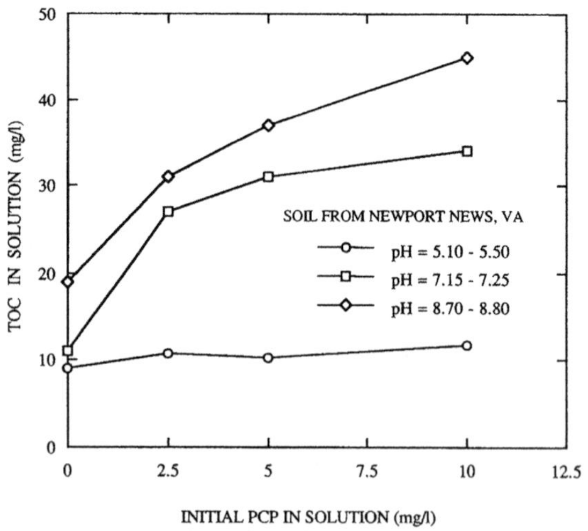

Fig. 5. Release of soil organics resulting from PCP-soil contact at different pH values

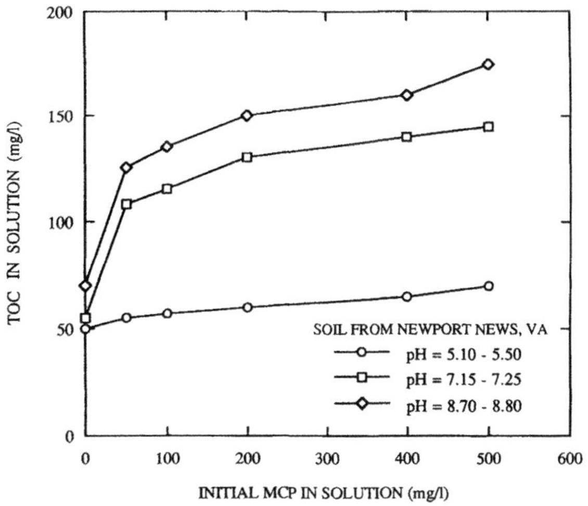

Fig. 6. Release of soil organics resulting from MCP-soil contact at different pH values

at three different pH levels. The pH was adjusted by the use of HCl. At pH values below 7, relatively small releases of soil organics could be observed (Figs. 5 and 6). When the pH was raised above 7, higher levels of soil organics were released into solution and their concentration was directly influenced by chlorophenol concentrations. The TOC concentrations were measured after centrifugal separations at 8000 g.

4 Discussion

The study showed that PCP and MCP, in aqueous solutions with soil, caused the release from the soil surface of soil organics and increased colloidal stability. The colloidal stabilization observed in the experiments was clearly influenced by the concentration and nature of the soil organic matter in solution. At higher concentrations of soil organics the separation of the colloids achieved either by settling or by centrifugation was improved.

Galil and Novak (1995) showed in similar conditions that most of the PCP remained in the aqueous phase, in association with the soil organics released by PCP, tentatively identified as fulvic acids. Sorption of organic solutes, especially HOCs, is inversely related to solubility. Enhanced solubility, which in this case could be caused by fulvic acids from the soil surface, would lead to a decrease in sorption. Water solubility enhancement by dissolved humic and fulvic acids from the soil have been determined for selective organic pollutants by Chiou et al. (1986), who reported significant solubility enhancements of HOC by dissolved humic materials from soil. In a later study, Chiou et al. (1987) showed that the enhancement effect varies with the molecular composition of the humic material.

Enhancement solubility of HOC by humic acid solutions was used by Abdul et al. (1990) to remove organic contaminants from hydrogeologic systems. The effects on solubility and sorption of organic chemicals upon addition of more organic material to an aqueous solution were defined by Rao et al. (1991) as cosolvency. Robinson (1990) and Crane (1992) found substantial sorption of chlorinated phenols to dissolved humic acids at pH 7.

Separations based on gel chromatography reported by Galil and Novak (1995) showed that all the PCP found in the aqueous phase was connected with the fulvics, creating a complex of higher molecular weight and without any "free" PCP in solution

In the present study, the two soils responded quite differently to exposure to PCP and MCP. Although the unique soil characteristics which account for these differences were not identified, the history and nature of the soils would suggest that differences are likely. The Eagle Point soil was from a river deposit and had been subject to glaciation. The pH was above 7 and organic content was high. The Newport News soil was at a lower pH, highly weathered and likely the result of oceanic deposition. For the Eagle Point soil, exposure to water would release recently deposited and loosely bound organic matter, much of which may have resulted from the river flooding. The Newport News organics would be expected to be more highly weathered and tightly bound. The exposure to higher pH water, in the absence of chlorophenols, released less organic matter than from the Eagle Point soil.

The binding of soil organic matter to soil surfaces is thought to be due to ionic interactions between positively charged surface oxides and negatively charged organic matter (Wershaw 1993). It is possible that the negatively charged phenolate ions attracted and desorbed with the soil organic matter, so that soil organic matter was released, resulting in the increase of TOC in solution.

The release of turbidity differed for the two soils. For the Eagle Point soil, the deposition of river deposits would result in a loose association between silts, clay and soil organic matter. With the more highly weathered Newport News soil, the clays are more likely to be bound or trapped within the soil organic matter and be released along with the humic and fulvic acids. Visual observations of the Newport News soil indicated that additions of either PCP or MCP always resulted in a release of turbidity, however at PCP levels above 2.5 mg/l and MCP above 50 mg/l, substantial recoagulation occurred over the 24-h settling period.

5 Summary

It was found that PCP and MCP in aqueous solution released colloids and low molecular soil organics, such as fulvic acids, from the soil creating a mobile complex. The affinity of the colloids to this complex varied with the type of soil and chlorophenol concentration.

This release of soil organics and colloids may provide a mechanism of enhanced transport of contaminants in both ground and surface water systems.

References

Abdul SA, Gibson LT, Rai ND (1990) Use of humic acid solution to remove organic contaminants from hydrogeologic systems. Environ Sci Technol 24:328–333
Chiou TC, Malcolm LR, Brinton IT, Kile ED (1986) Water solubility enhancement of some organic pollutants and pesticides by dissolved humic and fulvic acids. Environ Sci Technol 20:502–508
Chiou TC, Kile ED, Brinton IT, Malcolm LR, Leenheer JA (1987) A comparison of water solubility enhancements of organic solutes by aquatic humic acids. Environ Sci Technol 21:1232–1234
Choi J, Aomine S (1972) Effects of the soil on the activity of pentachlorophenol. Soil Sci Plant Nutr 16:255–260
Choi J, Aomine S (1974a) Adsorption of pentachlorophenol by soils. Soil Sci Plant Nutr 20:135–144
Choi J, Aomine S (1974b) Mechanisms of pentachlorophenol adsorption by soils. Soil Sci Plant Nutr 20:371–379
Crane CE (1992) Sorption of pentachlorophenol to humic acids and subsequent effects on biodegradation and solvent extraction. MSc Thesis, Virginia Polytechnic Institute and State University, Blacksburg, Virginia
Davies BE (1974) Loss-on-ignition as an estimate of soil organic matter. Soil Sci Soc Am Proc 38:150–151
Galil NI, Novak JT (1995) Pentachlorophenol – induced release of soil organics. Water Res 29:1533–1540
Johnson RL, Brilliante SM, Isabelle LM, Houck JE, Pankow JF (1985) Migration of chlorophenolic compounds at the chemical waste disposal site at Alkali Lake Oregon. 2. Contaminant distribution, transport and retardation. Groundwater 23:652–666
Lee LS, Rao PSC, Nkedi-Kizza P, Delfino JJ (1990) Influence of solvent and sorbent characteristics on distribution of pentachlorophenol in octanol-water and soil-water systems. Environ Sci Technol 24:654–661
McCarthy JF, Zachara JM (1989) Substransport of contaminants. Environ Sci Technol 23:496–502
McGinnis GD, Borazjani H, McFarland LK, Pope DF, Strobel DA (1989) Characterization and laboratory soil treatability studies for cresolate and pentachlorophenol sludges and contaminated soil. US Environmental Protection Agency, Ada, Oklahoma, EPA/600/S2–88/055
Patterson RJ, Liebscher HM (1987) Laboratory simulation of pentachlorophenol behaviour in an alluvial aquifer. Water Pollut Res J Con 22:147–155
Pinal R, Rao PSC, Lee LS, Cline PV (1990) Cosolvency of partially miscible organic solvents on the solubility of hydrophobic organic chemicals. Environ Sci Technol 24:639–547
Rao PSC, Lee LS, Wood AL (1991) Solubility, sorption and transport of hydrophobic organic chemicals in complex mixtures. US Environmental Protection Agency, Ada, Oklahoma EPA/600/M-91/009
Robinson KJ (1990) Bioavailability of chlorinated phenol sorbed to dissolved organic matter. PhD Thesis, Virginia Polytechnic Institute and State University, Blacksburg, Virginia
Schellenberg K, Leunberger C, Schwarzenbach RP (1984) Sorption of chlorinated phenols by natural sediments and aquifer materials. Environ Sci Technol 18:652–657
Schnitzer M (1978) Humic substances: chemistry and reactions. In: Schnitzer M, Khan SU (eds) Soil organic matter. Elsevier, New York, pp 1–64
US Environmental Protection Agency (1986) Method 8040: phenols. Wershaw RL (1993) Models for humus in soil an sediments. Environ Sci Technol 27:814–816

Part III

Biological Aspects

10 Indigenous Soil Microflora Versus Introduced Microorganisms in Subsurface Bioremediation

R. Armon and T. Arbel

Contents

1 Introduction

Microorganisms are the main biocatalysts involved in bioremediation processes. Their ability to degrade a large variety of chemical compounds advocate their use in bioremediation applications. Soil microbial content can vary according to soil type, depth, water content and organic load. In general, mollisol fertile soil types may contain 10^8–10^9 bacteria/g, as well as other microorganisms such as: actinomycetes (10^7–10^8/g) and fungi (10^5–10^6/g) (Alexander 1977; Lynch 1988). Despite the larger number of bacteria in soil, fungi may play an equal or greater role in soil microbiology as indicated by their roughly equivalent biomass (bacteria: 300–3000 wet kg/ha, actinomycetes: 300–3000 wet kg/ha and fungi: 500–5000 wet kg/ha) (Stotzky 1972). The present review will focus, however, on bacteria and their potential to biodegrade organic pollutants intrinsically or by augmentation.

Soil contamination with highly toxic organic and inorganic pollutants (through unsaturated and saturated zones) may not only alter the number of microbial flora but also its biological diversity. Therefore, from the bioremediation aspect it is

Environmental & Water Resources Engineering, Faculty of Civil Engineering, Technion, Haifa 32000, Israel

H. Rubin/N. Narkis/ J. Carberry
Soil and Aquifer Pollution

possible that some bioaugmentation will be necessary in order to increase biological activity for degradation of recalcitrant contaminants. This review includes discussion on different aspects of subsurface bioremediation with emphasis on the advantages and disadvantages of intrinsic or bioaugmented bioremediation of non-aqueous phase liquids (NAPLs).

2 Soil Microorganisms

2.1 Bacterial Variety

A large variety of bacterial genera have been isolated from the soil subsurface, among them are: *Achromobacter*, *Cytophaga*, *Flavobacterium*, *Micrococcus*, *Nocardia*, *Gallionella*, *Leptotrix*, *Naumannilla*, *Siderocapsa*, *Siderococcus*, *Beggiatoa*, *Thiotrix*, *Toxothrix*, *Crenotrix*, *Sphaerotilus*, *Agrobacterium*, *Caulobacter*, *Clostridium*, *Hyphomicrobium*, *Microcyclus*, *Planctomyces*, *Prosthecomicrobium*, *Arthrobacter*-like spp., *Pseudomonas*, *Micrococcus*, *Chromobacterium* and *Bacillus* spp. (Dobbins et al. 1992). Out of this large variety of bacteria, many were found to possess the capability to biodegrade and mineralize NAPL pollutants as already reviewed elsewhere (Smith 1994).

2.2 General Processes of Bacterial Biodegradation

Biodegradation may occur in two or more steps: biotransformation and/or mineralization. Biotransformation is basically an incomplete biodegradation of a complex organic substance to a more simple and stable by-product. The process of biotransformation often accompanies cometabolism of substrates which are regularly utilized by bacteria for growth and energy (Schmidt and Alexander 1985). Complete biodegradation of contaminants to inorganic or mineral elements is termed mineralization. Some organic pollutants present in natural waters, are biotransformed when present at high concentration and mineralized at low concentrations (Wang et al. 1984).

3 NAPLs

Soil pollution of the unsaturated zone can be categorized in four phases: (1) gas or air phase – vapour in the pore spaces; (2) adsorbed phase – adsorbed to subsurface particles; (3) aqueous phase – dissolved in water; and (4) liquid phase – non-aqueous phase liquids (NAPLs). This last, immiscible, phase of pollutants can be divided into two groups according to their density vs water: light non-aqueous phase liquids (LNAPLs) and dense non-aqueous phase liquids (DNAPLs) (Huling and Weaver 1991). Examples of LNAPLs are petroleum hydrocarbons, and examples of DNAPLs are chlorinated solvents. Among the important NAPLs discussed in the present review and often found in contaminated unsaturated zones are: benzene, toluene, ethylbenzene and xylene *ortho*, *meta* and *para* mixtures, commonly abbreviated as BTEX.

3.1 BTEX and Microbial Metabolic Pathways

Under aerobic conditions (usually persistent in well aerated subsurface unsaturated zones), biodegradation of individual compounds such as benzene, toluene, ethylbenzene and xylene (BTEX) has been confirmed by many field and laboratory studies (Gibson et al. 1968, 1974; Worsey and Williams 1975; Rifai et al. 1995; Wiedemeier et al. 1995). For example, benzene biodegradation follows the formation of catechol and further cleavage of the aromatic ring (through *meta* or *ortho* cleavage of the ring) (Fig. 1).

The catabolic pathway of benzene degradation is called the *tod* pathway and that for toluene and xylene *tol* pathway (Lee et al. 1995). Nevertheless, a mixture of benzene, toluene and xylene cannot be successfully mineralized simultaneously because of the formation of 3,6-dimethylcatechol from *p*-xylene, which is a dead-end product in the *tod* pathway (Gibson et al. 1974). This metabolic obstacle was recently successfully overcome by construction of a hybrid strain of *P. putida* TB105 (by introduction of the constructed plasmid RSF1010, encoding toluene dioxygenase into *P. putida* mt-2 which contains TOL plasmid pWWO) able to mineralize benzene, toluene and p-xylene simultaneously without metabolic intermediate accumulation (Lee et al. 1995).

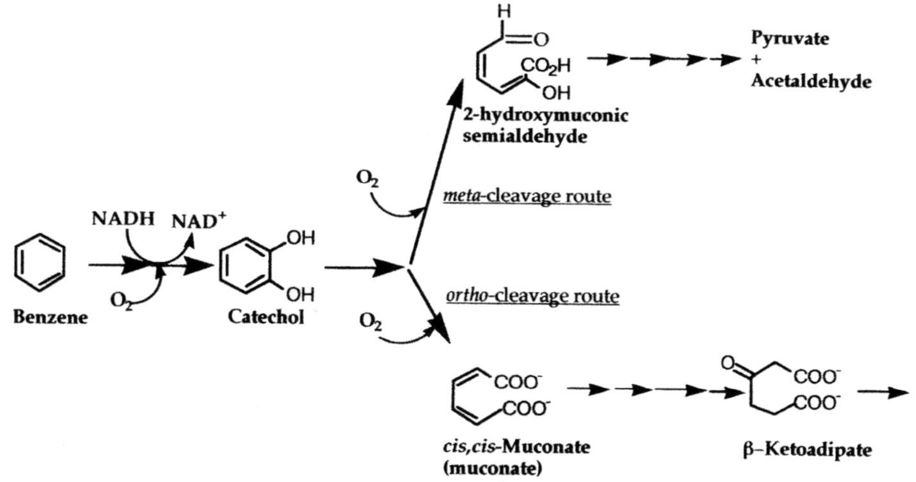

Fig. 1. The catabolic pathways of benzene by microorganisms. (After Smith 1990)

4 Important Factors in Successful Bioremediation

The binding property of the pollutant to soils or sediments is one of the most important factors controlling bioremediation. For example, toluene has been reported to be reversibly bound to soil in loose and tight forms. The degradation of toluene is enhanced when it is available in soil pore liquid, following desorption (Robinson et al. 1990).

Physical contact between bacteria and the pollutant is a second important factor. In aquatic environments, the contact is much faster due to chemical mixing and diffusion. In soil environments, following inoculation, mixing of soil particles is very much necessary in order to bring microorganisms in close contact with pollutants. This activity is not always possible in deep soil layers, therefore leaving desorption as the main factor for microorganism–pollutant direct contact.

Insufficient oxygen in the subsurface will favour anaerobic degradation of pollutants, which will generally occur at a much slower rate (Green and Trett 1989). Therefore, an external supply of oxygen to the subsurface will considerably enhance biodegradation. This factor was experimentally confirmed by the efficiency of bioventing remediation processes (DuPont 1993).

Another important factor in soil bioremediation is nutrient availability, in order to enhance bacterial activity and potential cometabolism. For biodegradation to occur, nutrients should be assimilated in N:C, P:C or S:C mass ratios corresponding to those in the biomass. Therefore, in poor soil found in semi-arid or arid zones, nutrients must be added to increase bioremediative activity.

Solubility of pollutants in water is also very important. For example, benzene solubility is 1790 mg/l; toluene, 579 mg/l; ethylbenzene, 135 mg/l and xylene (*p*-221 mg/l, *m*-260 mg/l and *o*-215 mg/l). The poor water solubility of these cytotoxic compounds can be overcome by several means. Rezessy-Szabó et al. (1987) used an organic solvent such as dibuthyl phthalate (not attacked by their bacterium) to dissolve benzene and cultivate *Pseudomonas* sp. 50 at much higher benzene concentrations. Schwartz and Bar (1995) used β-cyclodextrin that forms inclusion compounds with hydrophobic organics such as toluene. Toluene toxicity decreased substantially with the addition of β-cyclodextrin and allowed its degradation by a *P. putida* strain at concentrations up to 10 g/l. Finally, many microorganisms produce emulsification agents or biosurfactants which disperse the insoluble pollutant and increase the contact area (Green and Trett 1989).

Microbial growth is strongly related to temperature. Variation of this important factor can significantly alter the biodegradation rate. The optimum temperature for hydrocarbon degradation is 20–35 °C, although it occurs less at 0 °C and at as high as 70 °C (Green and Trett 1989). For example, aerobic biodegradation of four marker hydrocarbons was approximately similar at 0 °C and 27 °C in slurry samples taken from a lake with a past history of oil pollution (Cooney et al. 1985). In unpolluted slurry samples the low temperature was found to significantly limit biodegradation, indicating the existence of different microbial populations at the two locations.

Other important factors involved in successful biodegradation of pollutants and related to the main factors listed above are: environmental pH, redox conditions, pollutant concentration and presence of predator organisms.

4.1 Intrinsic Bioremediation

Intrinsic bioremediation describes the biological activity of indigenous microorganisms on different pollutants found in soil and groundwater. Intrinsic bioremediation has been explored for more than two decades and successful results were published by many authors (Borden et al. 1994; Wilson et al. 1994). The advantages and disadvantages of intrinsic bioremediation are discussed later and summarized in Table 1. In general, it can be stated that indigenous microorganisms are less susceptible to genetic manipulations and the only control available is by supplying different nutrients and surface active compounds. It can be stated that "you get what you have". Nevertheless, indigenous microorganisms comprise a large variety of species which may compensate for each other to provide complete biodegradative enzymatic pathways.

4.2 Augmented Bioremediation

Augmented bioremediation describes the process in which laboratory grown microorganisms are added to the soil in order to biodegrade contaminants. The

Table 1. Summary of advantages and disadvantages of two bioremediation processes: intrinsic and bioaugmentation

	Intrinsic Bioremediation		Bioaugmentation	
	Advantage	Disadvantage	Advantage	Disadvantage
Induction		*	*	
Soil history		*	*	
Survival	*			*
Transport	*			*
Pollutant recalcitrance		*	*	
Multiple degradation capability		*	*	
Bacterial concentration		*	*	
GEM	*			*[a]
Cometabolism		*	*	
Cost	*			*[b]
Die-off	*			*
Adaptation	*			*
Micro-encapsulation		*	*	

[a] Considering the restriction on environmental spread of GEM.
[b] Short-term.

application is necessary on polluted sites that otherwise would require years for the completion of a satisfactory biodegradation process. In overall terms, bioaugmentation leads to a higher degree of confidence in remediation success, and for certain sites has been shown to save time and money over alternative approaches. In the early years of practice, bioaugmentation failed to enhance bioremediation due to lack of knowledge of microbial ecology and physiology of the introduced microorganisms. This failure reminds us of the early years in agricultural practice when *Rhizobium* genus was inoculated to enhance nitrogen fixation in leguminous crops. The inability of introduced bacteria to compete with indigenous *Rhizobium* was overcome dramatically by understanding the host-bacterium specificity, soil and climatic conditions, and the biochemistry involved. This lesson should be also applied to bioaugmentation and successful results will follow.

4.3 Transport of Microorganisms Through Unsaturated Porous Media

One of the important factors in bioaugmentation of contaminated soil with laboratory grown bacteria is transport through the porous matrix. Wan et al. (1994) clearly showed the importance of the gas-water interface on transport of microorganisms through unsaturated porous media (Fig. 2).

Their results clearly show that microorganisms (*Pseudomonas* and *Arthrobacter*) are retained in the porous medium by their preferential sorption onto the gas-water interface. The more hydrophobic the bacteria, the higher retention by the gas-water interface. Even hydrophilic bacteria such as *P. cepacia* (Fig. 2a) were sorbed onto the gas-water interface, altering its movement in the porous medium. In order to mobilize the gas-water interface with previously sorbed bacteria into the porous medium, the authors suggested increasing the shear stress. The study performed by Wan et al. (1994) clearly points out the significance of gas-water interface sorption as the main factor in microorganism distribution in the subsurface matrix. This aspect should be carefully considered when bioaugmenting contaminated sites with laboratory-grown bacteria.

5 GEMs Used in Bioremediation

Advances in genetic engineering provides the means for construction of improved microorganisms. The ability to transform microorganisms by genetic manipulations (plasmid inclusion, gene transfer, etc.) into biodegraders provides endless possibilities. However, a major opposition was raised by the scientific community against genetically engineered microorganisms (GEMs), based mainly on our lack of knowledge on GEM behaviour in the environment, the main threat being the release of an uncontrollable factor into the environment, and loss of biological control. Recently, Ronchel et al. (1995) have published an exciting manipulation of *P. putida* as biologically contained bacteria and its application in bioremediation

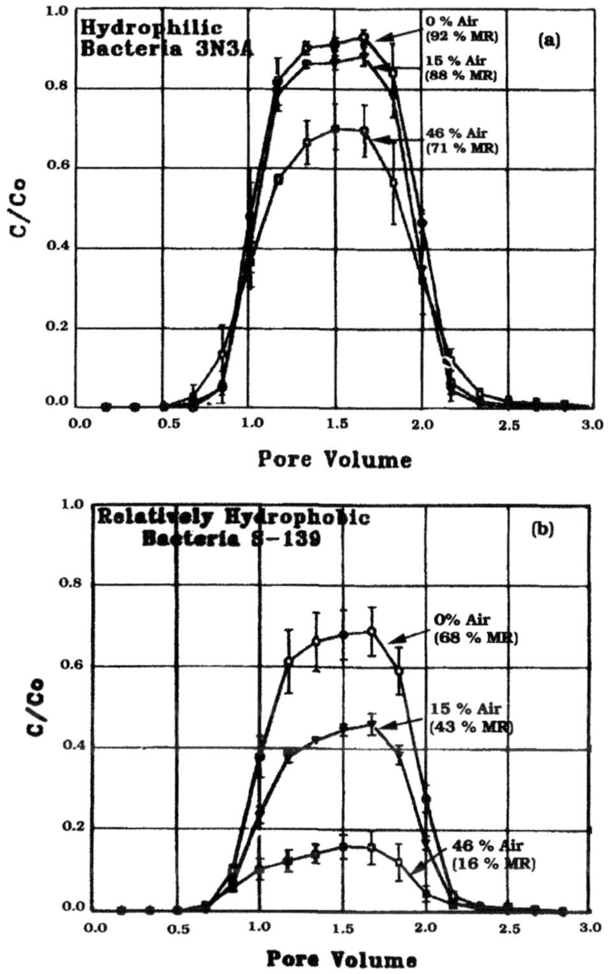

Fig. 2. (a) Breakthrough curves of hydrophilic *Pseudomonas cepacia* strain 3N3A and (b) hydrophobic *Arthrobacter* strain S-139 from columns with different air saturations. % *MR* Percentage of total cell mass recovery. (After Wan et al. 1994)

(Fig. 3). The authors used a *P. putida* strain that bears the TOL plasmid pWWO which allows growth on alkylbenzoates (3-methylbenzoate this study) and plasmid pCC102 which carries the *xyl* S2 gene (regulator) and the P_m::*lac*I fusion (P_m is the promotor for the *meta* pathway of the TOL plasmid) as a regulatory cassette. The killing cassette element consisted of a fusion of P_{lac} promoter to the *gef* gene in a mini-Tn5 element donor. The expression of *gef* protein results in the insertion of this protein into the cell membrane, causing the collapse of the membrane potential and death. The constructed elements work as follows: in the presence of alkylaromatics (3MB), the regulatory *xyl* S gene (expressed constitutively) becomes active, promoting transcription from P_m which leads to synthesis of the *lac*

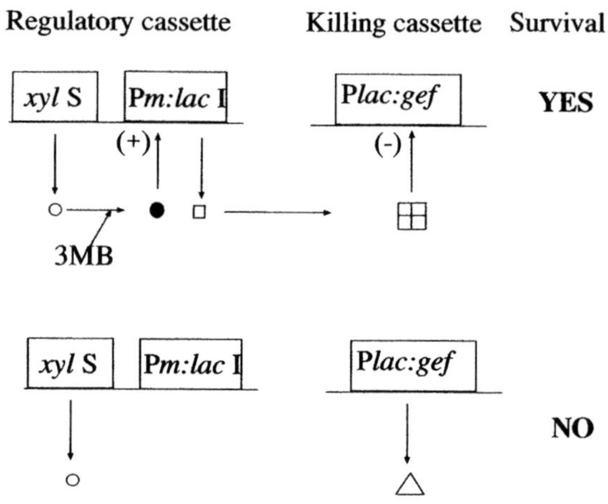

Fig. 3. Regulatory and killing cassette constructed in *P. putida* that allows its growth only in the presence of 3-methylbenzoate (3MB). (After Ronchel et al. 1995)

I repressor; that repressor forms a tetramer which prevents expression of the killing cassette (P_{lac}: $_{gef}$), and therefore the bacteria survives. In absence of alkylaromatics as substrate, the (Pl_{lac}: $_{gef}$) is activated causing death of the bacterial cells. This system is illustrated in Fig. 3. In 0.1% (wt/wt) *m*-methylbenzoate amended fulvisol soil, the biologically contained strain showed significant growth for 8 days (from 10^5 to 10^8 CFU/g) and then a slight decline up to 1 month. In unamended soil the decrease in the contained strain was significant by two orders of magnitude during the 1 month of the experiment. This pioneer experiment clearly shows the potential of using self-containing GEMs in bioremediation processes.

6 Summary and Discussion

Table 1 summarizes the advantages and disadvantages of intrinsic (indigenous bacteria) and bioaugmented bioremediation. In general, bioaugmentation has more advantages as the biodegrading microorganisms do not have to be induced and already have the capability to degrade contaminants (by preliminary selection). They can be genetically built to biodegrade multiple contaminants (Lee et al. 1995) and also to contain self-destruction elements in order to prevent undesired genetic exchange with indigenous microorganisms, or those widespread in the environment (Ronchel et al. 1995). Two major disadvantages of bioaugmentation are: transport in porous media and survival. Bioaugmentation with laboratory grown bacteria in the unsaturated zone is problematic due to their irreversible sorption to the gas-water interface and lack of mobility (Wan et al. 1994). Survival of laboratory strains in the environment is inferior as shown by many studies, yet it can be increased by augmenting the numbers of bacteria. To accomplish success in bioaugmentation it will be necessary to understand the ecology, physiology and biochemistry of laboratory grown microorganisms in soil and water environments. Conceivably, the best example for such an approach is the article of A. M. Chakrabarty (a pioneer in this field) on *Microbial degradation of toxic chemicals: evolutionary insights and practical considerations* (Chakrabarty 1996). In his article, Chakrabarty describes the work of Kellog and Chaterjee, who isolated from a toxic dump site a *P. cepacia* strain AC1100 (now *Burkholderia cepacia*), able to grow on 2,4,5-trichlorophenoxyacetic acid. Following different selection steps, the authors isolated a bacterium able to biodegrade 2,4,5-trichlorophenoxyacetic acid at concentrations from 1000 to 20000 ppm in 1 to 6 weeks in contaminated-soil samples. They also found that this strain is fragile and has a rapid die-off in the absence of the pollutant. In addition, this bacterial strain produces a surface-active compound which emulsifies the hydrophobic 2,4,5-trichlorophenoxyacetic acid. This classical experiment points to the endless possibilities, still to be explored, that may advance our knowledge and allow better bioremediation practices.

References

Alexander M (1977) Introduction to soil Microbiology, 2nd edn. John Wiley, New York

Borden RC, Gomez CA, Becker MT (1994) Natural bioremediation of a gasoline spill. In: Hinchee RE, Alleman BC, Hoeppel RE, Miller RN (eds) Hydrocarbon Bioremediation. Lewis, Boca Raton, pp 290–295

Chakrabarty AM (1996) Microbial degradation of toxic chemicals: evolutionary insights and practical considerations. ASM News 62:130–137

Cooney JJ, Silver SA, Beck EA (1985) Factors influencing hydrocarbon degradation in three freshwater lakes. Microb Ecol 11:127–137

Dobinns DC, Aelion CM, Pfaender F (1992) Subsurface, terrestrial microbial ecology and biodegradation of organic chemicals: a review. CRC Crit Rev Environ Control 22:67–136

DuPont RR (1993) Fundamentals of bioventing applied to fuel contaminated sites. Environ Prog 12:45–53

Gibson DT, Koch JR, Kallio RE (1968) Oxidative degradation of aromatic hydrocarbons by microorganisms. I. Enzymatic formation of catechol from benzene. Biochemistry 7:2653–2662

Gibson DT, Mahadevan V, Davey JF (1974) Bacterial metabolism of *para*- and *meta*-xylene: oxidation of the aromatic ring. J Bacteriol 119:930–936

Green J, Trett MW (1989) The fate and effects of oil in fresh water. Elsevier Applied Science, New York, 338 pp

Huling SG, Weaver JW (1991) Dense non-aqueous phase liquids. Ground Water Issue Pap, EPA/540/4-91-002. Robert S Kerr Environmental Research Laboratory, Ada, Oklahoma

Lee J-Y, Jung K-H, Choi SH, Kim H-S (1995) Combination of the tod and the tol pathways in redesigning a metabolic route of *Pseudomonas putida* for the mineralization of a benzene, toluene, and p-xylene mixture. Appl Environ Microbiol 61:2211–2217

Lynch JM (1988) The terrestrial environment. In: Lynch JM, Hobbie JE (eds) Micro-organisms in action: concepts and applications in microbial ecology. Blackwell, Oxford, pp 103–131

Rezessy-Szabó JM, Huijberts GNM, Bont JAM (1987) Potential of organic solvents in cultivating micro-organisms on toxic water-insoluble compounds. In: Laane C, Tramper J, Lilly MD (eds) Biocatalysis in organic media, Proc Int Symp, Wageningen, The Netherlands, 7–10 Dec 1986. Elsevier, Amsterdam, pp 295–302

Rifai HS, Borden RC, Wilson JT, Ward CH (1995) Intrinsic bioattenuation for subsurface restoration. In: Hinchee RE, Wilson JT, Downey DC (eds) Intrinsic bioremediation. Battelle Press, Columbus, Ohio, pp 1–30

Robinson KG, Farmer WS, Novak JT (1990) Availability of sorbed toluene in soils for biodegradation by acclimated bacteria. Water Res 24:345–350

Ronchel MC, Ramos C, Jensen LB, Molin S, Ramos JL (1995) Construction and behaviour of biologically contained bacteria for environmental applications in bioremediation. Appl Environ Microbiol 61:2990–2994

Schmidt SK, Alexander M (1985) Effects of dissolved organic carbon and second substrates on the biodegradation of organic compounds at low concentrations. Appl Environ Microbiol 49:822–828

Schwartz A, Bar R (1995) Cyclodextrin-enhanced degradation of toluene and p-toluic acid by *Pseudomonas putida*. Appl Environ Microbiol 61:2727–2731

Smith MR (1990) The biodegradation of aromatic hydrocarbons by bacteria. Biodegradation 1:191–206

Smith MR (1994) The physiology of aromatic hydrocarbon degrading bacteria. In: Ratledge C (ed) Biochemistry of microbial degradation. Kluwer, Dordrecht, pp 347–372

Stotzky G (1972) Activity, ecology and population dynamics of microorganisms in soil. CRC Crit Rev Microbiol 2:59–137

Wan J, Wilson JL, Kieft TL (1994) Influence of the gas-water interface on transport of microorganisms through unsaturated porous media. Appl Environ Microbiol 60:509–516

Wang Y, Subba-Rao RV, Alexander M (1984) Effect of substrate concentration and organic and inorganic compounds on the occurrence and rate of mineralization and cometabolism. Appl Environ Microbiol 47:1195

Wiedemeier TH, Swanson MA, Wilson JT, Kampbell DH, Miller RN, Hansen JE (1995) Patterns of intrinsic bioremediation at two US air force bases. In: Hinchee RE, Wilson JT, Downey DC (eds) Intrinsic bioremediation. Battelle Press, Columbus, Ohio, pp 31–52

Wilson JT, Pfeffer FM, Weaver JW, Kampbell DH, Wiedemeier TH, Hansen JE, Miller RN (1994) Intrinsic bioremediation of JP-4 jet fuel. In: Proc EPA Symp on intrinsic bioremediation of ground water. US Environmental Protection Agency, EPS/540/R-94/515, 189 pp

Wolters N, Schwartz W (1956) Untersuchungen über Vorkommen und Verhalten von Mikroorganismen in reinem Grundwasser. Arch Hydrobiol 51:500–510

Worsey MJ, Williams PA (1975) Metabolism of toluene and xylenes by *Pseudomonas putida* (arvilla) [sic] mt-2: evidence for a new function of the TOL plasmid. J Bacteriol 124:7–13

11 Rate-Limiting Steps in the Microbial Degradation of Petroleum Hydrocarbons

E. Rosenberg, S. Navon-Venezia, I. Zilber-Rosenberg and E. Z. Ron

Contents

1 Introduction

Effective bioremediation of petroleum pollution relies heavily on a fundamental understanding of how microorganisms grow on hydrocarbons. Table 1 outlines the requirements for biodegradation of petroleum hydrocarbons.

The two essential characteristics that define the hydrocarbon-oxidizing microorganisms are:

1. Membrane-bound, group-specific oxygenases.
2. Mechanisms for optimizing contact between the microorganism and the water-insoluble hydrocarbon.

Because the oxygenases are membrane-bound, never extracellular, it follows that the microorganism must make direct contact with the water-insoluble substrate. The fact that the oxygenases are group-specific, e.g., some work on specific fractions of alkanes, others on specific cyclic alkanes or aromatics, etc., means that only a mixture of different microorganisms can efficiently degrade crude oil and petroleum fractions. Fortunately, hydrocarbon oxidizers are located in virtually all natural areas, although with large variations in cell concentration (Rosenberg 1991). In general, there is a correlation between the quantity of heptane-extractable hydrocarbons in soil or water and the concentration of hydrocarbon-

Department of Molecular Microbiology and Biotechnology George S. Wise Faculty of Life Sciences Tel Aviv University, Ramat Aviv, Israel 69978

H. Rubin/N. Narkis/ J. Carberry
Soil and Aquifer Pollution

Table 1. Requirements for biodegradation of petroleum

A.	Microorganisms with:
	Hydrocarbon-oxidizing enzymes
	Ability to adhere to hydrocarbons
	Emulsifier-producing potential
	Mechanisms for desorption from hydrocarbons
B.	Water
C.	Oxygen
D.	Utilizable nitrogen source
E.	Utilizable phosphorus source

oxidizing bacteria. In this review we will summarize our studies on the physical interactions of microorganisms with hydrocarbons and the development of a new biotechnology for the bioremediation of petroleum pollution.

The low solubility of hydrocarbons in water, coupled with the fact that the first step in hydrocarbon degradation involves a membrane-bound oxygenase, makes it essential for bacteria to come in direct contact with the hydrocarbon substrates. Two general biological strategies have been proposed for enhancing contact between bacteria and water-insoluble hydrocarbons: (1) specific adhesion/desorption mechanisms, and (2) emulsification of the hydrocarbon.

In order to understand the special cell-surface properties of bacteria that allow them to grow on hydrocarbons, it is useful to consider the dynamics of petroleum degradation in natural environments. Following an oil spill in the sea, the hydrocarbons rise to the surface and come into contact with air. Some of the low molecular weight hydrocarbons volatilize; the remainder are metabolized relatively rapidly by microorganisms which take up soluble hydrocarbons. These bacteria do not adhere to oil and do not have a high cell-surface hydrophobicity (Rosenberg and Rosenberg 1985). The next stage of degradation involves microorganisms with high cell-surface hydrophobicity, which can adhere to the residual high molecular weight hydrocarbons.

The requirements for oxygen and moisture is not a problem in oil spills in aquatic environments. However, on land, oxygen and water are often rate-limiting. If oil has not penetrated too deeply into the ground, then watering and tilling are often advantageous. These problems have been discussed theoretically by Atlas (1991) and Song et al. (1990) who have provided some recent useful data on the subject.

The major limitation in the biodegradation of hydrocarbons on land and water is an available source of nitrogen and phosphorus. In theory, approximately 150 mg of nitrogen and 30 mg of phosphorus are consumed in the conversion of 1 g of hydrocarbon to cell material. The nitrogen and phosphorus requirements for maximum growth of hydrocarbon oxidizers can generally be satisfied by ammonium phosphate. Alternatively, these requirements can be met with a mixture of other salts, such as ammonium sulfate, ammonium nitrate, ammonium chloride, potassium phosphate, sodium phosphate and calcium phosphate. When ammonium salts of strong acids are used, the pH of the medium generally decreases with growth. This problem can often be overcome by using urea as the nitrogen source. All of these compounds have a high water solubility which reduces their effectiveness in open systems because of rapid dilution. In principle, this problem can be solved by using oleophilic nitrogen and phosphorus compounds with low C:N and C:P ratios. For example, it was found that a combination of paraffinized urea and octyl phosphate was able to replace nitrate and inorganic phosphate, respective (Atlas and Bartha 1973).

In the last few years, we have developed a new technology for petroleum bioremediation based on a controlled-release, hydrophobic nitrogen and phosphorus fertilizer, referred to as F-1 (Rosenberg et al. 1992). A summary of the use of this technology to bioremediate an oil-contaminated beach is presented, following a discussion of the physical interactions of microorganisms and hydrocarbons.

2 Adhesion

2.1 Growth of Bound and Free *Pseudomonas* UP-2 on n-Tetracosane

Pseudomonas UP-2, isolated from a petroleum polluted environment, is able to grow on a large variety of paraffins, ranging from C_{12} to above C_{26} (Horowitz et al. 1975). Microscope observations revealed binding of the bacterial cells to both liquid and solid hydrocarbons during the growth phase. In order to study the adhesion of strain UP-2 to hydrocarbons, n-tetracosane ($C_{24}H_{50}$) with a melting point of 54 °C, was chosen as the model substrate because UP-2 grew well on n-tetracosane and the solid hydrocarbon was relatively easy to separate quantitatively from the bulk aqueous phase. The technique that was developed was based on filtering the culture at different stages of growth through Whatman 1 filter paper, thereby separating the large n-tetracosane particles (with bound cells) from the continuous phase (including unbound cells). Growth and metabolic activities of the cells were assayed by the sensitive method of radioactive labelling with orthophosphate.

The accumulative incorporation of ^{32}Pi into free and bound UP-2 during growth on n-tetracosane is summarized in Fig. 1. During the early stages of growth (up to 15 h), most of the label was incorporated into bound cells, whereas at later stages of growth most of the counts were associated with unbound cells. The fact that the radioactive bound cells reached a maximum at 20 h and thereafter remained constant could be explained by assuming that the surface of the n-tetracosane became saturated with a layer of cells. This was consistent with microscopic observations. However, the origin of the labelled free cells was not clear. Were they able to grow in the unbound state, even though n-tetracosane is insoluble in water, or did they grow and incorporate radioactivity while they were bound to the hydrocarbon and subsequently were released into the bulk phase?

In order to understand the mechanism by which free cells became radioactive, cultures were pulse-labelled for 5 min at 11, 14 and 37 h, and the free and bound

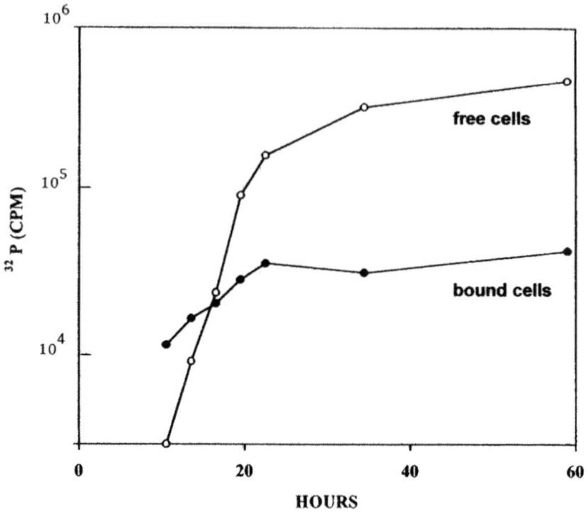

Fig. 1. Growth of *Pseudomonas* UP-2 on n-tetracosane ($C_{24}H_{50}$). The culture was grown in the presence of $^{32}PO_4$. At timed intervals, samples were removed and the tetracosane-bound and free cells separated by filtration through filter paper. Data calculated from Zilber et al. (1980)

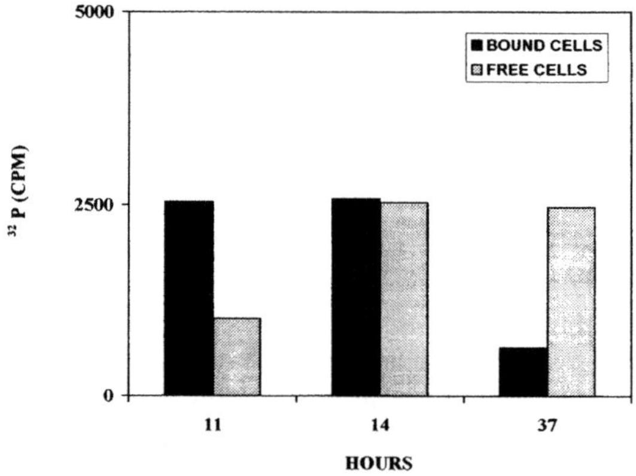

Fig. 2. Pulse-labelling of *Pseudomonas* UP-2 growing on n-tetracosane. The experiment was performed as described in Fig. 1 except the $^{32}PO_4$ was added for 5 min to cultures that had been grown for 11, 14 and 37 h. After determination of the total filterable $C_{24}H_{50}$, the filtrate was centrifuged to pellet bacterial cells and the distribution of hydrocarbon between the pellet and supernatant fractions determined. Data calculated from Zilber et al. (1980)

cells separated by filtration (Fig. 2). Again, it was seen that most of the label was incorporated into bound cells during the exponential growth phase (11 h) and into free cells during the early stationary phase (37 h). Thus, the model that cells grow only when in direct contact with the bulk hydrocarbon phase could not be correct.

The paradoxical situation in which free cells were incorporating considerable amounts of phosphate in a process that requires energy, even though tetracosane

Table 2. Utilization of filterable $C_{24}H_{50}$ by *Pseudomonas* UP-2[a]

Sample	$C_{24}H_{50}$ (ng/ml)		
	Total	Pellet	Supernatant
Control (no bacteria)	15		
24-h culture (t = 0)	330	280	49
24-h culture (t = 4 h)	102	53	49
24-h culture (t = 20 h)	55	13	42

[a] A culture was grown in the presence of n-tetracosane ($C_{24}H_{50}$) for 24 h and then filtered to remove insoluble hydrocarbon. The amount of n-tetracosane present in the filtrate was determined by gas chromatography immediately after filtration and following 4 and 20 h of additional incubation. After determination of the total filterable $C_{24}H_{50}$, the filtrate was centrifuged to pellet becterial cells and the distribution of hydrocarbon between the pellet and supernatant fractions determined. Data calculated from Zilber et al. (1980).

is insoluble in water, was resolved by a chemical analysis of the filtrate (Table 2). After 24 h of growth of strain UP-2 on n-tetracosane, the culture was filtered to remove bulk hydrocarbon. Analysis of the filtrate showed that it contained 330 ng/ml of n-tetracosane, whereas the control had 15 ng/ml. Peake and Hodgson (1967) reported that the solubility of n-tetracosane in water was 17 ng/ml. Furthermore, if a sample of the filtrate was centrifuged to pellet the free cells, 85% of the hydrocarbon was associated with the cells. When the remainder of the filtrate was incubated, the amount of n-tetracosane decreased from 330 to 102 and 55 ng/ml after 4 and 20 h, respectively. The fraction that was degraded was specifically the cell-bound hydrocarbon. The simplest explanation for this phenomenon is that cells adhere to the bulk n-tetracosane, grow for a period of time at the hydrocarbon/water interface, dissolve considerable amounts of the hydrocarbon into their membrane, and then come off the bulk n-tetracosane into the medium, carrying with them the hydrocarbon. The cell-associated hydrocarbon then serves as an energy source for the free cell for motility and growth (e.g., incorporation of radioactive Pi).

2.2 Cell-surface Hydrophobicity and the Requirement for Adhesion for Growth of *Acinetobacter calcoaceticus* on Hexadecane

Interest in the mechanism which enables direct contact between hydrocarbon-degrading cells and their water-insoluble liquid alkane substrates led us to develop a rapid quantitative assay for the hydrophobic interaction of cells with liquid hydrocarbons (Rosenberg and Rosenberg 1985). This method is based on the degree of adherence of cells to various liquid hydrocarbons following a brief period of mixing. After the mixing, the two phases are allowed to separate and the

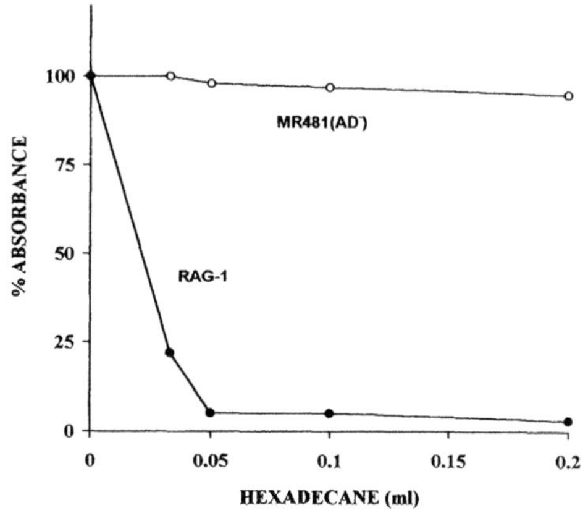

Fig. 3. Adhesion of *Acinetobacter calcoaceticus* RAG-1 and MR481 to hexadecane. (Rosenberg and Rosenberg 1981)

turbidity of the bottom aqueous phase is determined. If the cells have a high cell surface hydrophobicity they will adhere to the hydrocarbon droplets and rise to the surface, thereby causing a decrease in the turbidity of the aqueous phase. An example of this simple procedure is shown in Fig. 3. *A. calcoaceticus* RAG-1 is a hydrocarbon-degrading bacterium with a high cell surface hydrophobicity. The addition of 0.05 ml hexadecane to a turbid suspension of these cells resulted in a 97% decrease in turbidity in the aqueous phase. This simple, rapid and inexpensive method has been widely used by microbiologists and has been referred to as the MATH (microbial adhesion to hydrocarbons) test.

One of the advantages of the MATH test is that it allows for the ready isolation of non-adhering mutants. For example, when *A. calcoaceticus* RAG-1 was mixed with hexadecane and the phases allowed to separate, the aqueous phase was enriched in non-adhering mutants. One of these mutants, referred to as MR-481, exhibited no significant affinity for hydrocarbons (Fig. 3). Electron microscopy of strain MR-481 revealed that it lacked thin fimbriae (ca. 3.5 nm diameter) on the cell surface (Rosenberg et al. 1983). The wild type (RAG-1) possessed numerous thin fimbriae. Revertants of MR-481 selected for adhesion to hexadecane were isolated. All of the revertants contained thin fimbriae. Furthermore, adhesion of RAG-1 cells to hydrocarbons was greatly reduced after a shearing treatment to remove fimbriae. The data demonstrate that the thin fimbriae are the major factor responsible for adhesion of *A. calcoaceticus* RAG-1 to hydrocarbons.

The isolation of a mutant strain of *A. calacoaceticus* RAG-1 that is deficient in its ability to adhere to hydrocarbons permitted an examination of the role of adhesion in the growth of *A. calacoaceticus* RAG-1 on hexadecane. As seen in Fig. 4, the non-adherent mutant strain MR-481 failed to grow on hexadecane under conditions which supported growth of wild-type cells. Two trivial explanations for the inability of MR-481 to grow on hexadecane, insufficient oxygen and a faulty hydrocarbon-metabolizing enzyme system, were ruled out by the following data:

Fig. 4. Growth of *A. calcoaceticus* RAG-1 (•) and MR481 (○) on hexadecane. (Rosenberg and Rosenberg 1981)

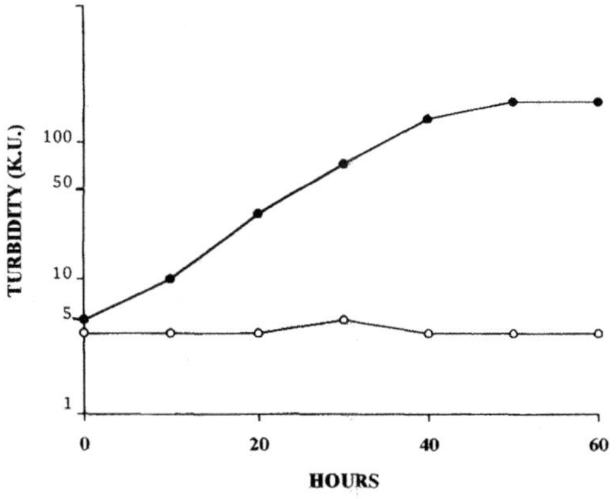

(1) MR-481 and RAG-1 cells grew with similar kinetics on acetate medium under conditions identical to those employed for growth on hexadecane; (2) MR-481 was able to grow on hexadecane medium after the addition of an emulsifier; and (3) MR-481 cells were able to grow on solid medium supplemented with hexadecane vapors as the sole carbon and energy source. We conclude from these data that adhesion is a prerequisite for the growth of RAG-1 on liquid hexadecane.

3 Bioemulsifiers of *Acinetobacter*

A summary of the properties of the bioemulsifiers of three strains of *Acinetobacter* is given in Table 3. The major component of RAG-1 emulsan is a heteropolysaccharide containing D-galactosamine, an amino uronic acid, and an uniden-

Table 3. Comparison of emulsion-stabilizing polymers: RAG-1 emulsan, BD4 emulsan and alasan

	Emulsan RAG-1	Emulsan BD4	Alasan
Producing bacterium:	*A. calcoaceticus* RAG-1	*A. calcoaceticus* BD4	*A. radioresistens* KA53
Chemical properties:	Polysaccharide containing fatty acid esters	Polysaccharide: protein complex	Polysaccharide with N-linked alanine: protein complex
Substrate specificity:	Requires both aliphatic and aromatic hydrocarbons	Similar to RAG-1	Long chain aliphatics and aromatics containing at least ten carbon atoms
Stability to heat and alkali:	Unstable	Unstable	Activated by heat; stable to alkali

tified amino sugar in approximately equal amounts (Zuckerberg et al. 1979). Fatty acids, comprising 15% of the emulsan dry weight, are attached to the polysaccharide backbone via O-ester linkages (Belsky et al. 1979).

Studies of the emulsifying activity of the RAG-1 emulsan (Rosenberg et al. 1979b) led to the development of a simple, quantitative emulsifying assay based on the linear relationship between the amount of emulsan present and the turbidity of the oil-in-water emulsion formed with an excess of hydrocarbon substrate. Emulsan was effective at low concentrations (0.01%–0.001%), representing emulsan-to-hydrocarbon ratios of 1:100–1:1000 and showed considerable substrate specificity (Rosenberg et al. 1979a). RAG-1 emulsan did not emulsify pure aliphatic, aromatic, or cyclic hydrocarbons; however, all mixtures that contained an appropriate mixture of an aliphatic and an aromatic (or cyclic alkane) were emulsified efficiently.

A. calcoaceticus BD4, initially isolated and characterized by Taylor and Juni (1961), produces a large polysaccharide capsule. The polysaccharide component was obtained by deproteinization and purification of the BD4 capsule (Kaplan et al. 1985). The polysaccharide and protein components had no emulsifying activity by themselves, but mixing the protein and polysaccharide fractions led to a reconstitution of the emulsifying activity (Kaplan et al. 1987).

Recently, we reported data on a new bioemulsifier, referred to as alasan (Navon-Venezia et al. 1995), produced by a strain of *Acinetobacter radioresistens*. Alasan efficiently stabilizes emulsions of pure long-chain alkanes and aromatics containing ten or more carbons. In general, the higher the molecular weight of the hydrocarbon, the more readily it was emulsified by alasan. In contrast, RAG-1 and BD4 emulsans require a mixture of aliphatics and aromatic hydrocarbons for emulsification. Alasan is a complex of an anionic polysaccharide and protein and has a molecular weight of approximately 1×10^6. The polysaccharide component of alasan is unusual in that it contains covalently bound alanine. Deproteinization of alasan caused a loss in most of the emulsifying activity. An interesting property of alasan is its activation by high pH and high temperature.

From purely theoretical considerations, emulsification of water-insoluble hydrocarbons is necessary for efficient biodegradation (Table 4). For example, a large drop of oil (r = 1 cm) would take 8000 h to be degraded, assuming that growth took place at a maximum rate, but only at the oil/water interface. If the oil was emulsi-

Table 4. Contribution of emulsification of water-insoluble substrates to biodegradation

Volume (cm^3)	Radius (μm)	Surface (μm^2)	Biodegradation rate[a] (h)
1.0	10000	3×10^8	8000
1.0	1000	3×10^9	800
1.0	10	3×10^{11}	8
1.0	2	1.5×10^{12}	1.6

[a] Assuming the surface is covered with microorganisms and the biodegradation rate is 1 g/24 h for 10^{11} cells.

fied (e.g., $r = 2\,\mu m$), then the much larger surface area would allow for many more microorganisms to be involved, and the biodegradation could theoretically occur in 1.6 h. Clearly, the production of bioemulsifiers would be an advantage to the producing organism.

Another potential advantage of bioemulsifier production by bacteria is in desorption from the substrate/substratum. This concept is supported by several lines of experimental data. First, bioemulsifier excretion into the medium by *Acinetobacter* species occurs at the end of the growth phase when cells become starved. Second, purified emulsans remove bacterial cells from hydrocarbons and other hydrophobic surfaces (Rosenberg et al. 1983). Third, emulsan minus mutants adhere to hydrocarbons, but fail to desorb (I. Pines, unpubl. data).

Studies on the growth, surface properties and production of emuslans have led to a model of growth on water-insoluble hydrocarbons (Fig. 5) that comprises the following steps:

1. Individual cells adhere to the surfaces of large oil drops. Adhesion of RAG-1 is mediated by the presence of hydrophobic fimbriae which are distributed over the entire surface of the bacterium.
2. The bacteria grow as a biofilm at the hydrocarbon/water interface, extracting water-soluble minerals, nitrogen and phosphate from the aqueous phase and

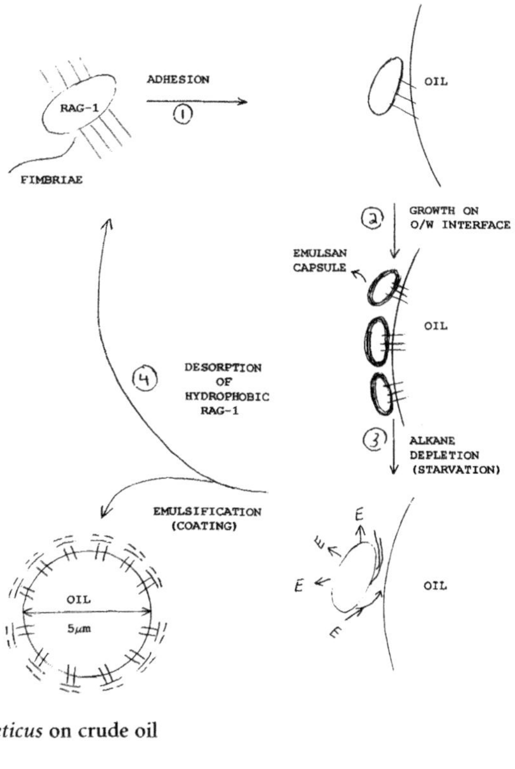

Fig. 5. Model for the growth of *A. calcoaceticus* on crude oil

oxidizing n-alkanes to carboxylic acids, which provide the carbon and energy source. During the growth phase, the cells accumulate emulsan on their surfaces in the form of a capsule. As the encapsulated cells cover the oil droplets, they reduce the interfacial tension between the hydrocarbon and the water. This results in the emulsification of the oil and an increase in the oil-water surface area for growth.

3. At the end of the growth phase, the individual oil droplets that have been colonized by the bacteria become depleted of n-alkanes. These droplets become enriched in nonutilizable aromatic and cyclic hydrocarbons. When the n-alkane content of the colonized droplets becomes depleted, the cells become starved.
4. The starved cells release their emulsan capsules. The extracellular emulsan attaches avidly to the oil droplet, thereby displacing the cells to the aqueous phase. Each "used" droplet is then covered with a stable monomolecular film of emulsan. The hydrophilic outer surface of the emulsan-coated oil prevents reattachment of the free bacteria. The released capsule-deficient bacteria are extremely hydrophobic and readily adhere to fresh hydrocarbon substrates.

4 Satisfying the Nitrogen and Phosphorus Requirements

Microorganisms which degrade hydrocarbons are widely distributed and can be found in virtually all natural areas (Rosenberg 1991). Thus, the rate limiting step in biodegradation of hydrocarbons is generally not the lack of appropriate microorganisms. The major limitation for the biodegradation of hydrocarbons on land and water is usually an available source of nitrogen and phosphorous (Atlas and Bartha 1973; Atlas 1991; Pritchard et al. 1992; Rosenberg and Ron 1996). In the laboratory, these requirements can generally be satisfied by salts, such as ammonium sulfate, ammonium nitrate and ammonium phosphate. However, all of these compounds have a high water solubility which reduces their effectiveness in open systems because of rapid dilution. Moreover, since the added salts are used up by all microorganisms, they become even more limiting for hydrocarbon degraders. In addition, it should be noted that the addition of soluble salts increases the mobile nitrogen and phosphorus in the soil, creating an additional environmental problem.

To overcome these problems we have developed a new hydrophobic nitrogen source, referred to as F-1 (Rosenberg et al. 1992; Rosenberg and Ron 1996) that also contains insoluble phosphorus (18% N, 10% P). F-1 has low water solubility and adheres to oil and other hydrophobic organic pollutants. The spontaneous degradation of F-1 in nature is very slow, and it is not utilized by most microorganisms.

Enriching for bacteria capable of degrading oil while using F-1 as a nitrogen source led to the identification of several bacterial strains that utilize F-1. These bacteria contain a cell-bound, inducible enzyme referred to as F-1-ase. The system comprising of F-1 and the special bacteria has been studied in the laboratory, and in field experiments. It has also been used successfully in open systems for treating petroleum pollution in soil and water. In this section we review the properties of F-1 in terms of its applicability for treating oil polluted soil and summarize data on the treatment of a beach polluted with several hundred tons of heavy crude oil.

4.1 Development of F1/Bacteria Technology

In order to determine the percentage of F-1 that sticks to soil, beach sand was treated with the nutrients and then washed with water. The water was collected and the level of nitrogen washed from the soil was determined. For comparison, the same treatment was performed with two commercial products used for bioremediation which use soluble nitrogen sources. The control was an ammonium salt. The results indicate that when F-1 was used only 3% of the nitrogen was washed into the water, in contrast to the other three preparations that lost 67–81% of the nitrogen. A further experiment demonstrating that F-1 binds to hydrocarbons and is not diluted into water is summarized in Table 5. Using F-1 as the nitrogen and phosphorus source enabled strains RL4, RL3, RT and a mixed culture of F-1 utilizing strains to grow on crude oil, even after replacing the aqueous phase. On the other hand, when water soluble nutrients (AS medium) was used, the washing procedure removed the N and P sources and thereby prevented growth of the strains.

Table 5. Growth of bacteria on crude oil and F-1 following removal of water soluble nutrients[a]

Strain	Nutrient	Cell density (cfu/ml)		Turbidity (K.U.) (t = 120 h)
		(t = 0)	(t = 120 h)	
RL4	F-1	5×10^5	9×10^7	550
RL3	F-1	8×10^5	9×10^7	860
RT	F-1	3×10^5	6×10^7	490
Mixed culture	F-1	3×10^5	1×10^8	900
RL4	AS	6×10^5	4×10^3	22
RL3	AS	4×10^5	2×10^6	25
R5	AS	4×10^5	4×10^5	20
Mixed culture	AS	5×10^5	1×10^6	23

[a] Data calculated from Rosenberg et al. (1992). F-1 medium contained 0.2% fertilizer (equivalent to 0.05% nitrogen). AS medium contained 0.1% ammonium sulfate (equivalent to 0.2% nitrogen) and 10 mM phosphate buffer. Both media contained 0.5% crude oil as a carbon source. Each 20 ml medium was mixed well and then allowed to stand for 5 min allowing the crude oil and the aqueous phase to separate. The aqueous phase was then removed as completely as possible and replaced with 20 ml fresh seawater. The procedure of removal of the remaining water soluble nutrients was repeated once more. Each flask was then inoculated with 40 μl of a pure or a mixed culture and incubated for 120 h at 20 °C with shaking. Viable counts were determined after inoculation (time 0) and at 120 h.

In order to find out whether bacteria that utilize F-1 can release into the medium intermediates that can be used by other bacteria, an experiment was performed to test for the existence of cross-feeding. A minimal medium containing 0.5% oil and 0.1% F-1 was divided into three flasks. The flasks were inoculated with bacteria of strain RL3, capable of utilizing F-1 (flask 1), with bacteria of strain *A. calcoaceticus RAG-1* that does not use F-1 as a nitrogen source (flask 2) and with a mixture of RL3 and *A. calcoaceticus RAG-1* (flask 3). The three flasks were incubated with aeration at 30 °C for 48 h, and the number of cells in each flask was determined by viable count. Since *A. calcoaceticus RAG-1* is resistant to ampicillin, its colonies could be distinguished from those of RL3 on plates containing the antibiotic. The results of this experiment indicate that the number of cells of strain RL3 (flask 1) increased more than 1000-fold, while there was no change in the cell number of *A. calcoaceticus RAG-1* (flask 2), confirming our previous findings that *A. calcoaceticus RAG-1* is incapable of using F-1 as a nitrogen source. The growth of *A. calcoaceticus RAG-1* was not stimulated by the presence of strain RL3 in flask 3, demonstrating that strain RL3 did not supply nitrogen-containing intermediates that could cross feed *A. calcoaceticus* RAG-1. These data suggest that the nitrogen product of the cell bound F-1-ase activity is efficiently taken up by strain RL3 and is not available for other bacteria.

To determine the capability of the system to degrade the various components of crude oil, sand was contaminated with 2000 ppm of weathered Alaska North Slope (ANS) crude oil and treated with F-1 and bacteria. The oil was extracted before and during treatment and analyzed in GC/MS by the National Environmental Technology Applications Center (NETAC), Pittsburgh, Pennsylvania, according to their standard procedure. The results indicated that the C_{15}-C_{33} fraction, that was the most abundant at day 0 was completely removed after one week. The C_{34}-C_{35} fractions were degraded to lesser degrees, about 20% in one week and about 70% in 4 weeks. During this time period, the concentrations of naphthalene, C_1-C_4 naphthalenes, fluorene, C_1-C_3 fluorenes, dibenzothiophene, C_1-C_3 dibenzothiophenes, anthracene, phenanthrene, C_1-C_3 phenanthrenes, flouranthene, pyrene, C_1-C_2 pyrenes, crysene and C_1-C_2 crysene were degraded to an undetectable level.

4.2 Bioremediation of Beach Sand

In August 1991, approximately 100 tons of heavy crude oil were accidentally spilled about 3 km north of Zvulon Beach, between Haifa and Akko in Israel. The oil-contaminated sand was collected into piles and subsequently spread over the beach.

The F/1 bacteria technology was chosen to clean the sand. In a small field trial ($50\,m^2$) the experimental plot, treated with F-1/bacteria, the degradation of oil reached 30% on the fourth day.

The biodegradation continued, reaching 50% on day 9 and 85% on day 25, when the experiment was concluded. The control plot showed a 18% degradation by day 9 which remained relatively constant until day 25.

Table 6. Bioremediation of hydrocarbon-contaminated sand at Haifa Beach, winter 1991–1992[a]

Day	Treated (% degradation)	Control (% degradation)
10	20	10
20	41	20
30	50	22
40	60	18
50	65	20
70	70	10
90	75	0
120	88	0

[a] Approximately 30 000 m^2 of contaminated beach were treated with 15 g F-1 per kg sand and 2000 l of the "mixed culture" (7×10^8 cells/ml). The area was plowed twice weekly.

Gas chromatographic analyses of the alkane fractions demonstrated the C_{20}-C_{32} fraction, which was the most abundant at day 0, was degraded to the extent of 94%, whereas the C_{14}-C_{18} and C_{36}-C_{40} fractions were degraded to lesser degrees, 80 and 74%, respectively. Visual examination of the beach sand following the treatment, in addition to the analytical data, indicated that the F-1/bacteria technology was applicable for bioremediation of the remaining sand.

In the large-scale cleanup, 30 000 m^2 of polluted sand was treated in the winter of 1992, which was an unusually hard winter for Israel, temperatures being around 5–10 °C for a couple of months. The results of the bioremediation treatment (Table 6) indicated that even under these conditions 88% of the oil was degraded after 4 months, whereas no significant degradation was measured in control plots.

Acknowledgments. This work was supported in part by the Pasha Gol Chair for Applied Microbiology, the Fogarty International Center for Advanced Study in the Health Sciences, N.I.H. and Makhteshim Chemicals Ltd.

References

Atlas RM (1991) Microbial hydrocarbon degradation-bioremediation of oil spills. J Chem Technol Biotechnol 52:149–156

Atlas RM, Bartha R (1973) Stimulated biodegradation of oil slicks using oleophilic fertilizers. Environ Sci Technol 7:538–541

Belsky I, Gutnick DL, Rosenberg E (1979) Emulsifier of *Arthrobacter* RAG-1: determination of emulsifier-bound fatty acids. FEBS Lett 101:175–178

Horowitz A, Gutnick DL, Rosenberg E (1975) Sequential growth of bacteria on crude oil. Appl Environ Microbiol 30:10–19

Kaplan N, Rosenberg E, Jann B, Jann K (1985) Structural studies on the capsular polysaccharide of *Acinetobacter calcoaceticus* BD4. Eur J Biochem 152:453–458

Kaplan N, Zosim Z, Rosenberg E (1987) *Acinetobacter calcoaceticus* BD4 emulsan: reconstitution of emulsifying activity with pure polysaccharide and protein. Appl Environ Microbiol 53:440–446

Navon-Venezia S, Zosim Z, Gottlieb A, Legmann R, Carmeli S, Ron EZ, Rosenberg E (1995) Alasan, a new bioemulsifier from *Acinetobacter radioresistens*. Appl Environ Microbiol 61:3240–3244

Peake E, Hodgson GW (1967) Solubilities of solid n-alkanes. J Am Oil Chem Soc 44:696–701

Pritchard PH, Mueller JG, Rogers JC, Kremer FV, Glaser JA (1992) Oil spill bioremediation: experiences, lessons and results from the Exxon Valdez oil spill in Alaska. Biodegradation 3:315–335

Rosenberg E (1991) Hydrocarbon-oxidizing bacteria. In: Ballows A (ed) The prokaryotes. Berlin Heidelberg New York, pp 441–459

Rosenberg E, Ron EZ (1996) Bioremediation of petroleum contamination. In: Crawford RL, Crawford DL (eds) Bioremediation: principles and applications. Cambridge University Press, Cambridge, pp 100–124

Rosenberg E, Zuckerberg A, Rubinowitz C, Gutnick DL (1979a) Emulsifier of *Arthrobacter* RAG-1: isolation and emulsifying properties. Appl Environ Microbiol 37:402–408

Rosenberg E, Perry A, Gibson DT, Gutnick DL (1979b) Emulsifier of *Archrobacter* RAG-1: specificity of hydrocarbon substrate. Appl Environ Microbiol 37:409–413

Rosenberg E, Gottlieb A, Rosenberg M (1983) Inhibition of bacterial adherence of hydrocarbons and epithelial cells by emulsan. Infect Immun 39:1024–1028

Rosenberg E, Legmann R, Kushmaro A, Taube R, Adler E, Ron EZ (1992) Petroleum bioremediation: a multiphase problem. Biodegradation 3:337–350

Rosenberg M, Rosenberg E (1985) Bacterial adherence at the hydrocarbon–water interface. Oil Petrochem Pollut 2:155–162

Song HG, Wang X, Bartha R (1990) Bioremediation potential of terrestrial fuel spills. Appl Environ Microbiol 56:652–656

Taylor WH, Juni E (1961) Pathways for biosynthesis of a bacterial polysaccharide. I. Characterization of the organism and polysaccharide. J Bacteriol 81:688–693

Zilber IK, Gutnick DL, Rosenberg E (1980) ^{32}P-incorporation and growth of *Pseudomonad* UP-2 on tetracosane. Appl Environ Microbiol 40:1086–1093

Zuckerberg A, Diver A, Peeri Z, Gutnick DL, Rosenberg E (1979) Emulsifier of *Arthrobacter* RAG-1: chemical and physical properties. Appl Environ Microbiol 37:414–420

12 Biodegradation of Coal Tar Constituents in Aquifer Sediments

E. J. Bouwer[1], W. Zhang[2], L. P. Wilson[1], and N. D. Durant[1]

Contents

1 Introduction

In situ bioremediation is an important technology for the cost-effective treatment of contaminated soils and groundwater (NRC 1993). Trial-and-error methods of implementing this complex process at a field scale are inefficient and costly. Therefore, it is important to conduct laboratory studies to establish feasible microbial reactions and to develop reliable engineering models that can analyze in situ options prior to field testing. Specifically, it is important to establish the appropriate chemical conditions required for biodegradation of the contaminants and to assess the relative importance of mass transfer (bioavailability) versus kinetic (biodegradation control) effects.

[1] Department of Geography and Environmental Engineering, The Johns Hopkins University Baltimore, Maryland 21218, USA
[2] Department of Civil and Environmental Engineering, Lehigh University, Bethlehem, Pennsylvania 18105, USA

H. Rubin/N. Narkis/ J. Carberry
Soil and Aquifer Pollution

This paper addresses the importance of chemical conditions and mass transfer effects to in situ bioremediation, particularly as they apply to a coal tar-contaminated site presented as a case study. The site was formerly used as a manufactured gas plant (MGP), and is presently contaminated with mono- and polycyclic aromatic hydrocarbons (PAHs). As part of an ongoing bioremediation feasibility study, an experiment was conducted to assess the ability of the sediment bacteria to mineralize ^{14}C-labeled forms of primary coal tar constituents (benzene, naphthalene, and phenanthrene), under site (intrinsic) conditions and under enhanced conditions. Modeling and experimental studies were combined to determine the influence of mass transfer rate (desorption rate) on the biodegradation performance. The first half of this chapter describes the experimental techniques and results of the biodegradation evaluation for the primary coal tar constituents dissolved in the groundwater at the site. In the second half of the paper we develop a theoretical model to describe the sorption-limited biodegradation that is characteristic of many hydrophobic aromatic compounds in sediment-water systems.

2 Site Description

The MGP site is contaminated with coal tar and petroleum hydrocarbons including benzene, toluene, ethylbenzene, xylene, naphthalene, acenaphthylene,

Table 1. Groundwater quality at the site of a former manufactured gas plant

Analyte	Range	Typical	Background
pH	3 to 7	6	6.3
Redox potential (mV)	−192 to 470	−20	0
Biological oxygen demand (mg/l)	<12 to 110	25	(a)
Chemical oxygen demand (mg/l)	<50 to 610	200	<50
BTEX (mg/l)	ND to 43	1	ND
Naphthalene (mg/l)	ND to 49[b]	0.5	ND
PAH-naphthalene[c] (mg/l)	ND to 1	0.05	ND
Total organic carbon	<0.5 to 43	10	(a)
Dissolved oxygen (mg/l)	<0.1 to 5	0.5	~2[d]
Nitrate (mg/l)	<0.5 to 4.6	<0.5	<0.5
Ammonia (mg N/l)	<0.05 to 83	20	(a)
Fe (II) (mg/l)	0.1 to 2350	50	16
Sulfate (mg/l)	<5 to 7300	500	4
Carbon dioxide (mg/l)	10 to 700	200	(a)

ND, not detected.
[a] Not measured.
[b] Upper bound value exceeds solubility due to presence of NAPL in aqueous sample.
[c] EPA priority pollutant PAH minus naphthalene.
[d] Dissolved oxygen measured by probe; may read higher than Winkler titration method or micro-oxygen electrode (Microelectrodes, Inc.).

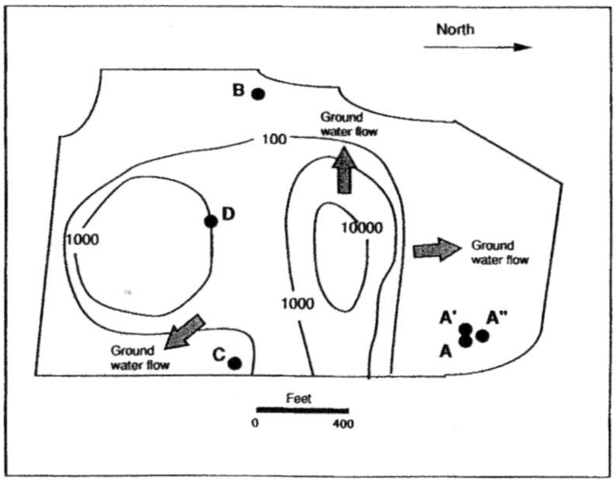

Fig. 1. Sampling borehole locations *A*, *A'*, *A"*, *B*, *C*, and *D*, groundwater flow direction, and naphthalene plume configuration (μg/l) in the upper aquifer (15–21 m). (Durant et al. 1995)

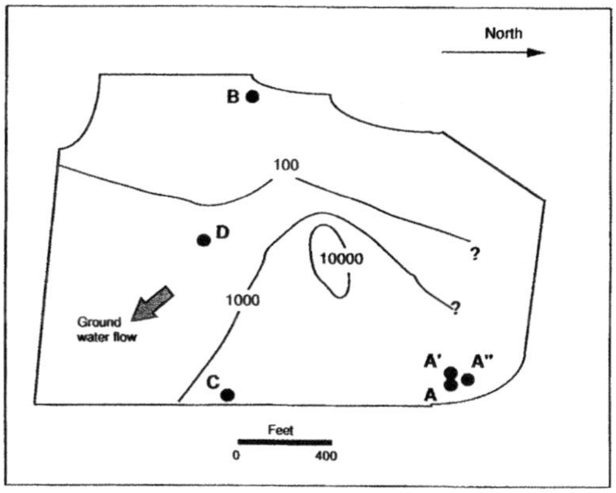

Fig. 2. Sampling borehole locations *A*, *A'*, *A"*, *B*, *C*, and *D*, groundwater flow direction, and naphthalene plume configuration (μg/l) in the lower aquifer (24–28 m). (Durant et al. 1995)

acenaphthene, fluorene, phenanthrene, anthracene, fluoranthrene and pyrene. Ranges for the aqueous concentration of these compounds, electron acceptors and other groundwater quality parameters measured in groundwater collected from various locations throughout the site are summarized in Table 1. Most of the contamination is contained within two sand and gravel aquifers at the 72 acre site (Figs. 1 and 2). Sulfate and ferric iron are the primary electron acceptors available in the subsurface, with only low concentrations of oxygen and/or nitrate detected at most locations.

3 Methods

3.1 Aquifer Material Sampling

Aquifer sediment samples were collected from six separate boreholes (Figures 1 and 2) in order to meet the following objectives: (1) to compare microbial activity from both contaminated and uncontaminated sediments; (2) to investigate the spatial distribution of microbial activity across the site; and (3) to investigate the extent to which amendments of O_2, N, and P affect biodegradation of aromatic compounds by the sediment bacteria. Sediment cores were obtained with a split spoon sampler and mud-rotary drilling. Cores were aseptically subsampled with a truncated sterile 60-ml syringe. Rhodamine-WT dye was added to the mud (5 mg/l) in order to trace mud intrusion into aquifer cores; sediments containing the dye as measured by fluorometry were assumed to be contaminated with non-indigenous bacteria and therefore were excluded from use in biological assays. Additional details on the sampling procedure are provided by Barranco et al. (1994) and Durant et al. (1995). Sediments from boreholes A, A′, B, C, and D were examined in the intrinsic assay, while only sediments from borehole A″ were analyzed in the enhanced assay.

3.2 Intrinsic Mineralization Assay

In this intrinsic mineralization experiment, aquifer sediments were combined with site ground water to assay the ability of the aquifer bacteria to degrade coal tar consituents in the absence of nutrient and electron acceptor amendments. The study entailed the addition of various radiolabeled (^{14}C) test compounds to batch sediment-water microcosms and the entrapment of $^{14}CO_2$ (using KOH) produced from mineralization. ^{14}C-benzene, ^{14}C-naphthalene, ^{14}C-phenanthrene, and ^{14}C-fluorene were selected for investigation because they are common constituents in the coal tar-contaminated aquifer. The sediments were also evaluated for their ability to mineralize ^{14}C-acetate and ^{14}C-*p*-hydroxybenzoate. Acetate was of interest because it was presumed that ^{14}C-acetate mineralization would provide an effective means of screening for microbial activity. ^{14}C-*p*-hydroxybenzoate was assayed to determine whether mineralization of this putative intermediate of aromatic hydrocarbon degradation would indicate the presence of bacteria capable of degrading benzene and PAHs.

Each microcosm consisted of a 15-ml serum bottle that contained 5 g of sediment and 3 ml of filter-sterilized groundwater. Groundwater was obtained from monitoring wells adjacent to the sediment sampling locations and from depths which corresponded to the sediment sample depth. Each microcosm received 0.07 μCi of radiolabeled and 1 ppm of unlabeled substrate. A sterile 2-ml glass vial containing 0.5 ml of 1 N KOH was placed inside each microcosm to capture the $^{14}CO_2$ generated from the mineralization of the aromatic substrate. The microcosms were sealed with Teflon-faced, rubber-butyl septa and aluminum crimp caps. The microcosms were incubated at 22 °C for 4 to 6 weeks. Anaerobic microcosms were prepared and incubated inside an anaerobic glove box (Coy) with an atmosphere consisting of 97:3% $N_2:H_2$. The majority of sediments were incubated under aerobic conditions because the well monitoring data for most of the locations examined indicated a prevalence of microaerophilic conditions. Anaerobic microcosms were prepared for sediments where ground water data indicated anaerobic conditions.

The microcosms were sampled using a syringe to extract and replenish the KOH in the $^{14}CO_2$ trap through the septa which were then resealed with silicone cement. ^{14}C activity was quantified by scintillation counting on a Beckman LS 3801 liquid scintillation counter. For each sample and test compound, triplicate active microcosms were prepared as well as duplicate killed control microcosms (with 1 ml of 0.5 N $HgCl_2$ and 1 ml of 2×10^{-3} N NaN_3 added to the microcosm supernatant). The difference in the amount of ^{14}C trapped in KOH in the control microcosm and the active microcosms represents the amount of mineralization of the test compound due to biotic processes. Biotic mineralization of the target compounds was assessed as the difference between mineralization observed in the average of the active microcosms less the average of the associated controls. The assay is a conservative estimate of biodegradation capability because it does not quantify the amount of parent compound converted to cells and intermediates.

3.3 Enhanced Mineralization Assay

Based on the results of the intrinsic mineralization assay, sediment samples to be used in the enhanced assay were collected from borehole A″ (located adjacent to borehole A) in order to assess the ability of the bacteria from this location to biodegrade aromatic compounds under enhanced conditions (supplemental electron acceptor and nutrients). Several aerobic and anaerobic samples were studied including sediments taken from 23.4 m which closely corresponds to the depth (22.2 m) where sediments with bacteria capable of both aerobic and anaerobic naphthalene biodegradation were identified in the intrinsic mineralization assay.

Using a design similar to the intrinsic assay, sediment microcosms were prepared to screen for the ability of indigenous subsurface microorganisms to degrade radiolabeled benzene, naphthalene, toluene and phenanthrene under enhanced aerobic conditions (supplemental O_2, NH_4^+, and PO_4^{3-}) and naphthalene and toluene under enhanced anaerobic denitrifying conditions (supplemental NO_3^- and PO_4^{3-}). Each sediment microcosm consisted of a 20-ml serum bottle containing 5 g of sediment and 3 ml of sterilized stock solution containing 128 mg/l ammonium (added as NH_4Cl) and 59 mg/l phosphate (consisting of KH_2PO_4, K_2HPO_4, and $Na_2HPO_4 \cdot 7H_2O$). In order to maximize oxygen availability, the stock solution used in the aerobic microcosms was purged with pure O_2 for 1 h, resulting in an initial dissolved oxygen concentration of 37 mg/l. The stock solution used in the preparation of anaerobic denitrifying microcosms contained 59 mg/l phosphate, 100 mg/l nitrate instead of ammonium and 0.5 mg/l sodium dithionate (oxygen scavenger), and was purged with nitrogen for 1 h prior to introduction of the stock into the anaerobic glove box.

The stock solution added to each microcosm was amended with ^{14}C-labeled and unlabeled benzene, toluene, naphthalene or phenanthrene. Microcosms containing benzene, naphthalene or phenanthrene received an average of 0.07 μCi/microcosm of ^{14}C-labeled compound and 1 mg/l of unlabeled compound. Microcosms containing toluene received an average of 0.03 μCi/microcosm of ^{14}C-labeled toluene and 50 mg/l of unlabeled toluene. Microcosms were sealed and sampled for $^{14}CO_2$ evolution in the same way as described in the intrinsic assay. Microcosms were prepared in triplicate along with duplicate killed control microcosms containing 8.2 g/l $HgCl_2$ (a biocide). Aerobic microcosms were incubated at constant temperature (22 °C) and anaerobic microcosms were prepared and incubated in an anaerobic glove box (Coy) at room temperature.

4 Results

4.1 Intrinsic Mineralization Assay

The results of the mineralization assay suggest that bacteria capable of aerobically biodegrading benzene and PAHs are present at a variety of locations throughout the subsurface of the former MGP site (Table 2). Although the majority of sediments analyzed could not degrade these compounds during four week incubations, benzene-, naphthalene-, and phenanthrene-degrading bacteria were detected at a variety of depths. Benzene mineralization ranged from 4–42%, naphthalene mineralization ranged from 8–55%, and phenanthrene mineralization ranged from 4–23% in the active microcosms (≥95% confidence level). No significant removal of fluorene was observed. The most extensive biodegradation of benzene, naphthalene and phenanthrene was observed in a sample from 3 m in borehole C. The in situ dissolved oxygen concentration at this location and sample depth was among the highest measured for this site. High oxygen enhances the growth of bacteria as well as their ability to biodegrade aromatic ring compounds. The presence of oxygen and substrate allows for the induction of oxygenase enzymes which facilitate the breakdown of the highly stable benzene ring. The high levels of oxygen in this portion of the site probably encouraged the proliferation of more metabolically active microbial populations.

The most widespread mineralization of naphthalene was observed throughout boreholes A and A′. Relatively high levels of mineralization of *p*-hydroxyben zoate, a structural analog of naphthalene biodegradation intermediates, was also observed. Both of these boreholes exhibited high viable and total cell counts (Durant et al. 1995), suggesting a link between biomass density and biodegradation capacity.

Two samples from 13.1 and 22.2 m from borehole A′ exhibited statistically significant naphthalene mineralization (8% ± 1 and 13% ± 2, respectively) under

Table 2. Microbial mineralization of test compounds over four weeks. (None of the samples from borehole D exhibited significant mineralization, and were therefore excluded from presentation in this table.)

Site	Sample depth	Sediment type	Percent mineralized (standard deviation in parentheses)			
			Acetate	p-*Hydroxy*	*Naph*	*Phen*
A	2.7	Silty sand	27 (5)	5	2 (3)	0 (3)
	8.8	Sandy gravel	12 (3)	6	0 (3)	0 (3)
	14.9	Sand	20 (2)	27	8 (6)	11 (11)
	18.0	Coarse sand	28 (13)	–	13 (3)	–
	18.0 MIC		23	10	–	–
	21.0	Gravel	24 (3)	30	20 (4)	–
	21.0 MIC		16	19	1	7
	24.7	Gravel	26 (12)	20	11 (12)	–
	24.7MIC		21	19	–	12.1
			Acetate	*Fluorene*	*Naph*	*Phen*
A	2.7	Silty clay	–	0	29 (5)	2 (1)
	5.8	Fine sand	70 (2)	0	1 (0)	2 (1)
	13.1	Sandy gravel	63 (5)	1 (2)	25 (7)	5 (5)
	13.1 AN		57 (4)	–	8 (1)	0 (2)
	16.8	Sand	47 (25)	–	29 (1)	9 (10)
	22.3	Gravel	–	–	8 (8)	–
	22.3 AN		52 (4)	–	13 (2)	–
			Acetate	p-*Hydroxy*	*Naph*	*Phen*
B	1.1	Sandy silt	4 (2)	6	0 (3)	0 (3)
	13.7	Gravel	0 (2)	–	2 (3)	–
	14.9	Sand	5 (2)	6	5 (3)	8 (3)
	18	Coarse sand	6 (2)	2	2 (3)	9 (4)
	18 MIC		25	0	7	4
	21	Gravel	1 (2)	5	2 (3)	12 (4)
	22.2	Saprolite	0 (3)	0	0 (3)	0 (3)
	22.2 MIC		0	0	0	0
			Acetate	*Benzene*	*Naph*	*Phen*
C	3.3	Sand	48 (13)	42	55 (5)	23 (3)
	12.5	Silty clay	6	0	17	0
	16.7 AN	Clayey silt	21	0	0	0
	17.4	Clayey silt	0 (2)	3	0	–
	22.2	Sand	0 (6)	0	0	1 (1)
	27.0	Gravel	–	–	2 (1)	–

AN, Anaerobic; MIC, microaerophilic; –, no data; *p*-Hydroxy, *p*-hydroxybenzoate; Naph, naphthalene; Phen, phenanthrene.

anaerobic conditions. Although the operative electron acceptor was not identified, significant levels of nitrate (13 mg/l) and sulfate (24 mg/l) were present in the groundwater used to prepare these microcosms. Aerobic acetate mineralization in these samples was very high (57% ± 4 and 52% ± 4, respectively), indicating the occurrence of facultative anaerobic bacteria. Other researchers have observed anaerobic biodegradation of naphthalene (Mihelcic and Luthy 1988; Grbic-Galic et al. 1991).

Statistically significant ($p < 0.05$) ^{14}C-acetate mineralization was observed in nearly half the sediments tested (Table 2). The presence of bacteria capable of mineralizing ^{14}C-acetate, however, did not always correspond with the presence of bacteria capable of mineralizing the aromatic compounds. While ^{14}C-acetate appears to be an effective indicator of microbial activity in aquifer sediments, our results suggests that it cannot be used to screen for the presence of bacteria capable of mineralizing benzene and PAHs. Mineralization of *p*-hydroxybenzoate may be a more effective tool for screening sediments for this purpose, but our results were inconclusive.

The extent of mineralization for aromatic substrates was typically limited (<50%), and this observation may be due to a number of factors, including volatilization in the microcosm headspace (especially true for benzene), a poor availability of oxygen in the microcosm supernatant, and poor substrate bioavailability due to sorption to the aquifer solids (especially true for naphthalene and phenanthrene). Limited degradation of PAHs is common for sediment–water microcosm studies (Madsen et al. 1991). Hot methanol extractions of some randomly selected sediments from our mineralization assay confirm that a significant portion of the ^{14}C-phenanthrene was sorbed, or otherwise not bioavailable in the microcosms (Durant 1993). Ancillary experiments on aerobic ^{14}C-naphthalene in the presence and absence of sediment have also observed sediment to significantly limit the extent of mineralization due to sorption and reduced bioavailability (Zhang et al. 1995).

The results of the intrinsic mineralization studies demonstrate that active microorganisms exist in the subsurface sediments at the site under both aerobic and anaerobic conditions. Some sediments contain bacteria capable of mineralization of PAHs, and in the case of naphthalene, biodegradation was observed under both aerobic and anaerobic conditions. This degradation was feasible without the use of nutrient (N and P) amendments.

4.2 Enhanced Mineralization Assay

Many samples from borehole A″ exhibited extensive mineralization of toluene, benzene, naphthalene, and/or phenanthrene in the presence of added nutrients and oxygen during eleven weeks of incubation. Sediments from 23.4 m demonstrated the largest mineralization of naphthalene and phenanthrene; limited mineralization of toluene was observed (Table 3). Toluene mineralization may have been limited due to toxicity effects. High levels of unlabeled toluene (50 mg/l) were used in the study to deliver a sufficient quantity of ^{14}C-toluene to each microcosm. The extent of mineralization for these incubations was typically higher than those observed for the intrinsic assays for sediments from boreholes A and A′, thus demonstrating the enhancing effect of oxygen and nutrient supplements on mineralization. The enhanced assays were incubated longer than the intrinsic

Table 3. Percent substrate mineralized (converted to $^{14}CO_2 \pm SD$) in oxygen and nutrient-amended microcosms under aerobic conditions after 77 days (borehole A″ sediments)

Depth (m)	Toluene	Benzene	Naphthalene	Phenanthrene
11.8	44 ± 10	51 ± 37	64 ± 20	55 ± 6
21	21 ± 5	62 ± 11	35 ± 49	52 ± 7
23.4	0 ± 1	27 ± 5	91 ± 4	85 ± 1

Table 4. Intrinsic versus enhanced mineralization of naphthalene by the MGP site aquifer bacteria

Study	Sample borehole and depth (m)	Condition	Compound	Extent of mineralization (%)
Intrinsic	A′ 13.1	Aerobic	Naphthalene	25 ± 7
Enhanced	A″ 11.9	Aerobic	Naphthalene	64 ± 20
Intrinsic	A′ 22.3	Aerobic	Naphthalene	8 ± 8
Enhanced	A″ 23.4	Aerobic	Naphthalene	91 ± 4
Intrinsic	A′ 22.3	Anaerobic[a]	Naphthalene	13 ± 2
Enhanced	A″ 23.4	Denitrifying[b]	Naphthalene	16 ± 5

[a] Approximately 13 mg/l nitrate and 25 mg/l sulfate were present in the groundwater used to prepare these microcosms.
[b] Microcosms amended with 100 mg/l nitrate.

assays, but since most (≥80%) of the mineralization occurred within the first 3 weeks, the difference in the incubation lengths for the two assays was not of significant importance.

Sediments from the shallower depths (11.8 and 21 m) were able to mineralize toluene although not as extensively as benzene, naphthalene and phenanthrene. These data may suggest that the aquifer bacteria proximal to borehole A″ may be better acclimated to biodegradation of unsubstituted aromatic hydrocarbons (i.e., benzene, naphthalene, and phenanthrene).

Under anaerobic denitrifying conditions, only the sediment sample from 23.4 m exhibited significant levels of mineralization of naphthalene (16 ± 5% mineralization after 80 days of incubation). To confirm denitrification, nitrate and nitrite (a product of denitrification) were measured by ion chromatography. This analysis revealed that 95% of nitrate was consumed relative to abiotic controls after 80 days with a concomitant conversion of 76% of the nitrate to nitrite. Since subsequent mineralization of naphthalene was not observed once 95% of the nitrate was consumed, these data suggest that an absence of available NO_3^- may have limited further degradation. The extent of denitrification observed, however, exceeded the stoichiometric requirements for the amount of naphthalene added (1 mg/l). It is possible that degradation of methanol, which was used as a carrier solvent for the addition of ^{14}C-naphthalene, served as an additional electron donor for denitrification.

The results of the enhanced mineralization assay confirmed that facultative anaerobic bacteria exist in certain sediments from the site and are capable of naphthalene mineralization under aerobic and denitrifying conditions. Addition of oxygen and nutrients enhanced the extent of naphthalene mineralization in sediments collected from adjacent locations (A′ and A″) in the subsurface at the site (Table 4).

5 Influence of Mass Transfer on Bioavailability in Aquifer Sediments

5.1 Introduction to Bioavailability

The laboratory intrinsic mineralization assay demonstrated that the indigenous site bacteria are capable of aerobic mineralization of benzene, toluene, naphthalene, and phenanthrene, and limited anaerobic mineralization of naphthalene. Enhanced mineralization work also demonstrated that the biodegradation of these pollutants by the aquifer bacteria could be enhanced through supplementation with oxygen and nutrients (N and P). For some sediments, stimulated anaerobic biodegradation may be achieved through the addition of nitrate. Data collected in this work and elsewhere, however, suggest that regardless of electron acceptor and nutrient enhancements, sediments can limit the extent of biodegradation of naphthalene and phenanthrene by sorption and reduced bioavailability. This is especially true in aquifers, where hydrophobic compounds are exposed to sediments for much longer periods than those examined in this work.

Prolonged exposure to sediments allows hydrophobic compounds such as naphthalene and phenanthrene to sorb strongly to intraparticle organic carbon. Therefore, the coal tar constituents at the site are likely to be less available for microbial utilization because they reside on the surface or in the pores of soil particles. The remainder of this chapter describes a theoretical and experimental study that was conducted to assess the influence of mass transfer rate on the overall biodegradation performance.

There must be a close association between a microorganism and contaminant for biodegradation to occur. The contaminant must be available for uptake and utilization by the microorganism (Mihelcic et al. 1993). Microorganisms and hydrophobic organic pollutants are distributed among the solid, liquid, and gas phases within the subsurface. In systems where a pure organic phase (NAPL) is present, bioavailability will be controlled, in part, by the rate at which compounds dissolve from the NAPL. In locations where free NAPL is absent, sorptive processes will ultimately control bioavailability.

Many organic contaminants of interest tend to sorb onto soil such that only a small fraction of the compound may actually be in the bulk water phase. Over long contact time, the sorbing pollutants slowly diffuse into the inorganic and organic solid matrix and may also form bound residues. Most evidence indicates the uptake of compounds by bacteria proceeds via the liquid phase. Consequently, a process such as sorption or volatilization that reduces the solution concentration tends to reduce the biotransformation rate. Furthermore, the accumulation of contaminants in fissures and cavities within subsurface solids renders them inaccessible to microorganisms and their enzymes. These processes decrease the bioavailability. The important conclusion from the influence of mass transfer in terms of reduced bioavailability is that the overall reaction rate (in the absence of NAPL) is controlled by the desorption rate and not by the activity of the degrading microorganisms. The practical effect of such slow diffusion from within soil aggregates and other kinetic limitations to desorption is a decrease of the rate of removal of the contaminant, thereby increasing the time required to achieve cleanup.

Sorption of contaminants tends to separate the direct contact between microorganisms and contaminants which is necessary for biodegradation to occur. For soils, sediments, and groundwater aquifers where the solid fraction and surface area are high, sorption is the primary process limiting bioavailability. Effects of sorption on biodegradation in the subsurface can be classified into concentration effects and desorption rate limitations as discussed below.

5.2 Concentration Effects

A direct impact of sorption on biodegradation is the reduction of organic compounds in the bulk water phase. As sorbed chemicals are effectively protected from direct biodegradation, microbial growth must rely on organic substrate(s) in the bulk water phase. The rate of bulk water uptake and metabolism of the organic compound is generally given by the Monod relationship:

$$\frac{dC}{dt} = -\frac{kXC}{K_s + C}, \tag{1}$$

where C is the bulk water concentration of the organic compound, t is time, k is the maximum specific biodegradation rate, X is the concentration of microorganisms in water, and K_s is the substrate concentration when the growth rate is at half the maximum. Sorption reduces the bulk water concentration and as a result will prolong the time to degrade a given amount of organic pollutant as compared with a system free of sediments or soil where sorption will not occur. At very low contaminant concentrations, often in the range of μg to ng per liter, insufficient energy and carbon may be available for microbial growth and maintenance. Rittmann and McCarty (1980) defined a critical concentration (C_{min}) at which microbial growth is just balanced by decay:

$$C_{min} = \frac{K_s b}{Yk - b}, \tag{2}$$

where Y is the biomass yield coefficient and b is the microbial decay coefficient. If sorption diminishes the concentration below C_{min}, biodegradation will decrease or even stop with time because there will be net decay of biomass.

On the other hand, at high concentrations, many organic compounds become toxic to microorganisms. One expression for the biodegradation rate is the Haldane equation which is given below (Andrews, 1968):

$$\frac{dC}{dt} = -\frac{kXC}{K_s + C + \frac{C^2}{K_i}}, \tag{3}$$

where K_i is the inhibition constant. Here sorption can reduce the bulk liquid concentration to lessen toxic inhibition and increase microbial growth and biodegradation.

5.3 Desorption Rate Limitations

Due to geometrical and mass transfer restrictions, most bacteria are present in the external surface of soil particles and in the bulk water. Only limited biological activities exist within the intraparticle pores (Jones et al. 1993). Decontamination of soils and sediments may involve three steps: (1) dissolution from the organic phase; (2) desorption of previously sorbed contaminant; and (3) biotic or abiotic transformation of the contaminant in the aqueous phase. In the absence of free NAPL, the apparent biodegradation rate in a solid-water system can be controlled either by the desorption or the biodegradation rate.

The influence of mass transfer on biodegradation can be characterized by a single parameter:

$$\frac{dC}{dt} = -B_f k_b C. \tag{4}$$

Here, B_f, termed the bioavailability factor, is used to account for the impact of exchange of the organic compound between soil and water, or sorption, on biodegradation rates. The objective of the following analysis is to derive B_f as function of fundamental sorption and biodegradation mechanisms.

In a soil–water slurry, the mass balance for an organic compound can be written as:

$$\frac{d}{dt}\left[VC + mS\right] = -Vf_{(C)}. \tag{5}$$

Here, V is the volume of the bulk water, m is the mass of the solid material, and S is the organic concentration in the solid phase. The left-hand side of this equation represents the change of organic compound with time. $f_{(C)}$ is the sink term which occurs only in the bulk water. If biodegradation is the only mechanism for removal of the organic compound from the system and is assumed to follow first-order kinetics, then

$$f_{(C)} = -k_b C, \tag{6}$$

where k_b is the first-order rate coefficient for biodegradation. The ratio of m to V is termed the soil/water ratio, $R_{s/w}$:

$$R_{s/w} = \frac{m}{V}. \tag{7}$$

Substitution of Eqs. (6) and (7) into Eq. (5) yields Eq. (8) below:

$$\frac{dC}{dt} + R_{s/w}\frac{dS}{dt} = -k_b C. \tag{8}$$

The Eq. 8 mass balance in a soil–water system can be rearranged to yield Eq. 9:

$$\left(1 + R_{s/w}\frac{dS}{dC}\right)\frac{dC}{dt} = -k_b C. \tag{9}$$

The following expression results when Eq. (9) is solved for dC/dt:

$$\frac{dC}{dt} = -\frac{k_b}{1 + R_{s/w}\dfrac{dS}{dC}}C. \tag{10}$$

B_f in Eq. (4) is thus defined as:

$$B_f = \frac{1}{1 + R_{s/w}\dfrac{dS}{dC}}. \tag{11}$$

The B_f can be quantified by selecting an explicit expression for dS/dC. There are two possible cases: (1) fast sorption/desorption, and (2) slow sorption/desorption.

(1) Fast sorption. If sorption is fast compared to biodegradation, sorption will be at equilibrium so that the following equation applies:

$$\frac{dS}{dC} = K_d, \tag{12}$$

where K_d is the soil–water distribution coefficient assuming a linear sorption isotherm. The corresponding B_f using Eq. (11) becomes:

$$B_f = \frac{1}{1 + K_d R_{s/w}}. \tag{13}$$

Here, B_f is analogous to the retardation factor used to model movement of a sorbing solute in porous media (Freeze and Cherry 1979).

(2) Slow sorption. When the rate of biodegradation is much faster than the sorption rate, microorganisms can effectively degrade available organic compound as soon as it desorbs from the solid phase. Consequently, the rate of biodegradation is approximately equal to the rate of desorption:

$$m\frac{dS}{dt} = Vk_bC. \tag{14}$$

One approximation for the mass exchange between soil and water (sorption/desorption) is a first-order reaction:

$$\frac{dS}{dt} = k_m(K_dC - S). \tag{15}$$

Here, k_m is the first-order sorption rate coefficient. The "driving force" for sorption is ($K_dC - S$) in terms of the solid phase concentration. Combining Eqs. (14) and (15), the following expression results for dS/dC:

$$\frac{dS}{dC} = K_d\left(1 + \frac{k_b}{k_mK_dR_{s/w}}\right). \tag{16}$$

The corresponding B_f for slow sorption [substitution of Eq. (16) into Eq. (11)] becomes:

$$B_f = \frac{1}{1 + K_dR_{s/w}\left(1 + \frac{k_b}{k_mK_dR_{s/w}}\right)}. \tag{17}$$

The four fundamental parameters, k_b, k_m, K_d, and $R_{s/w}$, can be combined into a dimensionless parameter, Φ, which is termed the Thiele modulus:

$$\Phi^2 = \frac{k_b}{k_mK_dR_{s/w}}. \tag{18}$$

The expression for B_f incorporating the Thiele modulus becomes:

$$B_f = \frac{1}{1 + K_dR_{s/w}\left(1 + \Phi^2\right)}. \tag{19}$$

Φ is a parameter for the kinetic effect of sorption on biodegradation, an analogy to the widely used parameter in chemical engineering for the effect of diffusion (mass transfer) on catalytic reactions in porous media (Satterfield 1970; Levenspiel 1972; Aris 1975; Ruthven 1984). The Thiele modulus (Φ) can be rearranged as shown in the following expression to reflect the ratio of characteristic times for sorption and biodegradation:

$$\Phi = \sqrt{\frac{k_b}{k_mK_dR_{s/w}}} = \sqrt{\frac{\dfrac{1}{k_mK_dR_{s/w}}}{\dfrac{1}{k_b}}} \tag{20}$$

$$\Phi = \sqrt{\frac{\text{characteristic time for sorption}}{\text{characteristic time for biodegradation}}}. \tag{21}$$

The practical value of the B_f model stems from its simplicity and ease of use. The B_f provides an estimate of the impact of sorption on biodegradation to be used as a substitute for extensive numerical calculations. Therefore, the overall rate of biodegradation in a soil/water slurry is determined by the two factors: B_f and k_b [Eq. (4)]. B_f is evaluated from the extent and rate of sorption. k_b is determined by the metabolic capability of microorganisms and environmental factors such as temperature. As applied here, the computation of k_b assumes that an excess of electron acceptor is present (i.e., biodegradation is not limited by electron acceptor availability). The time to reduce the bulk water concentration in half is given by:

$$t_{50\%} = \frac{\ln 2}{B_f k_b}. \tag{22}$$

A B_f value of 1.0 means there is no influence of sorption on biodegradation, and this situation corresponds to a well-mixed liquid culture system without sorbent. As the B_f decreases below 1.0, biodegradation will be significantly limited by sorption and the persistence of the chemical increases.

The relationship between the Thiele modulus (Φ) and the value of B_f as a function of the K_d for the organic contaminant is illustrated in Fig. 3. The curves in Fig. 3 were obtained using Eq. (19) and $R_{s/w} = 5\,\text{kg/l}$. Measured K_d values for many organic compounds of interest in the subsurface are greater than 0.1 l/kg (Ball and Roberts 1991). From the curves in Fig. 3, K_d values >0.5 l/kg will yield B_f values that are less than 0.3 for biodegradation in the subsurface. Consequently, organic compounds which can be readily biodegraded in liquid culture with half-lives ranging from a few hours to a few days (large k_b) can be very persistent in soils and groundwater aquifers (curves with $B_f < 0.3$ in Fig. 3). The Thiele modulus (Φ) incorporates the effect of desorption rate on the biodegradation rate. As Φ

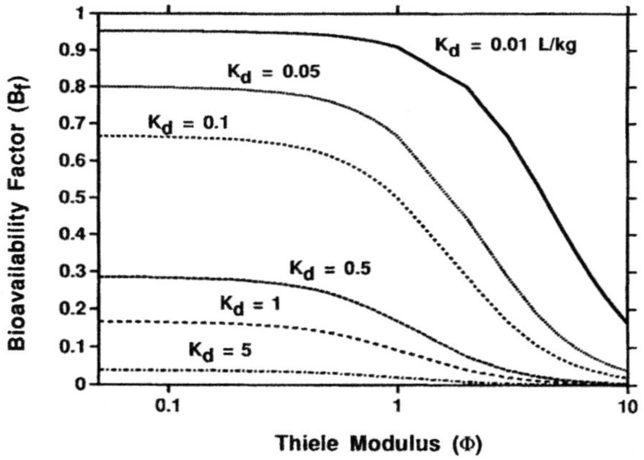

Fig. 3. Influence of Thiele modulus (Φ) and soil-water distribution coefficient (K_d) on the bioavailability factor (B_f). The curves were calculated from Eq. (19) with $R_{s/w} = 5\,\text{kg/l}$

Table 5. Calculated values for the Thiele modulus (Φ) and B_f for naphthalene and phenanthrene in the presence of MGP site aquifer sediments

Compound	$R_{s/w}$, kg/l	Φ	B_f
Naphthalene	5	0.94	0.28
Phenanthrene	5	0.64	0.11

increases, the B_f decreases which indicates that the overall biodegradation rate will be kinetically limited by the rate of desorption. Low bioavailability (small B_f) can be a major factor responsible for slow in situ bioremediation.

5.4 Application of the Bioavailability Model to MGP Site Aquifer Sediments

The modeling concepts described above were applied to estimate the extent of mass transfer control on biodegradation of naphthalene and phenanthrene for "average" sediments at the MGP site. Laboratory studies were conducted with composite MGP site sediment samples to evaluate the intrinsic biodegradation rates (k_b), the sorption partition coefficients (K_d), and the sorption rates (k_m) for naphthalene and phenanthrene. The organic carbon content for the sediment composite was 0.004%. Details of the sediment characteristics and batch methods to measure k_b, K_d, and k_m are presented elsewhere (Zhang 1995). From the values of k_b, K_d, and k_m determined by Zhang (1995), the calculated values of the Thiele Modulus (Φ) and B_f for naphthalene and phenanthrene appear in Table 5. The B_f values in Table 5 illustrate the importance of contaminant sorption. Given this model and composite sediment used, the aquifer sediments at the MGP site will slow naphthalene biodegradation by about 3.6 times ($1/B_f = 1/0.28$) and phenanthrene by slightly less than 10 times ($1/B_f = 1/0.11$) compared to biodegradation rates in the absence of aquifer sediment.

6 Concluding Remarks

Aquifer sediments recovered from a variety of locations at a coal tar-contaminated site contain bacteria capable of aerobically mineralizing benzene, toluene, naphthalene, and/or phenanthrene without an acclimation period. These bacteria appear to be distributed independent of depth in the aquifer, and occur both near and distal to contaminant sources. Substantial biodegradation was observed in the absence of nutrient amendments. This biodegradation could be stimulated by the addition of oxygen, nitrogen, and phosphate supplements. Acetate mineralization was an effective screening tool for aerobic and anaerobic microbial activity in the sediments, but could not be correlated to the presence of bacteria capable of degrading aromatic hydrocarbons. Limited anaerobic biodegradation of naphthalene was observed in certain sediments, and this anaerobic degradation appears to be linked to denitrification in some cases.

The results of mineralization experiments suggest that enhanced aerobic bioremediation would be an effective means of reducing the contaminant mass at the MGP site. Our work confirms that the aromatic hydrocarbon-degrading bacteria are present, and that the biodegradation rates can be improved. Empirical data and numerical simulations suggest, however, that the rate of bioremediation, intrinsic or enhanced, will be limited by contaminant sorption to the aquifer sediments. Modeling analyses (Φ and B_f) and batch experiments demonstrated that naphthalene and phenanthrene exhibit reduced extents and rates of biodegradation in the presence of aquifer sediments. Consequently, the mass transfer rate becomes an important control over the biodegradation rate. The practical effect of such slow diffusion from within soil aggregates and other kinetic limitations to desorption is a decrease of the rate of removal of the contaminant, thereby increasing the time required to achieve cleanup and the amount of chemicals that must be supplied to sustain and/or enhance microbial activity.

Acknowledgments. This work was supported in part by Cooperative Agreement ECD-8907039 between the National Science Foundation and Montana State University, and in part by the Hazardous Substance Management Research Center and the New Jersey Commission on Science and Technology headquartered at the New Jersey Institute of Technology.

References

Andrews JF (1968) A mathematical model for the continuous culture of microorganisms utilizing inhibitory substrates. Biotechnol Bioeng 10:707–723

Aris R (1975) The mathematical theory of diffusion and reaction in permeable catalysis. Clarendon Press, Oxford

Ball WP, Roberts PV (1991) Long-term sorption of halogenated organic chemicals by aquifer materials. I: equilibrium studies. Environ Sci Technol 25:1223–1237

Barranco FT, Kocornik JL, MacFarlane ID, Durant ND, Wilson LP (1994) Subsurface sampling techniques used for a microbiological investigation. In: Hinchee RE, Leeson A, Semprini L, Ong SK (eds) Bioremediation of chlorinated and polycyclic aromatic hydrocarbon compounds. Lewis, Ann Arbor, pp 474–478

Durant ND (1993) Screening for natural subsurface biotransformation of coal tar constituents at a former manufactured gas plant. Master's Thesis, Johns Hopkins University, Baltimore, pp 96–97

Durant ND, Wilson LP, Bouwer EJ (1995) Microcosm studies of subsurface PAH-degrading bacteria from a former manufactured gas plant. J Contam Hydrol 17:213–237

Freeze RA, Cherry JA (1979) Groundwater. Prentice-Hall, Englewood Cliffs

Grbic-Galic D, Henry SM, Godsy EM, Edwards E, Mayer KP (1991) Anaerobic degradation of aromatic hydrocarbons and aerobic degradation of trichloroethylene by subsurface microorganisms. In: Baker RA (ed) Organic substances and sediments in water, vol 3. Lewis, Chelsea, Micligan, pp 239–266

Jones WL, Dockery JD, Vogel CR, Sturman PJ (1993) Diffusion and reaction within porous packing media: a phenomenological model. Biotechnol Bioeng 41:947–956

Levenspiel O (1972) Chemical reaction engineering, 2nd edn. Wiley, New York

Madsen EL, Sinclair JL, Ghiorse WC (1991) In situ biodegradation: microbiological patterns in a contaminated aquifer. Science 252:830–833

Mihelcic JR, Luthy RG (1988) Degradation of polycyclic aromatic hydrocarbons under various redox conditions in soil-water systems. Appl Environ Microbiol 54:1182–1187

Mihelcic JR, Lueking DR, Mitzell R, Stapleton JM (1993) Bioavailability of sorbed- and separate-phase chemicals. Biodegradation 4:141–154

National Research Council (1993) In situ bioremediation when does it work? National Academy Press, Washington, DC

Rittmann BE, McCarty PL (1980) Model of steady-state biofilm kinetics. Biotechnol Bioeng 22:2343–2357

Ruthven DM (1984) Principles of adsorption and adsorption processes. John Wiley, New York

Satterfield CN (1970) Mass transfer in heterogeneous catalysis. MIT Press, Cambridge, Massachusetts

Zhang W (1995) Effect of sorption on bioavailability of hydrophobic organic contaminants: experimental and model studies. PhD Dept of Geography and Environmental Engineering, The Johns Hopkins University, Baltimore, Maryland

Zhang W, Bouwer E, Wilson L, Durant N (1995) Biotransformation of aromatic hydrocarbons in subsurface biofilms. Water Sci Technol 31:1–14

13 Reversible and Irreversible Binding of Organics to Soils and the Effect of Biodegradation

J. T. Novak[1], W. D. Burgos[2], and A. Bhandari[3]

Contents

1 Introduction

Remediation of contaminated soil is complicated by our incomplete understanding of processes which control sorption, desorption, and biodegradation of organic contaminants associated with soil surfaces. Because of the interest in intrinsic or natural remediation, knowledge of these processes is necessary if we are to understand the risks associated with this and other soil treatment strategies. One major area of concern is the potential for release of contaminants associated with soils that are not readily extracted during soil testing. Ball and Roberts (1991) have shown that for some soils, the time scale associated with adsorption and desorption can be of the order of several years. In addition, the apparently irreversible binding of organic contaminants to soils has been observed, especially for phenolic compounds, and this binding has generally been attributed to oxidative coupling wherein phenolics are incorporated into soil humic matter by chemical (Stone 1987) or biological (Bollag 1992) mechanisms. The magnitude and ultimate

[1] Dept. of Civil Engineering, Virginia Polytechnic Inst. and State University, Blacksburg, Virginia, 24061, USA
[2] Dept. of Civil and Environ. Engineering, The Pennsylvania State University, University Park, Pennsylvania, 16802, USA
[3] Dept. of Civil and Environ. Engineering, University of Michigan, Ann Arbor, Michigan, 48109, USA

H. Rubin/N. Narkis/ J. Carberry
Soil and Aquifer Pollution

fate of contaminants that are not easily extracted from soils are the main focus of this chapter.

The irreversible binding of phenolics in soils and activated carbon is well documented (Berry and Boyd 1985; Vidic and Suidan 1991). The covalent bonding of phenols to soil organic matter (SOM) may be catalyzed by metal oxide surfaces (Stone 1987), clay minerals (Wang et al. 1978), or by extracellular enzymes (Bollag 1992). Stone (1987) has described a mechanism by which phenolics are adsorbed then oxidized by manganese oxide particles, then the reactive oxidized organic can be covalently linked to soil humic matter by an oxidative coupling reaction. Biologically-mediated oxidative coupling of phenols by peroxidases, lacasses, and tyrosinases has been shown to produce cross-linked products. It has been suggested that fungal laccases play the most important role in linking phenols to soil humus (Sjoblad and Bollag 1981), because these reactions involve only the use of molecular oxygen. Peroxidases require H_2O_2, so while these may not be as important in natural systems, they could be relevant for engineered systems where hydrogen peroxide is added as a source of oxygen to support aerobic biological growth.

Also of interest in remediation processes is the fate of polyaromatic hydrocarbons (PAHs). Although the saturated PAHs are not thought to be "coupleable", their hydroxylated degradation intermediates may be. The rate of oxidative coupling of such intermediates compared to further biological oxidation is not known but Suflita and Bollag (1981) have shown that incubation of alpha-naphthol, a degradation intermediate of naphthalene, with fungal peroxidases produces polynapthols. Therefore, oxidative coupling of partially oxidized PAHs is probable but the degree to which this may occur is not known.

Separation of slowly desorbing materials from those irreversibly bound may be important in assessing risk, in certifying a site to be clean, or in devising remediation schemes. Furthermore, the use of oxidative coupling as a remediation method may be unacceptable to regulators because of the uncertainty of the eventual fate of these bound materials. Although oxidative coupling has been shown to occur through both biological (enzymatic) and chemical mechanisms in soils, the extent of these reactions and their relative importance in soils are not known.

It seems likely that oxidative coupling reactions may play a more important role in the disappearance of contaminants from soil than is generally thought. The techniques used to enhance in situ bioremediation processes are also likely to promote oxidative coupling and measurements of soil cleanliness are unable to distinguish between coupling, and complete degradation. In this study, we attempt to quantify the magnitude of irreversible binding reactions using readily coupleable phenolic compounds and to determine the extent to which degradation intermediates of polycyclic aromatics may also participate in oxidative coupling reactions.

2 Materials and Methods

2.1 Experimental Design

In order to attempt to determine the extent to which biological oxidative coupling and chemical oxidative coupling contribute to the irreversible binding of contaminants, the adsorption conditions were varied to separate the various processes. Experiments were conducted under oxic-biotic conditions where non-autoclaved soil under aerobic conditions was used; anoxic-biotic conditions where non-autoclaved soil under a nitrogen atmosphere was used; oxic-abiotic soils, autoclaved but in an ambient atmosphere; and for the phenols, anoxic-abiotic soils made up the remaining soil category. It was thought that in the oxic-biotic soils, both chemical and biological oxidative coupling would occur. In the anoxic-biotic and oxic-abiotic soils, oxidative coupling would be restricted, especially for the autoclaved soil where biological oxidative coupling would not be expected but the effect on chemical processes is unknown. For the oxic-abiotic experiments, soils were autoclaved three times over 5 days for 30 min at 121 °C at 20 psi. The anoxic-biotic experiments were conducted in a plexiglass glove box purged with nitrogen gas.

2.2 Materials

Radiolabeled chemicals (C^{14}) were purchased from the Sigma Chemical Co. and known amounts of the radioactive compounds were added to stock solutions of non-radiolabeled chemicals. Three replicates were used for each of the experimental conditions and concentrations.

Two soils (thermic, coated, Typic Quartzipsamments) collected from Surry County, Virginia, were selected after reviewing soil survey maps. These soils were selected to minimize the clay content and also to provide soil from the same parent material but with different organic contents. One of the soils, the high organic matter soil (HOM) was obtained from a forested area at a depth of 0 to 15 cm below the surface and had an organic content of 3.2%. The low organic matter soil (LOM) was collected from 15 to 30 cm below the surface in an adjacent field and had an organic content of 0.52%. The moist soils were passed through a US standard sieve No. 10 and stored at 4 °C. A more detailed description of these soils has been provided by Burgos (1995).

2.3 Adsorption/Desorption Experiments

2.3.1 Adsorption

Adsorption experiments were conducted at approximately 25 °C in glass 13 × 100 mm Teflon-capped test tubes by adding soil and then filling the tube to the top with the solution containing the test compound, and mixed on a shaker table for 2 d. A 2-day contact period was used to minimize biological reactions. At the end of the contact period, the tubes were centrifuged and 1 ml of solution transferred directly into a scintillation cocktail. The radioactivity was measured using a Beckman LSC 3510T. The soil-associated contaminant following 2 d of contact (Q_{EQ}) was calculated by subtracting the activity remaining in solution from the total activity added and using the data from soil-free tubes containing the radioactive solution as a control. Biotransformations during the 2-day sorption period were monitored using thin layer chromatography (TLC) and this was used to correct for minor biotic losses.

2.3.2 Desorption

Following sorption, the following procedure was used to determine the fate of the contaminant. After the 2-day contact period, tubes were removed from the shaker table and centrifuged at 1600 g. A 1 ml sample of the supernatant was analyzed for ^{14}C and an additional 4 ml of liquid removed. This was replaced with 5 ml of synthetic groundwater, remixed, allowed to stand for times ranging from 5 min to 24 h, then centrifuged again and the activity in the supernatant again counted. Depending on the organic compound studied, between 25 and 50 aqueous fill and draw extractions were carried out, followed by five to seven solvent extractions. For the phenols, dichloromethane was used as the solvent and for the naphthalene and naphthol, ethyl acetate was used. Initially, the water-extracted solutions were individually analyzed for ^{14}C but after the second extraction, the contents of each five or ten extractions were pooled for analysis. The amount of the organic removed by the combination of water and solvent extraction was considered to be extractable or reversibles Q_{REV}, while that which remained was considered to be non-extractable, Q_{NONEX}, according to Eq. (1).

$$Q_{EQ} = Q_{REV} + Q_{NONEX}. \tag{1}$$

The non-extractable contaminant was quantified by combustion for 4 min at 925 °C in a biological material oxidizer and collecting the $^{14}CO_2$. In addition, one set of tubes was first extracted with 0.1 N NaOH prior to combustion, in order to quantify the alkali soluble humic and fulvic acids. The difference between the combustable ^{14}C with and without alkali extraction was considered to be the humic and fulvic associated fraction. Because the oxidative coupling reactions are thought to bind organics to humic acids, it was useful to see if the irreversibly sorbed fraction was associated with this portion of SOM.

Two other types of experiments were used to further evaluate the binding mechanisms. For monochlorophenol, a set of experiments was conducted in which H_2O_2 was added to the tubes to see how this influenced irreversible binding. The amount added was the same as used by Farmer (1989) to provide a source of oxygen to the organisms for biodegradation without causing toxicity or altering soil properties. The second set of experiments were long-term incubation experiments using naphthalene in oxic-biotic soils. These experiments were originally to provide evidence for the production of biological intermediates but also gave information about the potential for conversion to and oxidative coupling of naphthalene oxidation products. In these studies, incubation times of up to 200 days were used.

3 Results and Discussion

3.1 Naphthalene/Naphthol Studies

Some of the desorption isotherms for napthalene contacted with soil under oxic-biotic conditions are shown in Fig. 1. Only oxic-biotic conditions were used for the study of napthalene because this was considered to be the condition which would be most likely to lead to oxidative coupling. As can be seen from the data for both the high and low organic matter soils, the combination of water and solvent extraction removed most but not all of the sorbed naphthalene. Because unsubstituted PAHs have not been shown to participate in oxidative coupling reactions, no irreversible binding was expected for naphthalene. In addition, TLC analysis of the liquid phase from the reactors after 2 days of incubation indicated that no intermediates were formed, and therefore the binding of degradation products at this short contact was not likely to have occurred. Therefore, it was concluded that the naphthalene which remained after water and solvent extraction was probably the result of mass-transfer limited processes and not irreversible binding.

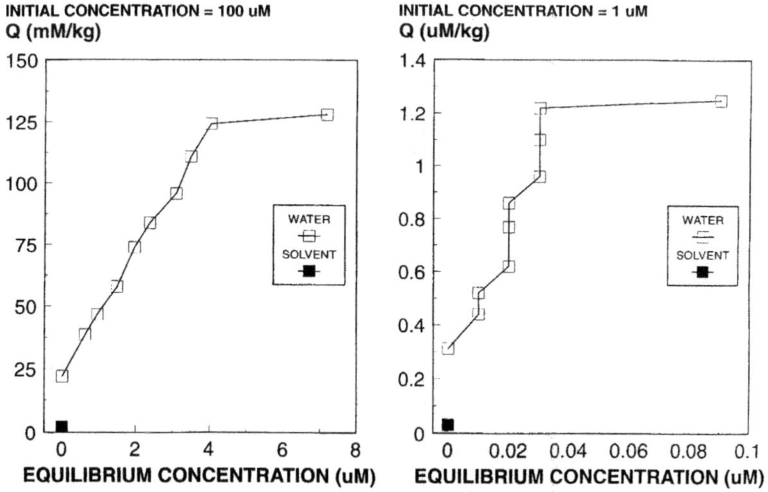

Fig. 1. Desorption isotherms for naphthalene at two initial sorption concentrations, 100 and 1 μM, where sorption occurred under oxic-biotic conditions. *Solid symbols* denote the fifth solvent extraction

For α-naphthol, the nonextracted material was thought to consist of both a portion retained by mass-transfer limitations ($Q_{TRAPPED}$) and an irreversibly bound fraction (Q_{BOUND}). The bound fraction is thought to be covalently bound to the soil and it is this fraction and the processes that control the rate and extent of this binding that is of interest in this study. For the naphthalene, because the Q_{BOUND} portion was thought to be near zero, the material that was not extracted was assumed to be trapped and this fraction was subtracted from the total unextracted fraction for soils where oxidative coupling was thought to occur, to provide an estimate of the oxidatively coupled portion (Q_{BOUND}). This distribution of nonextractable materials is shown in Eq. (2).

$$Q_{NONEX} = Q_{TRAPPED} + Q_{BOUND}, \quad (2)$$

where $Q_{TRAPPED}$ for alpha-naphthol is assumed to be equal to the value of Q_{NONEX} for naphthalene.

Aqueous and solvent extraction results for α-naphthol under oxic-biotic and anoxic-biotic conditions are shown in Fig. 2. The last data point in each desorption isotherm (solid symbol) corresponds to the final (fifth) solvent extraction. It can be seen from the data in Fig. 2 that a considerable fraction of the initial sorbed material remains unextracted following 50 water washes and 5 solvent extractions. The greatest amount of nonextractable material is associated with the oxic-biotic soil where 33–71% of the sorbed material is not extracted, depending on the initial contaminant concentration. For the anoxic-biotic soils, the bound fraction varied from 14 to 41%. The results of desorption experiments under oxic-abiotic conditions produced data very similar to the anoxic-biotic data so are not shown. In

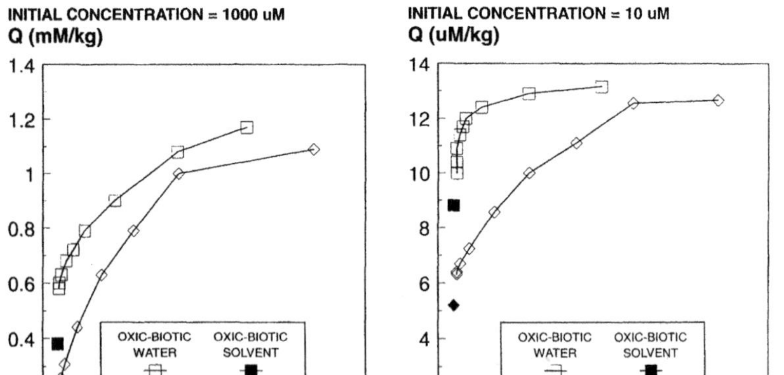

Fig. 2. Desorption isotherms for α-naphthol at two initial sorption concentrations, 1000 and 10 μM, where sorption occurred under oxic-biotic and anoxic-biotic conditions. *Solid symbols* denote the fifth solvent extraction

contrast, naphthalene retained only 2% of its adsorbed organic for high organic matter soil and 7% for low organic matter soil.

The difference between the biologically active oxic and anoxic soils is attributed by us to biological oxidative coupling. The Q_{BOUND} for the anoxic system is believed to be due to chemical oxidative coupling in the manner described by Stone (1987) and is catalyzed by metal oxide surfaces on the soil. For naphthol, approximately 40–60% of the oxidative coupling is believed to be due to biological processes. TLC analysis indicated that only 4.6% of the activity extracted from the high-organic soil resulted from transformation products of naphthol. Therefore, it appears that the biological coupling reaction under the conditions of these studies was kinetically favored over complete biological oxidation.

In Tables 1 and 2, some of the data for the desorption and extraction experiments are shown. Comparisons between the high and low organic matter content at three different initial concentrations of naphthol can be made. It can be seen that the major difference between the high and low organic matter soils is in the initial adsorption. Once adsorbed, a nearly constant percent of the adsorbed

Table 1. Desorption/extraction of naphthol sorbed to high organic matter soil

Adsorption conditions	2-Day aqueous equilibrium (μM/l)	2-Day adsorbed equilibria (μM/kg)	Remaining after solvent extraction (μM/kg)	Percent retained after solvent extraction
Oxic-biotic	153	1170	376	32
Oxic-biotic	7.18	128	69.4	54
Oxic-biotic	0.47	13.1	8.8	67
Anoxic-biotic	180	1130	184	16
Anoxic-biotic	10.4	124	38.4	31
Anoxic-biotic	0.83	12.6	5.19	41
Oxic-autoclaved	186	1110	152	14
Oxic-autoclaved	10.8	121	38.3	32
Oxic-autoclaved	0.84	12.6	3.83	30

Table 2. Desorption/extraction of naphthol sorbed to low organic matter soil

Adsorption conditions	2-Day aqueous equilibrium (μM/l)	2-day adsorbed equilibria (μM/kg)	Remaining after solvent extraction (μM/kg)	Percent retained after solvent extraction
Oxic-biotic	794	284	100	35
Oxic-biotic	68.9	42.9	24.2	56
Oxic-biotic	5.49	6.23	4.44	71
Anoxic-biotic	752	342	65.7	19
Anoxic-biotic	76.4	38.3	13.1	34
Anoxic-biotic	5.93	5.61	2.83	50
Oxic-autoclaved	827	239	33.8	14
Oxic-autoclaved	77.3	31.4	8.8	28
Oxic-autoclaved	7.59	3.32	0.78	24

material was bound to the soil. It can also be seen that the anoxic-biotic and oxic-abiotic binding was similar for most of the test conditions. Also, the component of binding that is assumed to be the result of biological processes is in the range of 50% although it is higher for the lowest initial concentration of contaminant.

Extraction by alkali to remove humic and fulvic acids from the soil always reduced the amount of bound material, but the removals showed no decernable trends, probably because this extraction procedure is not quantitative. However, the association of substantial fractions of contaminant with the humic and fulvic acids of the soils supports the idea that these materials have been incorporated into soil organic matter. The percent of bound activity extracted by alkali ranged from a high of nearly 75% to a low of 1%. The largest fraction of base-extractable naphthol was found for the material associated with the long-term contact data shown in Fig. 3. It can be seen that after 200 days of contact, about 75% of the nonextractable material was in the humic and fulvic fraction of the soil organic matter.

The data for the binding of naphthalene to the LOM and HOM soils under oxic-biotic conditions are shown in Table 3. It can be seen that the LOM soil retains a higher fraction of naphthalene than does the HOM soil. It appears that the LOM soil permitted better access by naphthalene to the mineral surface than did the HOM soil. Therefore, the contaminant concentration at the mineral surface would be higher allowing more material to enter the pores of the soil. It appears that mass transfer-limited processes may be more significant as the soil organic matter content declines.

Data for naphthalene was also obtained at extended contact times in order to see if naphthalene was degraded and to assess the fate of the degradation products. As shown in Fig. 3, when the contact time was extended to 200 days, more of the activity was found to be nonextractable. After 2 days of contact, 1.6% of the activity was not extractable. At 30 days, 6% was retained and after 200 days, nearly 12% of

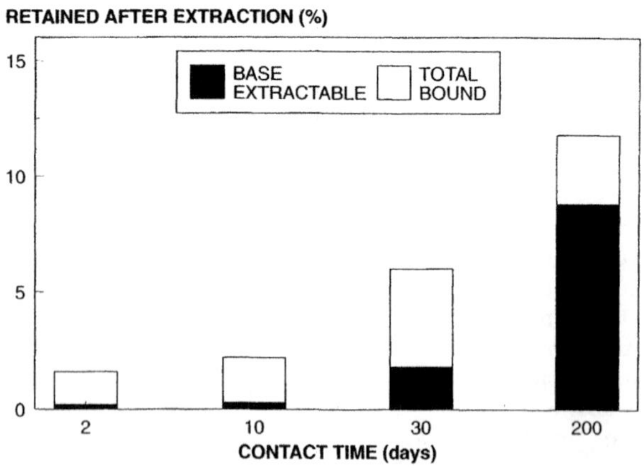

Fig. 3. Evidence for the apparent oxidative coupling of naphthalene oxidation products and the incorporation of this material into soil humic matter

Table 3. Desorption/extraction of naphthalene sorbed under oxic-biotic conditions (2-day contact period)

Soil organic content	2-Day aqueous equilibrium (μM/l)	2-Day adsorbed equilibrium (μM/kg soil)	Retained after water and solvent extraction (μM/kg soil)	Percent retained after water and solvent extraction
Low	59.2	56.4	4.8	8.5
Low	4.23	7.96	0.66	8.3
Low	0.46	0.75	0.06	8
High	7.13	128	2.1	1.6
High	0.83	12.7	0.31	2.4
High	0.09	1.26	0.03	2.4

the activity was considered to be irreversibly sorbed. For autoclaved soils with contact times up to 30 days, no additional binding was observed. TLC analysis of the supernatants from the 30- and 200-day reactors showed the presence of a more polar metabolite and we believe this metabolite, probably α-naphthol, was oxidatively coupled as it was produced. Evaluation of the supernatant from these reactors showed that more than 94% of the activity was naphthalene, indicating that intermediates were not accumulating in the reactors. Also, recovery of total activity in these systems was comparable to that of the other reactors where naphthalene was added, suggesting little evolution of $^{14}CO_2$.

Therefore, based on the extensive binding of α-naphthol after only 2 days of contact, especially under oxic-biotic conditions and on the increase in binding for long contact times which occur in conjunction with the production of degradation intermediates of naphthalene, it appears that oxidative coupling is a process competitive with complete biological mineralization. Furthermore, the conditions necessary to promote biodegradation of PAHs are also those which lead to partial mineralization and binding of PAHs.

3.2 Phenol/Chlorophenol Studies

Experiments similar to those used to characterize naphthalene and naphthol binding were also conducted to determine the binding characteristics of phenol, 4-monochlorophenol, 2,4,6-trichlorophenol and pentachlorophenol. The phenols are known to be used as building units for the enzyme-mediated synthesis of SOM or soil humus (Stevenson 1982) so it was expected that all of these compounds would oxidatively couple to some degree. Phenol and its chlorinated derivatives are important industrial compounds and commonly found as soil and groundwater contaminants. Therefore, these were of interest because they are common pollutants and also because they could be used as model compounds to investigate the influences of the octanol/water partition coefficient, degree of substitution, solubility, etc. Some of the characteristics of these compounds are shown in Table 4.

Table 4. Properties of the phenols used in this study

	Phenol	4-MCP	TCP	PCP
Molecular weight	94	128.5	197.5	266.5
Aqueous solubility (mg/l)	82 000 (15 °C)	27 100 (20 °C)	800 (25 °C)	14 (20 °C)
Log K_{OW}	1.47	2.39	3.38	5.01
pK_a	9.98	9.7	6.1	4.75

Table 5. Desorption/extraction of chlorophenols

Sorbate	2-Day adsorbed equilibrium (μM/kg soil)			Retained after water and solvent extraction (μM/kg soil)			Percent retained after water and solvent extraction		
	o-b	an-b	an-au	o-b	an-b	an-au	o-b	an-b	an-au
Phenol (0.53 mM)	196	201	159	106	49	21	54	24	13
Phenol (5.3 mM)	1376	1571	1382	587	301	171	43	21	12
4-MCP (78 μM)	54	56	55	7.7	3.5	1.3	14	6	2
4-MCP (78 μM)	405	417	427	47	41	20	12	10	5
2,4,6-TCP (51 μM)	67	65	62	17.9	9.7	7.7	27	15	12
PCP (38 μM)	55	55	54	9.3	4.5	2.4	17	8	4

o-b, Oxic-biotic; an-b, anoxic-biotic; an-au, anoxic-autoclaved.

Data for the adsorption, sequential water and solvent extraction (25 water washes followed by 7 solvent extractions) along with the percent of nonextractable activity are presented in Table 5. Recoveries were excellent, all above 90%, and generally above 95%. In these studies, reactors under anoxic-abiotic conditions were also included in the study matrix. By autoclaving and maintaining a nitrogen atmosphere, it was expected that little oxidative coupling would occur, so the unextracted material was used to estimate the fraction that would be retained in soil pores and be mass transfer-limited as opposed to permanently bound.

It can be seen in Table 5 that the adsorption conditions (oxic versus anoxic) had little effect on the amount of material adsorbed but had a major impact on the material irreversibly bound. As with naphthol, the greatest fraction of irreversibly bound material occurred with the oxic–biotic soils and approximately half of the irreversible binding appeared to be due to biological processes where the difference between oxic-biotic and anoxic-abiotic conditions is considered to be the contribution of biological processes. It can also be seen that the amount retained under anoxic-abiotic conditions, thought to be the mass transfer limited fraction in the soil pores, ranged from 2 to 12%, depending on the compound used and its concentration.

3.2.1 Chlorine Substitution

The degree of substitution of the phenols was expected to have a major impact on the oxidative coupling. However, when the data were plotted as a function of the initial solution concentration (Fig. 4), it appeared that the extent of binding was not greatly affected by the degree of substitution. Data for binding under oxic-biotic conditions are shown in Fig. 4A, for binding under anoxic-biotic conditions in Fig. 4B, and for anoxic-abiotic conditions in Fig. 4C. It can be seen from the data

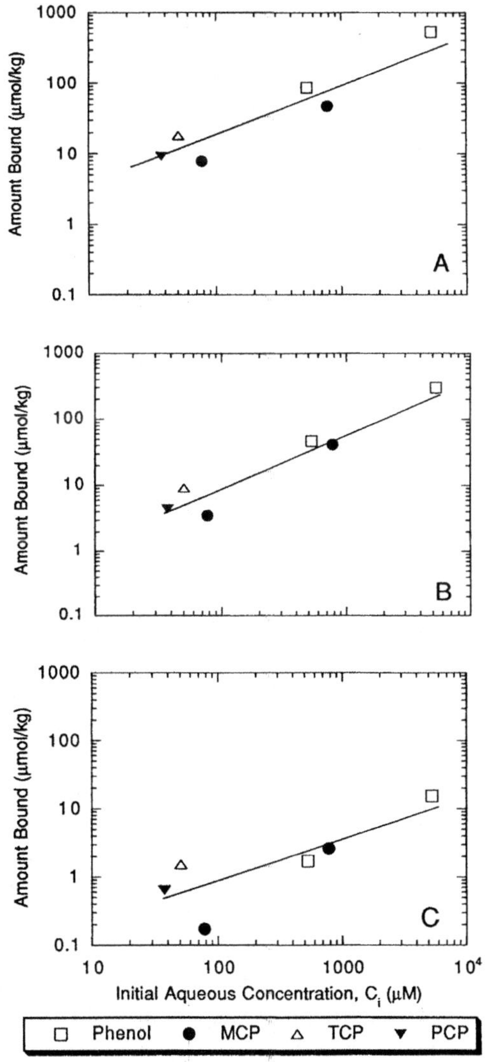

Fig. 4. Effect of the initial concentration of some phenols on the binding to soils under **A** oxic-biotic conditions; **B** anoxic-biotic conditions; **C** anoxic-abiotic conditions

in Fig. 4 that the fraction of material that is irreversibly bound increases with decreasing initial concentration and the biological contribution decreases as a percent of the coupled material as the initial concentration of the phenols increases.

The trends in the data in Fig. 4 appear to be the result of the combined effects of adsorption and "coupleability" of the compounds. The more highly substituted phenols such as PCP would be expected to be less subject to oxidative coupling (Berry and Boyd 1984), while phenol and MCP would be much more readily coupled (Bollag 1992). On the other hand, Ulrich and Stone (1989) showed that because of higher partitioning of the chlorophenols on manganese oxide as the degree of chlorine substitution increased, the amount of chemical oxidative coupling also increased and this was attributed to the stronger adsorption associated with the increase in chlorination. In this study, the higher sorption of the phenols with increased chlorine substitution probably resulted in increased binding. Therefore, it is likely that the pattern observed in this study is unique to this soil and somewhat different results could be found for other soils. What is important, however, is that we found that all of the phenols were oxidatively coupled and, over the concentration range from 0.01 to 1 mM, the extent of irreversible binding ranged from 10 to 50% of the sorbed organics with roughly half of this due to biological processes and half to chemical processes.

3.2.2 Biodegradability

In order to evaluate the biodegradability of the bound contaminants, microorganisms were grown using 4-monochlorophenol as the sole substrate and used to seed biocosms containing the monochlorophenol-contaminated soil that had first been extracted with water and solvent, then rinsed to remove any excess solvent. The objectives of these studies was to determine the biodegradability of nonextracted material from soils under different conditions. The degradable material was expected to be approximately equal to the material retained by mass transfer limitations. In Fig. 5, biodegradation, measured as the evolution of $^{14}CO_2$ from solvent extracted soils, one where sorption occurred under oxic–biotic conditions and the other under anoxic conditions, are compared to the monochlorophenol retained under oxic-biotic conditions. The soils that were subjected to biodegradation were first extracted with CH_2Cl_2 (seven times), washed with distilled water to remove excess solvent, then placed into biocosms and seeded with the acclimated organisms. It can be seen that the anoxic soil produced the least activity over 80 days and the amount of CO_2 produced was nearly the same as that found for the anoxic-abiotic soil, indicating that the material that was biodegraded was diffusion-limited but ultimately bioavailable. In contrast, the solvent extracted oxic-biotic soil produced more CO_2 than could be accounted for by mass transfer-limited material, suggesting that degradation of "irreversibly bound" material was occurring. It can be seen in Fig. 5 that not only was the diffusion-limited portion degraded but also, about 50% of the material thought to be oxidatively coupled was

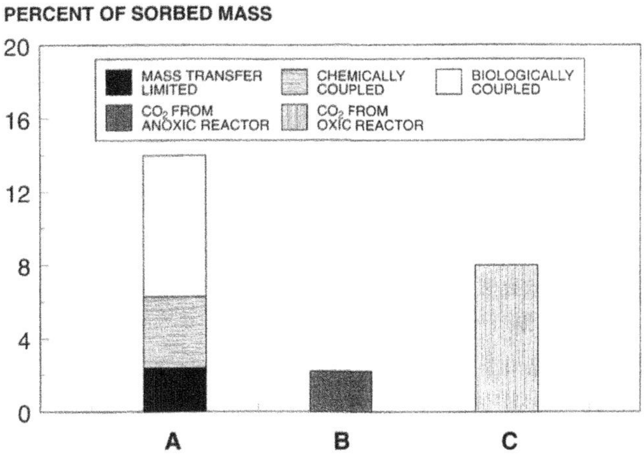

Fig. 5. Comparison of bound monochlorophenol to the amount mineralized following solvent extraction. Bound monochlorophenol was for oxic-biotic conditions. Mineralization was for two sorption conditions, oxic-biotic and anoxic-biotic

degraded. It is possible that the microorganisms were better at extracting the MCP than the solvent and therefore, the bound contaminant remaining at the end of biodegradation was the true irreversibly bound fraction. On the other hand, it is more likely that some of the bound MCP was being microbially "decoupled" and further degraded. Dec and Bollag (1990) also observed mineralization of dichlorophenol that had been coupled into synthetic humic acid. Dec and Bollag (1994) further showed that dechlorination could occur simultaneously with oxidative coupling so it is possible that some of the oxidatively coupled material could be more easily degraded once separated from the SOM. This suggests that coupling of certain toxic organics results in a transformation that also renders the compound less toxic so that even if decoupling occurred, the decoupled compound would be both more degradable and of less harm to the environment.

3.2.3 Effect of Hydrogen Peroxide

Hydrogen peroxide has been used as a source of oxygen for in situ bioremediation. Because hydrogen peroxide is known to be required for some enzymatic reactions that participate in oxidative coupling, it was of interest to quantify the effect of H_2O_2 on irreversible binding. In Fig. 6, data for the irreversible binding of 4-MCP in both the high and low organic matter soils under various conditions including in the presence of H_2O_2 are shown. It can be seen that the effect of peroxide is to increase the quantity of irreversibly bound material by about a factor of 4 in the oxic-biotic soil without peroxide and by a factor of 6 to 10 for the autoclaved soils. It can also be seen in Fig. 6 that the low organic matter soil binds more MCP than the high organic matter soil. This contrasts with the naphthol data shown in Table

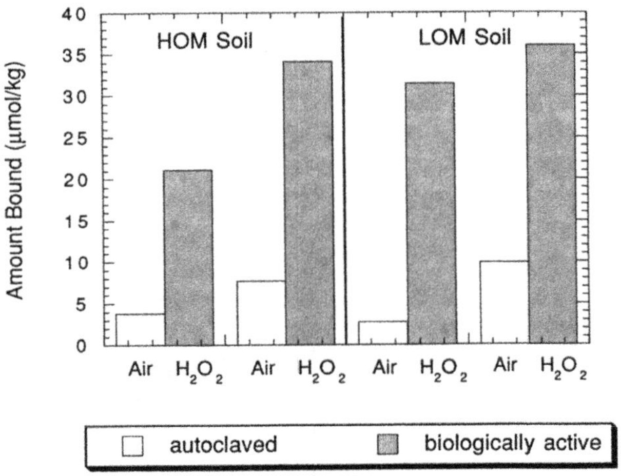

Fig. 6. The effect of hydrogen peroxide on the amount of irreversible binding for a low organic and a high organic matter soil

1 where binding to the high organic matter soil is two to six times greater than for the low organic matter soil.

Hydrogen peroxide can be expected to contribute to irreversible binding in several ways. H_2O_2 is an essential substrate for the peroxidase enzymes so can be expected to enhance biological coupling. It also can contribute to chemical binding by oxidizing surface metals to make them more reactive. Finally, oxygen supplied by decomposition of hydrogen peroxide can support the oxidative processes, and this would be expected to be especially important for the static systems used in this study and also in soils and groundwaters. The large increase in binding for the autoclaved soils suggests that the largest effect of hydrogen peroxide is on the chemical binding processes. These data show that addition of H_2O_2 to the nonautoclaved soil resulted in the binding of 63% of the sorbed MCP compared to approximately 15% in the oxic system without peroxide.

4 Environmental Significance

The results of this study indicate that surface soils contaminated with "coupleable" organics such as phenols, anilines or catechol-like compounds can be immobilized by covalent binding to the SOM. It has also been found that organics which undergo partial oxidation to coupleable intermediates can also be bound and at least for one compound, naphthalene, the coupling reactions are competitive with further oxidation. Further, it was found that, for nonautoclaved soil under aerobic conditions (oxic-biotic), binding of greater than 50% of the sorbed matter often occurred. Furthermore, addition of hydrogen peroxide resulted in an increase in the irreversible binding by as much as eight times. Even for the anoxic systems, irreversible binding was substantial and was thought to be due primarily to chemical oxidative coupling processes.

It remains to be decided if oxidative coupling can be considered to be an acceptable form of detoxification and remediation. It appears that the conditions that promote bioremediation of contaminated soils also optimize oxidative coupling. In this regard, it is likely that in many instances where bioremediation was used, at least some of the removal has been due to irreversible binding rather than mineralization. This is especially likely where hydrogen peroxide has been used. Most of the data for this study was obtained using a 2-day contact time in order to reduce the likelihood of biological degradation. For longer contact periods, more binding is likely and data from Young (1992) support this.

Decoupling of bound organics may be of concern because of the potential for reintroduction of these materials back into the environment. There are two factors that suggest that this may not be a problem. First, decoupling seems to be a biological process, and therefore it would be expected that organics would be degraded as they are decoupled. In addition, data in the literature suggest that coupling is also accompanied by detoxification, at least for the chlorinated phenols. Therefore, the coupled compounds may lose their toxicity as they become bound to SOM.

In this study, the experiments were limited to two soils and several organics. It is expected that the extent of oxidative coupling would vary considerably among soils, especially with regard to variations in the surface metal oxides of soils and the amount and type of soil organic matter. However, it is clear that the potential for oxidative coupling exists in soils and these binding reactions need to be considered in transport modeling, in the design and operation of engineered remediation systems, and also in the evaluation of natural or intrinsic remediation processes.

References

Ball WP, Roberts PV (1991) Long-term sorption of halogenated organic chemicals by aquifer material. 2. Intraparticle diffusion. Environ Sci Technol 25:1237–1249

Berry DF, Boyd SA (1984) Oxidative coupling of phenols and anilines by peroxidases: structure-activity relationships. Soil Sci Soc Am J 48:565–569

Berry DF, Boyd SA (1985) Reaction rates of phenolic humus constitutients and anilines during cross-coupling. Soil Biol Biochem 17:631–636

Bhandari A, Novak JT, Berry DF (1996) Binding of 4-monochlorophenol to soil. Environ Sci Technol 30(7):2305–2311

Bollag J-M (1992) Decontaminating soil with enzymes. Environ Sci Technol 26:1876–1881

Burgos WD (1995) Reversible and irreversible sorption of alpha-naphthol to soil. PhD Diss, Virginia Polytechnic Institute and State University, Blacksburg, Virginia

Dec J, Bollag J-M (1990) Detoxification of substituted phenols by oxidoreductive enzymes through polymerization reactions, Arch Environ Contam Toxicol 19:543–550

Dec J, Bollag J-M (1994) Dehalogenation of chlorinated phenols during oxidative coupling. Environ Sci Technol 28:484–490

Farmer WS (1989) A microcosm study of the biodegradability of adsorbed toluene by acclimated bacteria in soils. MS Thesis, Virginia Polytechnic Institute and State University, Blacksburg, Virginia

Michelcic JR, Luthy RG (1988) Degradation of polyaromatic hydrocarbon compounds under various redox conditions in soil-water systems. Appl Environ Microbiol 54:1182–1187

Sjoblad RD, Bollag J-M (1981) Oxidative coupling of aromatic compounds by enzymes from soil microorganisms. In: Paul EA, Ladd JM (eds) Soil biochemistry, vol 5. Marcel Dekker, New York

Stevenson FJ (1982) Humus chemistry: genesis, composition, reactions. Wiley, New York

Stone AT (1987) Reductive dissolution of Mn(III/IV) oxides by substituted phenols. Environ Sci Technol 21:979–988

Suflita JM, Bollag J-M (1981) Polymerization of phenolic compounds by a soil-enzyme complex. Soil Sci Soc Am J 45:297–302

Ulrich H-J, Stone AT (1989) Oxidation of chlorophenols adsorbed to manganese oxide surfaces. Environ Sci Technol 23:421–428

Vidic RD, Suidan MT (1991) Role of dissolved oxygen on the adsorptive capacity of activated carbon for synthetic and natural organic matter. Environ Sci Technol 25:1612–1618

Wang TSC, Li SW, Ferng YL (1978) Catalytic polymerization of phenolic compounds by clay minerals. Soil Sci 126:15–21

Young R (1992) Effect of organic matter and contact time on the sorption and bioavailability of chlorophenols. MS Thesis, Virginia Polytechnic Institute and State University, Blacksburg, Virginia

Part IV

Contaminant Modeling

14 On the Importance of Dimensionality in the Simulation of Dense Non-Aqueous Liquid Migration in the Subsurface

J. Guarnaccia and G. Pinder

Contents

1 Introduction

The simulation of dense non-aqueous phase liquid migration in the subsurface is computationally demanding. As a consequence it is common to reduce the dimensionality of the problem. Herein we investigate the impact of reducing the dimensionality of a DNAPL migration problem from three to two.

The problem to be considered is presented diagrammatically in Fig. 1. A source of DNAPL is introduced into the partially saturated soil at the land's surface. The liquid moves downward under the influence of gravity and capillarity through the unsaturated zone and into the saturated zone. Evaporation into the soil gas and dissolution into groundwater flowing around and through the DNAPL containing soil gradually reduces the amount of DNAPL present.

The movement of any fluid is impeded by the presence of other fluid phases. Thus, either air or DNAPL in the soil impedes the movement of groundwater. One would expect, therefore, that groundwater would tend to move around DNAPL-saturated soil. The three-dimensional simulation of DNAPL would, of course, permit this behavior to occur. However, in a two-dimensional representation of this phenomenon, the water is unable to move laterally around the DNAPL-saturated soil and tends to move through it. In the most extreme case of a DNAPL-saturated column of soil such as shown in Fig. 1, all groundwater is forced to flow

J. Guarnaccia and G. Pinder College of Engineering and Mathematics University of Vermont Burlington, Vermont 05405, USA

H. Rubin/N. Narkis/ J. Carberry
Soil and Aquifer Pollution

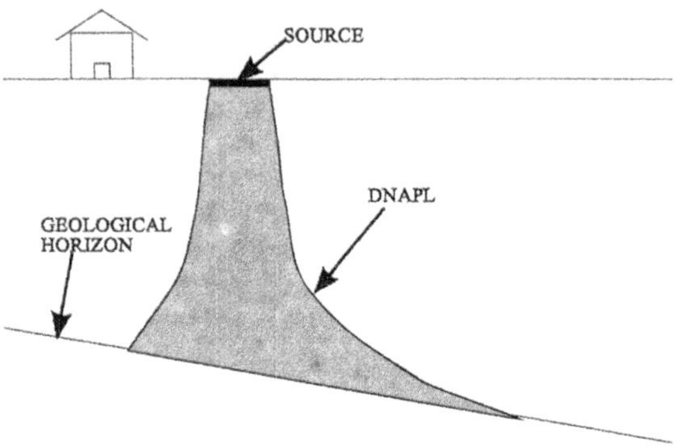

Fig. 1. Diagrammatic representation of DNAPL migration from a surface source

through the DNAPL-saturated column if it is to move from one side of the column to the other.

Three questions arise when one attempts to simulate DNAPL migration using a two-dimensional Cartesian model. Firstly, what is the impact of the reduction in dimensionality on the saturation distribution of the DNAPL. Secondly, how much of an impact does the DNAPL have on the flow of the groundwater. Finally, what is the impact of the reduction in dimensionality on the rate of dissolution of the DNAPL. In this chapter, we present a model three-phase flow problem designed to addresses the first two questions. A discussion of the results provides insight into answering question three.

2 Theory

The mathematical model used in this analysis has the following attributes:

1. Kinetic representation of the mass transfer between all three phases;
2. Hysteresis in the representation of the relative permeability-saturation-capillary pressure relationships, where residual phase distributions result from fluid entrapment processes;
3. Reduction in the residual NAPL saturation as it is dissolved into the aqueous phase and/or evaporated into the air phase;
4. Imbibition of additional NAPL when the system is in a state of residual, or less than residual, NAPL saturation (hysteresis).

The method of approximation employed is the collocation finite element method based upon cubic Hermite interpolation polynomials.

2.1 The Balance Equations

The flow and transport of DNAPL contaminants requires the solution of a set of equations describing fluid flow and another set describing contaminant transport. To effectively represent DNAPL migration during imbibition, drainage, residual emplacement, dissolution, and evaporation, these equations must be fully coupled. In addition, it is necessary to accommodate the removal of one or more flow equations as either the DNAPL, air or water phases approach an immobile residual state. The equations that describe the flow of DNAPL in the presence of water and gas are the following (see Guarnaccia and Pinder 1992):

$$\frac{\partial\left(\varepsilon S_\alpha \rho^\alpha\right)}{\partial t} - \nabla \cdot \frac{\mathbf{k} k_{r\alpha}}{\mu^\alpha}\left(\nabla p^\alpha - \rho^\alpha \mathbf{g}\right) = \rho^\alpha q^\alpha + \tilde{\rho}^\alpha, \qquad (1)$$

where S_α is the saturation of the α phase, $\mathbf{k}$ is the permeability, $k_{r\alpha}$ is the relative permeability of the α phase, ε is the porosity, ρ^α is the density of the fluid α, p^α is the fluid phase pressure, q^α is a source term, μ^α is the viscosity of the α phase, and $\tilde{\rho}^\alpha$ is the interphase mass-exchange term which accounts for dissolution and evaporation. In the case of NAPL flow in a partially saturated or saturated medium, α would represent water (w), gas (g), and NAPL (n).

The equation describing the transport of species i in phase α is given by:

$$\frac{\partial\left(\varepsilon S_\alpha \rho_i^\alpha\right)}{\partial t} + \nabla \cdot \left[\varepsilon S_\alpha \rho_i^\alpha \mathbf{v}^\alpha\right] - \nabla \cdot \left[\varepsilon S_\alpha \mathbf{D}^\alpha \nabla \rho_i^\alpha\right] = \rho_i^\alpha q^\alpha + \tilde{\rho}_i^\alpha, \qquad (2)$$

where ρ_i^α is the concentration of the ith species in the α phase, $\tilde{\rho}_i^\alpha$ is the species interphase mass exchange term, $\mathbf{v}^\alpha$ is the velocity vector, and $\mathbf{D}^\alpha$ is the dispersion tensor, a known function of velocity. We have not included adsorption in this expression because its influence is small relative to NAPL dissolution and evaporation. However, if required, this phenomenon can be easily added to the above expression.

The above equations are written in operator notation. Without further modification they are applicable to either two- or three-dimensional flow and transport. Here, we shall compare the solutions obtained using a fully three-dimensional and a two-dimensional Cartesian formulation.

2.2 The Relative Permeability-Saturation-Pressure Relationship

In the above equations, the relative permeability, $k_{r\alpha}$, is a function of saturation, and saturation, S_α, is in turn a function of the capillary pressure which is defined as the difference in pressure between the various fluid phases. A relative permeability-saturation-capillary pressure model (k–S–P model) has been derived in Guarnaccia and Pinder (1995) which has the following attributes:

1. It is capable of modeling the simultaneous flow of water, DNAPL and gas;
2. It is based on wettability considerations and two-phase data;

3. It will reduce to the appropriate two-phase and single-phase cases when appropriate;
4. It includes flow-path-history-dependent functionals (hysteresis);
5. It includes a mechanism for fluid entrapment during drainage (residual emplacement);
6. It employs capillary pressure scaling to accommodate variable soil and fluid properties (after Leverett 1941).

The current three-phase hysteretic *S–P* model (Guarnaccia and Pinder 1995) is a modification of the van Genuchten model (1980), where hysteresis is accommodated by employing two empirical evaluation procedures:

1. The procedure of Scott et al. (1983) which scales the main drainage and imbibition curves to fit between predefined endpoints;
2. A procedure to quantify trapped phase magnitudes as a function of flow-path history.

The current three-phase hysteretic *k–S* model extends the development of the two-phase non-hysteretic *k–S* model of Luckner et al. (1989) and the three-phase non-hysteretic *k–S* model of Parker et al. (1987), to include hysteresis due to fluid entrapment. The model defines residual saturation as that saturation at which the relative permeability is zero.

Because the development of the *k–S–P* model is lengthy and complex, it is not detailed in this chapter (see Guarnaccia and Pinder 1992 or 1995 for the functional form). An illustrative example of the hysteretic *k–S–P* model for a two-phase system is presented in Fig. 2, where the notation PDC, MDC, PIC, and MIC stands for primary and main drainage curves, and primary and main imbibition curves, respectively. Curves 3–6 are drainage and imbibition scanning curves. In addition, S_{wr} and S_{nr} are the residual wetting and non-wetting phase saturations, respectively.

2.3 The Interphase Mass-Transfer Parameter

The interphase mass transfer parameter, $\tilde{\rho}_i^\alpha$, is a very important function. Predictive flexibility is obtained by using a rate-limited kinetic relationship of the form (Imhoff et al. 1992a):

$$\tilde{\rho}_i^\alpha = C^{\alpha\beta}\left(K_i^\alpha - \rho_i^\alpha\right), \tag{3}$$

where $C^{\alpha\beta}$ [1/T] is a mass transfer rate coefficient for species i between phases α and β, and K_i^α is the equilibrium concentration of species i in the α phase. It has been shown in Imhoff et al. (1992a) for a two-phase water–NAPL system, that $C^{\alpha\beta}$ is functionally sensitive to the soil properties, NAPL saturation and fluid velocities:

$$C^{\alpha\beta} = \gamma_1\left(\varepsilon S_n\right)^{\gamma_2}\left|\mathrm{v}^{w}\right|^{\gamma_3}, \tag{4}$$

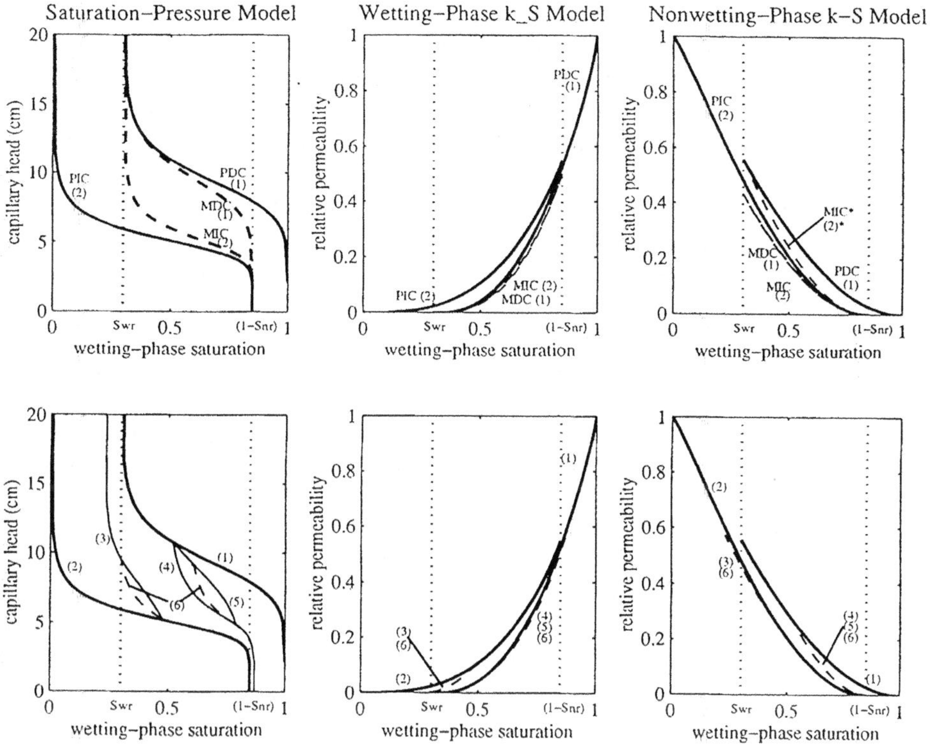

Fig. 2. Hysteretic relative permeability–saturation–pressure relationships

where γ_1, γ_2, and γ_3 are fitting parameters.

This model may also be utilized for the study of DNAPL volatilization mass transfer. Finally we note that whether we use a rate-limited or equilibrium approach for the volatilization of NAPL, we believe that it is essential that a mobile gas phase be included in the model. Considerable interest exists in this phenomenon from scientific, remediation, and investigative perspectives.

3 Numerical Solution

3.1 The General Concept

In Guarnaccia and Pinder (1995), we solved the above equations for a three-fluid system and one dissolved species. The procedure involves the approximation of Eqs. (1) and (2) using an implicit backward difference scheme for the time derivative. The resulting equations are subsequently linearized using Picard iteration. The dependent variables p^α, S_α, and ρ_i^α in (1) and (2) are approximated using cubic Hermite interpolation polynomials. The remaining variables that are space-dependent are approximated using linear Lagrange interpolation polynomials. The non-linear terms in these equations are evaluated at the nodal locations and are interpolated into the element using linear Lagrange polynomials.

The above equations are reduced to a set of algebraic expressions using *orthogonal collocation.* This is a method of weighted residuals wherein the weighting function is the Dirac delta function. The method is equivalent to driving the residual to zero at specified points in the elements. These points are known as collocation points. Four collocation points are used per element in two space dimensions and eight in three space dimensions. The complete development of the collocation procedure is beyond the scope of this chapter and the reader is referred to Lapidus and Pinder for an introduction to collocation as applied to partial differential equations such as those considered here (Lapidus and Pinder 1982).

The phase mass balance equations define the distribution of the phases, given the phase composition. The constituent or species balance equations define the distribution of DNAPL or other soluble components in the water and gas phases, given the phase distribution and the velocity field. In an attempt to minimize the overall computational effort, we choose to solve the balance equations by sequentially lagging in time certain dependent variables, such that the balance equations become uncoupled. This strategy minimizes the size of the system matrix for each solution step as the equations can be solved independently. The equations presented above provide the basis for the mode employed in this analysis.

The time and space discrete representation described above generates systems of linear algebraic equations which are solved using a conjugate-gradient approach.

4 Analysis

To examine the effect of model dimensionality on the simulation of DNAPL migration, a problem of the type presented in Fig. 1 has been solved using both a two- and three-dimensional model (2-D and 3-D). Consider the following experimental-scale multiphase flow experiment. A homogeneous, coarse sand, free-surface aquifer with bedding inclined to the horizontal and water flowing in an up-dip direction is initially at steady state flow conditions. At two known time intervals, a quantity of DNAPL is injected at a point near the surface. The rate, while assumed unknown, is sufficient to result in the following NAPL saturations near the source: maximum during injection, 0.24; residual after source removal, 0.12. The geometry, fluid and porous medium parameters, and initial, boundary and source conditions for the problem are presented diagrammatically in Fig. 3. In addition, the primary and main curves of the three-phase k–S–P model used in this simulation are provided in Fig. 4. The gas–NAPL and NAPL–water retention curves are computed from the gas–water retention curve scaled by the ratio of appropriate fluid interfacial tension pairs (Leverett scaling 1941). Note the short length scale of the S–P model.

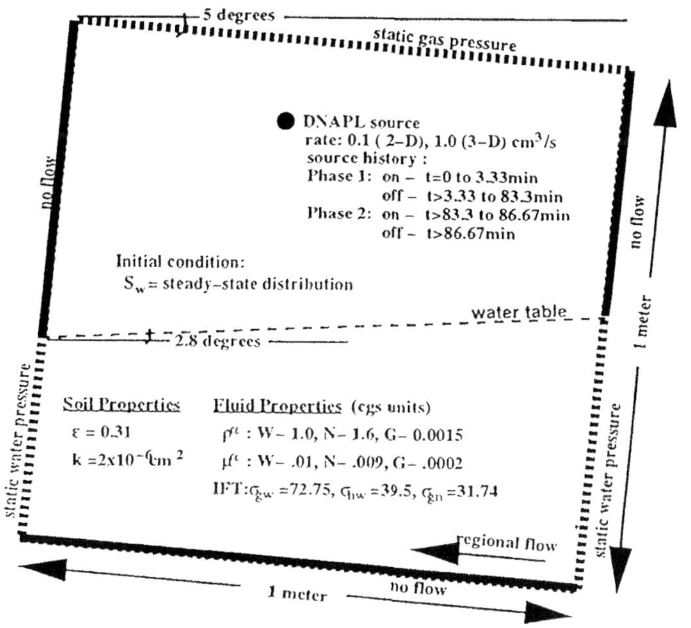

Fig. 3. Definition sketch of the DNAPL problem. Domain cross section is 20 by 20 elements uniform. Solved using both a two-dimensional (one element deep = 1 cm), and three-dimensional (6 elements deep = 30 cm) Cartesian coordinate system

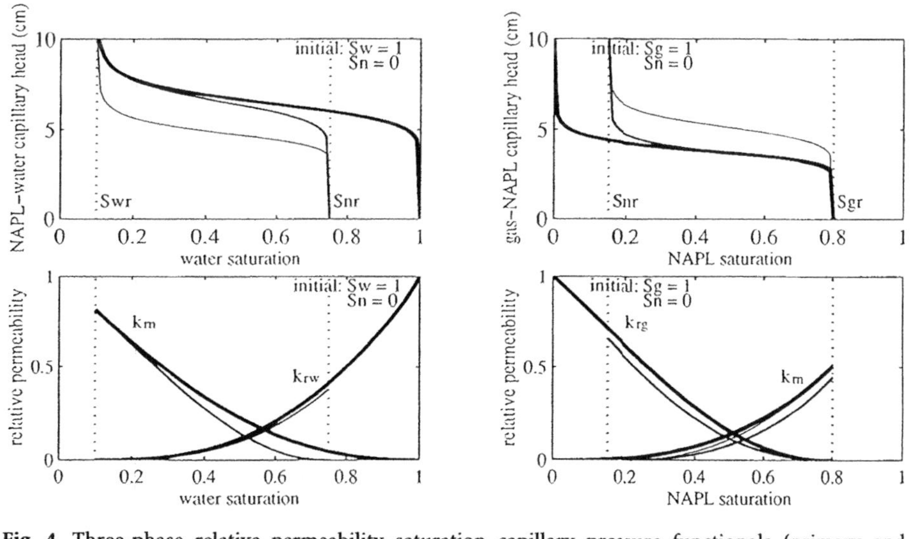

Fig. 4. Three-phase relative permeability-saturation-capillary pressure functionals (primary and main curves), for soil and fluid properties defined in the problem setup figure. The van Genuchten curve-fit parameters for a gas-water system are; $\eta = 11.434$, α (drainage) = 0.085/cm, α (imbibition) = 0.12/cm

The problem is set up such that there is symmetry on each side of the section. The 2-D and 3-D representations differ in the number of elements in the direction normal to the page (one and six, respectively), and in the DNAPL injection rate. To obtain the target saturations near the source, the source rate for the 2-D and 3-D problems is defined as 0.1 and 1.0 cm^3/s, respectively. The DNAPL saturation distribution and water velocity arrows computed for the 2-D and 3-D models is shown in Fig. 5 (T = 2.83 h, during the second DNAPL injection episode) and 6 (T = 3.67 h, near-residual NAPL saturation conditions). The figures show that the solutions are qualitatively similar in the vadose zone and differ significantly from each other in the liquid-saturated zone: the DNAPL front has traveled further in the case of the 2-D model. An explanation for this perhaps lies in the definition of the three-phase S-P model. The interfacial tension between the fluids considered for this problem results in the condition that the DNAPL front will be retarded as it approaches the capillary fringe (the DNAPL will pool at the interface until the DNAPL-water break-through pressure is reached). The 3-D model provides an additional dimension along which the DNAPL can spread at the fringe.

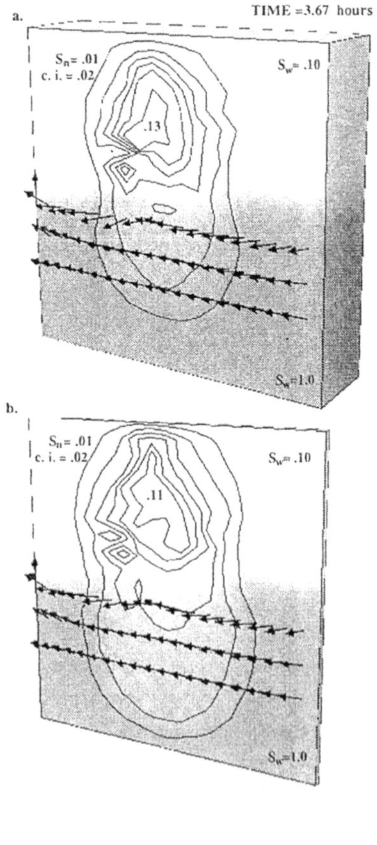

Fig. 5. Water (*shading*) and DNAPL (*contours*) saturation distributions along the center cross section of the domain at T = 2.83 h: **a** 3-D model solution; **b** 2-D model solution. In addition, water velocity *arrows* along three planar levels are shown (magnitude ≈ 0.005 cm/s)

Fig. 6. Water (*shading*) and DNAPL (*contours*) saturation distributions along the center cross section of the domain at T = 3.67 h: **a** 3-D model solution; **b** 2-D model solution. In addition, water velocity *arrows* along three planar levels are shown (magnitude ≈ 0.005 cm/s)

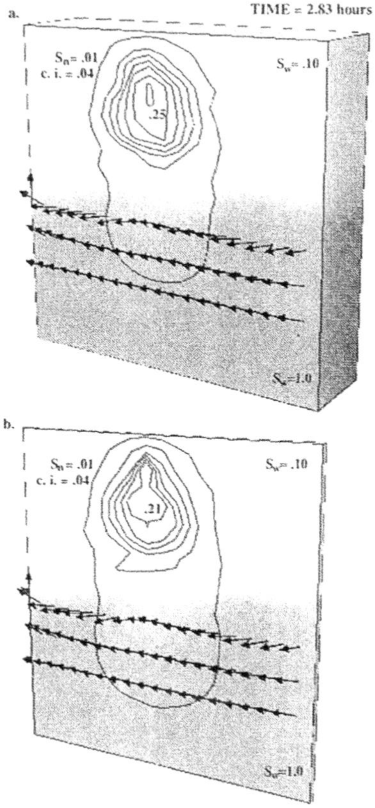

From this view of the domain it appears that the presence of DNAPL in the fluid saturated zone has little effect on the water velocity vectors. Low DNAPL saturations entering the aquifer result in low residual saturation values (less than 5%). Figure 7 shows the groundwater flow velocity field obtained using the two-dimensional and three-dimensional simulators.

5 Summary and Conclusions

The simulations of two and three phase DNAPL flow in two and three dimensions are compared. It is found, for the particular problem considered, that when similar saturations are defined at the source, the two- and three-dimensional models provide similar, although different, results. The groundwater flow velocity is affected by the dimensionality of the DNAPL plume but, for this problem, the effects are relatively small.

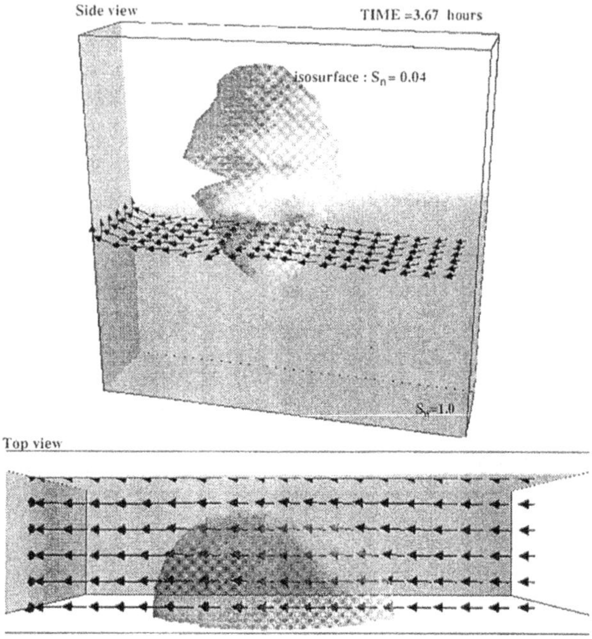

Fig. 7. Water velocity solution (*arrows*) shown along a planar section near the capillary fringe. The DNAPL distribution is depicted by the 4% isosurface. Two views of the same solution are shown

Acknowledgment. This research was supported by the US Department of Energy, Office of Energy Research.

References

Guarnaccia JF (1992) Two-phase flow, dissolution and transport in porous media: a numerical model using a collocation-based parallel algorithm Doctoral Diss, Princeton University, Princeton, New Jersey

Guarnaccia JF, Pinder GF (1992) A new two-phase flow and transport model with interphase mass exchange. Proc 9th Int Conf on Computational methods in water resources, Denver, Colorado, pp 281–288

Guarnaccia JF, Pinder GF (1995) NAPL: a mathematical model for the study of NAPL contamination in granular soils. Draft report submitted to US EPA RS Kerr Environmental Research Laboratory, Ada, Oklahoma, 88 pp

Imhoff PT, Jaffe PR, Pinder GF (1992a) An experimental study of the dissolution of trichloroethylene in saturated porous media. Doctoral Thesis, Princeton University, Princeton, New Jersey

Imhoff PT, Jaffe PR, Pinder GF (1992b) An experimental study of the dissolution of trichloroethylene in saturated porous media. Princeton University Water Resour Prof Rep WR-92-1

Lapidus L, Pinder GF (1982) Numerical solution of partial differential equations in science and engineering. John Wiley, New York, 677 pp

Leverett MC (1941) Capillary behavior in porous solids. Trans AIME 142:152–169

Luckner L, van Genuchten MT, Nielsen DR (1989) A consistent set of parametric models for the two-phase flow of immiscible fluids in the subsurface. Water Resour Res 25(10):2187–2193

Parker JC, Lenhard RJ, Kuppusamy T (1987) A parametric model for constitutive properties governing multiphase flow in porous media. Water Resour Res 23(4):618–624

Scott PS, Farquhar GJ, Kouwen N (1983) Hysteretic effects on net infiltration. In: Advances in infiltration. American Society of Agricultural Engineers, St Joseph, Michigan, pp 163–170

Van Genuchten MA (1980) Closed-form equation for predicting the hydraulic conductivity of unsaturated soils. Soil Sci Soc Am J 44:892–898

15 Modeling of Non-Aqueous Phase Liquid Migration and Recovery in Heterogeneous Aquifers

D. W. Waddill[1] and J. C. Parker[2]

Contents

1 Introduction

In recent years, researchers have attempted to incorporate soil variability into models that predict the fate of environmental pollutants in the subsurface, and the emphasis in modeling has shifted from a deterministic to a stochastic approach. Numerous authors (for example, Gelhar 1986; Jury et al. 1987; Dagan 1990; Russo 1991) have developed and reviewed the theory of subsurface stochastic flow and transport. Stochastic theory is based on the assumption that a heterogeneous soil property can be treated as a single sample, or realization, taken from a random field. Quantification of spatial variability is generally limited to estimating the mean and variance of the field variable, which is said to be second order stationary if its mean and variance are constant throughout the domain. In general, only one measurement of the field variable is available at a given point in space, but estimates of the mean and variance may still be obtained under the ergodic hypothesis (Lumley and Panofsky 1964; Sposito et al. 1986).

[1] Department of Civil Engineering 200 Patton Hall/Virginia Tech Blacksburg, Virginia 24061, USA
[2] Environmental Systems and Technologies, Inc. 2608 Sheffield Drive Blacksburg, Virginia 24060-8270, USA

H. Rubin/N. Narkis/ J. Carberry
Soil and Aquifer Pollution

Stochastic flow and transport in the water phase have been studied by numerous researchers (for example, Yeh et al. 1985; Russo 1991; Chen et al. 1994; Russo et al. 1994). However, the influence of soil heterogeneity on the subsurface flow of non-aqueous phase liquids (NAPLs or simply "oil") has received limited attention. Kaluarachchi (1996) has shown that simulated oil recovery is affected by treating the intrinsic permeability as a stochastic model input. In addition, the van Genuchten (1980) capillary curve shape parameters (α and n), and the residual water saturation (S_m) have a direct influence on free oil distribution and may affect recovery. Essaid et al. (1993) found that field samples at a crude oil spill showed wide variability in liquid retention curve characteristics, implying that α and n may be considered stochastic variables. In addition, the maximum residual oil saturations in the saturated (S_{or}) and unsaturated zones (S_{og}) have been shown to vary with soil type (Wilson et al. 1990). These parameters have a direct influence on oil trapping and may therefore influence oil migration and recovery.

The purpose of the present study was to use the multiphase flow model ARMOS (Kaluarachchi and Parker 1989; ES&T 1996) with enhancements derived by Waddill and Parker (1996) to evaluate the effects of soil heterogeneity on the infiltration, redistribution, and recovery of oil from the subsurface. Hypothetical, yet realistic, sandy loam soils were generated using the turning bands method (Mantoglou and Wilson 1982). Stochastic input parameters for the model included the saturated hydraulic conductivity (K_{sw}), α, n, S_m, S_{or}, and S_{og}. For comparison, model simulations were also performed using the geometric mean of the above soil parameters to produce an "equivalent" homogeneous porous medium. Homogeneous and heterogeneous results were compared for two basic scenarios. In the first, oil migration along the regional water table gradient was examined using statistical and spatial moment analyses, while in the second, oil recovery from a single pumping well was analyzed.

2 Model Description

The multiphase flow model solves for the elevations of the air–water table (Z_{aw}) and the air–oil table (Z_{ao}) at nodal points of a discretized areal domain. The elevation Z_{aw} is defined as the location of equal air and water pressures, while Z_{ao} is the location of equal air and oil pressures, and the oil–water table (Z_{ow}) is the elevation of equal water and oil pressures. In the absence of air pressure gradients, lateral flow of free oil occurs in response to gradients in Z_{ao}, while water flow depends on gradients in Z_{aw}. The governing equations for water and oil flow in an areal domain are given (Parker and Lenhard 1989) by

$$\frac{\partial V_w}{\partial t} = \frac{\partial}{\partial x}\left(-T_w \frac{\partial Z_{aw}}{\partial x}\right) + \frac{\partial}{\partial y}\left(-T_w \frac{\partial Z_{aw}}{\partial y}\right) + \bar{J}_w \tag{1a}$$

$$\text{and} \quad \frac{\partial V_o}{\partial t} = \frac{\partial}{\partial x}\left(-T_o \frac{\partial Z_{ao}}{\partial x}\right) + \frac{\partial}{\partial y}\left(-T_o \frac{\partial Z_{ao}}{\partial y}\right) + \bar{J}_o, \tag{1b}$$

where V_w and V_o are total water and oil volumes per horizontal area [L] at a point in the x-y plane; T_w and T_o are water and oil transmissivities [$L^2 T^{-1}$]; and $\tilde{J}_w$ and $\tilde{J}_o$ are vertically integrated source–sink terms [$L T^{-1}$] for water and oil. Further details of the governing equations, the finite element formulation, and the consitutive relationships for the multiphase flow model are given elsewhere (Kaluarachchi and Parker 1989; ES&T 1996).

If fluid levels are not changing rapidly, then the distributions of fluid pressures may be assumed to be in vertical equilibrium. Under this assumption, free oil saturation (S_{of}) can be calculated as a function of elevation (z) using the van Genuchten (1980) model as modified by Parker et al. (1987). Model parameters include the van Genuchten α and n, the water saturation at field capacity (S_m), soil porosity (ϕ), and fluid dependent scaling factors β_{ow} and β_{ao}. For each node in the computational grid, the model calculates free oil specific volume (V_{of}) as a function of the oil thickness (H_o) that would be observed in a monitoring well placed at the grid point (Parker and Lenhard 1989). Using the method described by Waddill and Parker (1996), the model accounts for hysteresis in the liquid saturation-pressure relations, which allows V_{of} to take on more than one value for a given H_o, depending on the wetting history of the soil. The quantity V_{of} represents the free oil volume per unit area of porous media, and the total free oil volume is obtained by summing V_{of} over the model domain.

Oil may be trapped in the unsaturated and saturated zones due to fluctuations in the fluid tables. The maximum unsaturated zone residual oil saturation (S_{og}) represents the irreducible oil saturation that follows gravity drainage from a previously high oil saturation. The model calculates the specific volume of unsaturated zone residual oil (V_{og}) by considering the vertical drainage rate of oil as a function of the oil relative permeability in the unsaturated zone (Corey 1986). The maximum saturated zone residual oil saturation (S_{or}) represents oil trapped by water imbibition into a zone of high oil saturation. As Z_{ow} rises, the model uses the minimum of S_{or} versus the current average oil saturation to compute the specific volume of residual oil in the saturated zone (V_{ot}). As with V_{of}, V_{og} and V_{ot} are computed at each grid point and summed to obtain the total volume of residual oil in the unsaturated and the saturated zones.

Transmissivity of the oil phase (T_o) is derived from vertical integration of Darcy's equation (Parker and Lenhard 1989) subject to the vertical equilibrium assumption, which implies that fluid flow in the horizontal direction is much greater than vertical flow. The model uses the van Genuchten (1980) model as refined by Kaluarachchi et al. (1990) to calculate the relative permeability to oil at elevation z. Subsequently, T_o may be integrated numerically as a monotonically increasing function of V_{of}. Oil transmissivity varies linearly with oil specific gravity (ρ_{ro}), and inversely with the ratio of oil to water viscosity (η_{ro}). The average velocity (v_o) of an oil plume may be expressed (Parker et al. 1994) by

$$v_o = M_o \frac{dZ_{ao}}{dx}, \tag{2}$$

where M_o is the oil mobility factor, defined as $M_o = T_o/V_{of}$. At high values of V_{of}, M_o reaches a constant maximum value defined by

$$M_o^{max} = \frac{\rho_{ro} K_{sw}}{\eta_{ro} \phi (1 - S_m)}, \tag{3}$$

which may be used to estimate plume migration rate (see Section 4.4).

3 Generation of Heterogeneous Soil Property Distributions

In all cases, the geometric mean soil properties were characteristic of a sandy loam soil (Carsel and Parrish 1988). Fluid properties were typical of gasoline (Table 1) and were treated as deterministic. Soil properties for the homogeneous case were also treated as deterministic, with ϕ, K_{sw}, α, n, and S_m taken from Carsel and Parrish (1988). The parameters α, n and S_m were adjusted using the method of Lenhard and Parker (1990) to represent the "quasi-static" fluid distributions that occur in the soil under field conditions. Values for S_{or} and S_{og} were estimated from S_m using

$$S_{or} = f_{or}(1 - S_m) \quad \text{and} \quad S_{og} = f_{og} S_m (1 - S_m). \tag{4}$$

The empirical factors f_{or} and f_{og} typically range between 0.2 and 0.5 (ES&T 1996), and the median value of 0.3 was chosen for these simulations.

For the heterogeneous soils, porosity was assumed constant at 0.41. All other soil parameters were treated as stochastic variables, such that a distinct set of soil parameters was used for each node in the domain. The geometric means of the stochastic parameters were identical to those used for the homogeneous case. The parameters K_{sw}, α, and n were assumed to be lognormally distributed, locally stationary, random space variables described by the following two-dimensional isotropic covariance function:

$$C(s_x, s_y) = \sigma^2_{\ln(i)} \exp\left[-\left(s_x^2 + s_y^2\right)/\eta^2\right], \tag{5}$$

Table 1. Fluid parameters and geometric mean soil parameters used by the multiphase flow model. The maximum and mimimum values show the range of each parameter in the hypothetical heterogeneous soil

Soil properties				Fluid properties
Parameter	Geometric mean	Minimum	Maximum	
K_{sw} (m day^{-1})	1.06	0.0221	45.5	$\rho_{ro} = 0.8$
α (m^{-1})	7.56	1.096	61.7	$\eta_{ro} = 0.6$
n	2.00	1.45	2.71	$\beta_{ao} = 3.2$
S_m	0.25	0.101	0.642	$\beta_{ow} = 1.5$
S_{or}	0.225	0.107	0.270	
S_{og}	0.055	0.027	0.075	

where s_x and s_y are the separation distances in the x and y-directions, η is the correlation scale, and $\sigma^2_{\ln(i)}$ is the variance of ln(i) for i = K_{sw}, α, or n. The turning bands method (Mantoglou and Wilson 1982), as implemented by Tompson et al. (1987), was used to produce the random fields of K_{sw}, α, and n.

Surveys of the literature (Gelhar 1986; Jury et al. 1987) have shown that the standard deviation of $\ln K_{sw}$ ($\sigma_{\ln K_{sw}}$) does not often exceed 1.0; thus $\sigma_{\ln K_{sw}} = 1.0$ was chosen to represent a relatively high degree of variability in K_{sw}. Based on the work of Carsel and Parrish (1988), $\sigma_{\ln\alpha}$ was chosen to be 0.5, and $\sigma_{\ln(n)}$ was set at 0.082. In soils with an overall horizontal scale similar to the present study, η has been shown to range from 14 to 39 m (Russo and Bressler 1981; Viera et al. 1981). In order for a discretized model domain to capture spatial correlation Ababou (1988) has recommended using at least four nodes per correlation scale. Moreover, to satisfy the ergodic hypothesis, Dagan (1990) recommends that the flow domain extend over a large number of correlation scales. Based on the above criteria, the correlation scale (η) was chosen to range from 3.66 to 6.1 m in specific cases that will be described below.

Carsel and Parrish (1988) concluded that for many soils, a high correlation (ρ) exists between K_{sw} and α. Such correlation exists if the average pore size controls the rate of flow rather than fractures or other large scale openings in the porous medium. Based on the reported value for the sandy loam soil (Carsel and Parrish 1988), K_{sw} and α were assumed to be correlated in most cases with $\rho = 0.85$. This degree of correlation was produced in the random K_{sw} and α fields by generating them together as two horizontal layers in the turning bands procedure. The parameter n was assumed to be uncorrelated with K_{sw} and α, so the distribution of n was produced by an independent run of the turning bands program in all cases.

The method of Lenhard and Parker (1990) was used to estimate the quasi-static S_m for each of the Soil Conservation Service textural classifications of soils described by Carsel and Parrish (1988). Using regression, S_m was subsequently derived as the following quadratic function of $\ln(K_{sw})$:

$$S_m = 0.00852x^2 - 0.0912x + 0.344, \tag{6}$$

where $x = \ln(K_{sw})$ and $R^2 = 0.98$. Thus, the S_m field was obtained from the generated K_{sw} field, and values for S_{or} and S_{og} (Table 1) were calculated from S_m using Eqs. (4).

4 Oil Migration Following a Hypothetical Spill

4.1 Description of the Test Case

The model domain for comparison of oil migration in a homogeneous versus a heterogeneous aquifer was 131 m in the x-direction by 62 m in the y-direction. The area was divided into a 9936 node mesh using 0.91 by 0.91 m elements, and the bottom of the aquifer was chosen as the datum (z = 0). The boundary condition for

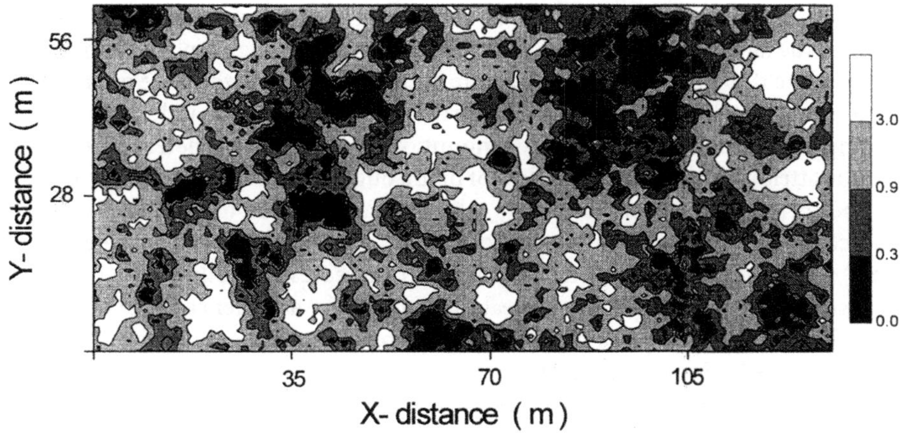

Fig. 1. Filled contour intervals for the heterogeneous distribution of K_s, with dimensions of the scale bar in m day^{-1}

the water table (Z_{aw}) was maintained at a constant head of 30.5 m at x = 0, and 27.9 m at x = 131, while the top (y = 62 m) and bottom (y = 0) boundaries were specified as no flow. Thus the average gradient in Z_{aw} was 2%, but variations in K_{sw} produced local irregularities in the heterogeneous soil. Significant quantities of oil never reached the boundaries of the domain in any of the simulations.

For each of the stochastic soil parameters, Table 1 shows the minimum and maximum value in the domain. Since α and n have relatively small standard deviations, these parameters show a narrower range of values than K_{sw}. The actual values for K_{sw} vary over several orders of magnitude, but due to autocorrelation produced by the turning bands procedure, the ratio in K_{sw} for any two adjacent nodes rarely exceeded 4:1. In addition, autocorrelation produces observable zones of high and low values in the spatial distribution of K_{sw} (Fig. 1).

4.2 Oil Plume Migration

For both the homogeneous and the heterogeneous soils, the oil plume was produced by using the model to simulate the release of 153.8 m^3 of oil at x = 22.9 m and y = 32.0 m over a 120-day period. Subsequently, the oil was allowed to redistribute for 1140 days, during which time the plume spread laterally and migrated in the direction of the regional water table gradient. The distributions of well oil thickness (H_o) at the end of infiltration and at various times during redistribution are shown for the heterogeneous soil (Fig. 2, left side) and the homogeneous soil (Fig. 2, right side). During redistribution in the homogeneous soil, the plume remains symmetrical about the axis y = 32 m, while in the heterogeneous soil, the plume is asymmetrical (Fig. 2). For the homogeneous soil, the plume does not reach the boundary; as the edge of the plume travels only 86 m in the x-direction and 21 m in the y-direction relative to its location at the end of infiltration. However, at 1260 days in the heterogeneous soil, the plume just reaches the right and bottom boundaries of the domain. Thus, relative to the homogeneous soil, the edge of the plume in the heterogeneous soil travels 26% further in the x-direction and 52% further in the negative y-direction. The large amount of spreading in the heterogeneous soil can be attributed to preferential flow of oil through areas of high conductivity. Due to the correlation between α and K_{sw}, these areas also tend to have high α, which indicates high average pore size. As α increases, the specific volume of free oil (V_{of}) that can occupy the soil at a given H_o also increases and allows for higher oil transmissivity. Thus the high permeability zones have a higher capacity to hold oil, and the effects of permeability and capacity combine to promote spreading of the plume in the heterogeneous soil.

The effect of soil heterogeneities on the free and trapped oil distributions is apparent in Fig. 3, which depicts the actual nodal values for H_o, V_{of}, V_{ot}, and V_{og} at 720 days, along with nodal values of H_o in the heterogeneous soil at 240 days. The values were taken along a transect through the center of the plume corresponding to the y = 32 m axis. The free and trapped oil specific volumes reveal a high degree of spatial variability in the heterogeneous case as compared with the homogeneous

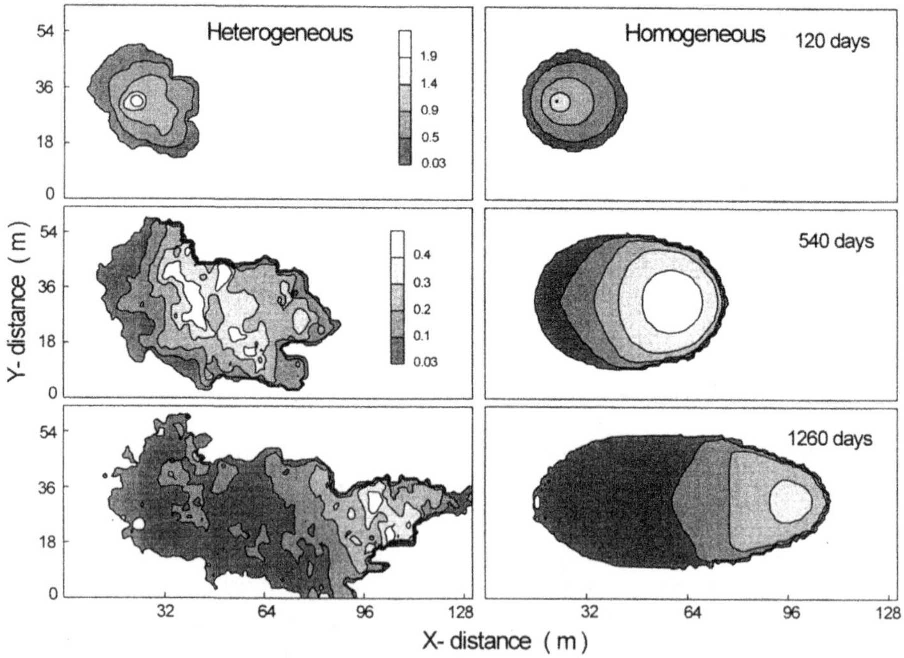

Fig. 2. Well oil thickness (H_o) following infiltration (t = 120 days), and at t = 540 and 1260 days during redistribution of the oil plume in the heterogeneous soil (*left side*), and the homogeneous soil (*right side*). The dimensions for H_o are in meters, and the scale bar for the 540-day contours also applies to the 1260-day contours

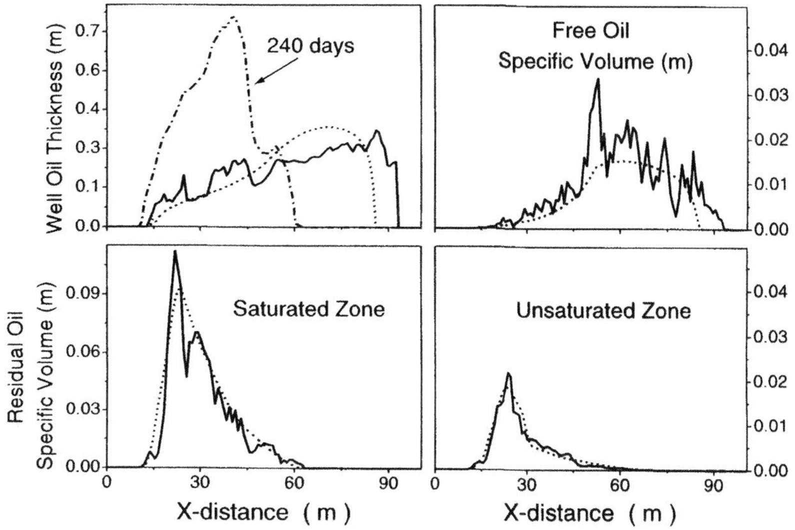

Fig. 3. Nodal values along the y = 32 m axis for well oil thickness (H_o), free oil specific volume (V_{of}), and residual oil specific volumes in the saturated (V_{ot}), and the unsaturated zone (V_{og}) at an elapsed time of 720 days for the homogeneous soil (*dotted line*) and the heterogeneous soil (*solid line*). For comparison, the nodal values of H_o in the heterogeneous soil are included at 240 days

soil. As free oil volume diminishes during the time between 240 and 720 days, nodal values of H_o become smaller and more variable. Thus, at later times, reduced oil flow allows the H_o distribution to be controlled more by soil heterogeneity than gradients in the fluid levels.

4.3 Statistical Analysis

For a given position on the y-axis, the coefficients of variation (CVs) of the well oil thickness and free oil specific volume were computed over the x-direction of the domain (Fig. 4). The CV is a normalized measure of variability and is calculated as the standard deviation of a group of data divided by the mean. For both soils, the CVs of V_{of} are higher than those of H_o, indicating that H_o is a more regular variable than V_{of}. This may be explained because H_o is the difference between two potentiometric surfaces (Z_{ao} and Z_{ow}) whose values must vary continuously throughout the domain. In contrast, V_{of} shows a nonlinear dependence on H_o and the stochastic parameters α, n, and S_m. Furthermore, hysteresis allows a single H_o to produce a range of V_{of} values, so large differences occur in V_{of} between adjacent nodes.

The CVs for both soils increase with time as the free oil volume decreases (Fig. 4). For the homogeneous soil, the values of all CVs are symmetrical about the axis y = 32 m, and the values are relatively uniform across the domain at 240 days.

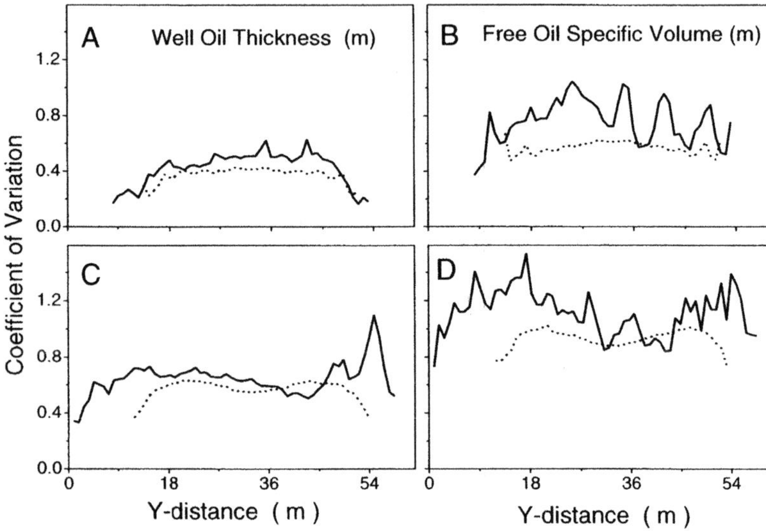

Fig. 4. Coefficients of variation (CVs) of the well oil thickness (**A, C**) and free oil specific volume (**B, D**) calculated for a given position on the y-axis by computing over the x-direction of the domain of the homogeneous (*dotted lines*) and heterogeneous soil (*solid lines*). Values are given at 240 days (**A, B**) and 1260 days (**C, D**)

However, by 1260 days, the CVs for both H_o and V_{of} (Fig. 4C,D) show a depression in the center of the domain where free oil is highest. The same effect occurs in the heterogeneous soil, although the curves are not symmetrical. For the heterogeneous soil, the variation in soil properties is reflected in the large and fluctuating values of the CVs, especially for V_{of}. At 1260 days, the CVs for H_o (Fig. 4C) are highest where there is little oil at large y-values of the domain (see Fig. 2). Thus the variability in the oil plume is greatest in areas of small V_{of}, while areas of large V_{of} tend to maintain a smoother distribution.

4.4 Analysis of Spatial Moments of the Oil Plume

The average, large scale movement of each oil plume was evaluated by calculating spatial moments of V_{of} at 5-day intervals throughout the simulation. At time t, the *ij*th spatial moments of V_{of} may be defined as

$$M_{ij}(t) = \int_{-\infty}^{\infty}\int_{-\infty}^{\infty} V_{of}(x,y;t)x^i y^j \, dx \, dy. \tag{7}$$

The zero order moment (M_{00}) provides an estimate of the total free oil volume in the domain, while the first order moments (M_{10} and M_{01}) can be used to locate the center of volume of the oil plume. Since oil density is assumed constant, the center of volume corresponds to the center of mass (x_c, y_c), which is computed as

$$x_c = M_{10}/M_{00} \quad \text{and} \quad y_c = M_{01}/M_{00}. \tag{8}$$

The locations of x_c and y_c for the homogeneous and the heterogeneous soils were compared during the entire 1260-day simulation (data not shown). Overall, the location of x_c was similar for both soils, indicating that soil heterogeneities do not have a large influence on the average movement of the entire plume in the direction of flow.

The average velocity of the oil plume in the x-direction (v_x) at time t was calculated as

$$v_x = \frac{\left(x_c^t - x_c^{t-\Delta t}\right)}{\Delta t}, \tag{9}$$

where the time interval Δt was a constant 5 days. For both soils, v_x increases sharply (Fig. 5) when infiltration stops at 120 days. After 120 days in the homogeneous soil, v_x decreases with time as oil trapping reduces the free oil volume and therefore the oil transmissivity. For the heterogeneous soil, v_x increases between 200 and 400 days (Fig. 5) as oil travels into zones of higher conductivity. Thus, average oil transmissivity undergoes a net increase in spite of the reduction in free oil volume during this period. From about 400 days until the end of the simulation, v_x for the heterogeneous soil exceeds that of the homogeneous soil. At the beginning of the release event, v_x for the heterogeneous plume is negative for several days, indicating that the center of mass of the plume actually moves against the direction of average water flow. This occurs because oil mounding directly under the spill causes gradients in Z_{ao} to point outward in all directions. In the heterogeneous soil, a zone of high K_{sw} and α exists just up-gradient of the spill, so large volumes of oil

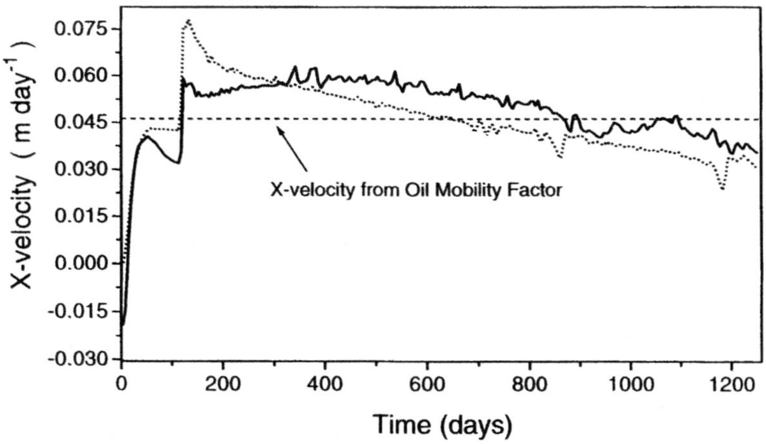

Fig. 5. Comparison of the velocity of the center of mass in the x-direction of the oil plume versus time for the homogeneous (*dotted line*) and the heterogeneous soil (*solid line*). The average velocity as calculated from the maximum oil mobility factor (M_o^{max}) is also depicted (*dashed line*)

move in the negative x-direction along the gradient in Z_{ao} and against the average gradient in Z_{aw}. Thus, when the spill volume is small and covers a limited number of correlation scales, plume velocity is sensitive to soil variability.

During the redistribution period from 120 to 1260 days, v_x for the homogeneous soil falls from 0.079 to 0.037 m day^{-1} (Fig. 5). During the same time period, v_x for the heterogeneous soil varies from 0.061 to 0.040 m day^{-1}. Thus, the average velocity of the oil plume is relatively stable, despite the decrease in free oil volume from 53.8 to 8.75 m^3 in the homogeneous soil and 53.8 to 10.3 m^3 in the heterogeneous soil. For comparison, the average x-velocity of the oil plume was also calculated using the maximum oil mobility factor (M_o^{max}) described in Section 2. Using Eqs. (2) and (3), and assuming that the gradient in Z_{aw} may be substituted for dZ_{ao}/dx, the average velocity of the oil plume is calculated as 0.046 m day^{-1}. This value lies in the middle of the range of velocities predicted by the transient analysis for both soils (Fig. 5). Early on, the plumes are moving at a higher velocity than predicted by M_o^{max}, probably due to the fact that the plume has not yet flattened out, so the gradient in Z_{ao} is larger than the gradient in Z_{aw}. Later, the reduction in free oil volume causes both plumes to move at lower velocities than predicted by M_o^{max}. Nevertheless, the discrepancy is not large, suggesting that the average oil velocity computed from M_o^{max} may be a useful approximation of the overall plume migration rate.

5 Oil Recovery from a Hypothetical Spill

5.1 Description of Test Case

The multiphase flow model was used to analyze oil recovery from a single pumping well in a homogenous versus three heterogeneous aquifers. The model domain for each case was 91 m in the x-direction by 90 m in the y-direction. The area was divided into a 9999 node mesh using 0.91 by 0.91 m elements, and the bottom of the aquifer was chosen as the datum ($z = 0$). Boundary conditions for the water table (Z_{aw}) were maintained at a constant head of 30.5 m at $x = 0$ and 30.0 m at $x = 91$. The top ($y = 90$) and bottom ($y = 0$) boundaries were specified as no flow. Thus the average gradient in Z_{aw} prior to pumping was 0.5%, but variability in K_{sw} produced local irregulatities in the gradient for the heterogeneous soil. Oil never reached the boundaries of the domain in any of the simulations.

In both the homogenous and the three heterogeneous cases (Table 2), the geometric mean soil properties and the fluid properties were identical to those used in the previous section. The correlation scale (η) was chosen as 6.1 m in case 1, and 3.66 m in cases 2 and 3. For cases 1 and 2, K_{sw} and α were assumed to be correlated with $\rho = 0.85$, while for case 3, K_{sw} and α were assumed to be uncorrelated (Table 2). As before, the parameter n was assumed to be uncorrelated with K_{sw} and α. Due to the difference in η, case 1 parameter values remain correlated over larger distances, so the ranges are smaller in case 1 than cases 2 and 3. The correlation scale also influenced the K_{sw} distribution (data not shown), as the

Table 2. Heterogeneous soil cases

	Correlation scale (η)	Correlation (ρ) between K_{sw} and α
	(m)	
Case 1	6.10	0.85
Case 2	3.66	0.85
Case 3	3.66	Uncorrelated

larger η of 6.1 m allowed areas of similar K_{sw} to be larger and more connected than when η was 3.66 m.

For each soil, the model predicted oil flow during three stages of the simulations. In the first stage, 202 m^3 of oil was allowed to infiltrate at $x = 37$ and $y = 45$ m over a 400-day period. In the second stage, the oil plume was allowed to redistribute for 200 days. In the third stage, oil was recovered by simulating oil and water pumping for 1400 days from a recovery well placed just down-gradient of the center of the plume at $x = 46$ and $y = 45$ m. The well was placed at this location in order to maximize product recovery with minimal pumping. Water pumping was simulated at a rate of 55 m^3 day^{-1}, which provided hydraulic containment of the oil plume with limited lateral spreading at early times. Oil pumping was sufficient to maintain zero oil thickness in the well throughout the simulations.

5.2 Distribution of the Oil Plume

The distribution of H_o was asymmetrical in all three heterogeneous soils, as the oil flowed preferentially into the randomly distributed zones of high K_{sw} and α. The edge of the oil plume moved furthest in case 1 with $\eta = 6.1$ m because the high η allowed high K_{sw} and α zones to be more connected than when $\eta = 3.66$ m. In all three heterogeneous cases, the edges of the plume spread further than they did in the homogeneous soil. However, plume spreading was limited in case 3 because K_{sw} and α were uncorrelated, so soil zones with high K_{sw} often had low α and therefore limited capacity to hold oil. This reduced capacity tended to restrict oil transmissivity, even in zones having high K_{sw}.

Statistical and spatial moment analyses revealed that the distributions of the oil plumes followed the same trends as described in Sections 4.3 and 4.4. Specifically, the CVs of V_{of} were higher than those of H_o, and all CVs tended to increase with time as the free oil volume decreased. By the end of pumping, x_c for all three heterogeneous soils had moved farther than x_c of the homogeneous soil; however, the difference was small, especially when $\eta = 3.66$ m. In all soils, oil recovery eventually led to limited contraction of the free oil area as compared to the decrease in free oil volume. For example, in heterogeneous case 2, pumping reduced the free oil area by only 27% in spite of a 99% reduction in free oil volume. Again, this reflects the fact that the oil has negligible mobility in soil zones where V_{of} is small, and complete recovery from these zones is difficult. Thus, in practice, monitoring wells may produce small but measurable H_o, even when free oil recovery has essentially disappeared.

5.3 Oil Recovery

The cumulative volumes of recovered oil and residual oil in the saturated and unsaturated zones are plotted versus time in Fig. 6 for the homogeneous soil and cases 1 and 2 of the heterogeneous soils. For all soils, the oil recovery rate is initially high (Fig. 6C), but diminishes as the free oil volume decreases. Total oil recovery is greatest for the homogeneous soil, intermediate for heterogeneous case 1 ($\eta = 6.1$ m) and least for heterogeneous case 2 ($\eta = 3.66$ m). This trend suggests that potential oil recovery diminishes with decreasing η and the accompanying decrease in continuity of soil properties throughout the domain. At the end of pumping, the homogeneous soil yielded 1.1% more oil than heterogeneous case 1, and 6.6% more oil than case 2. Thus, the overall difference in recovery due to the soil heterogeneities is small.

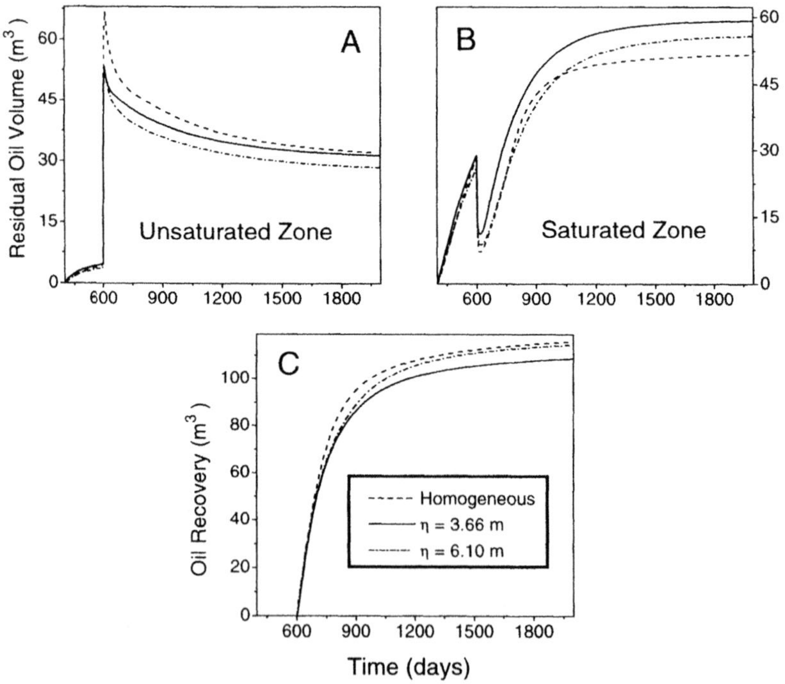

Fig. 6. Cumulative volumes of unsaturated zone residual oil (**A**), saturated zone residual oil (**B**), and oil recovered (**C**) versus time for the homogeneous soil (*dashed line*) and the heterogeneous cases 1 (η = 6.10 m, *dash-dotted line*) and 2 (η = 3.66 m, *solid line*)

Relative to the homogeneous soil, the heterogeneous soils trap more residual oil in the saturated zone (Fig. 6B) and less residual oil in the unsaturated zone (Fig. 6A). The onset of pumping draws down the water table and releases most of the saturated zone residual oil that had been trapped during the redistribution period from 400 to 600 days (Fig. 6B). The reduction in saturated zone residual oil is greatest in the homogeneous soil. In all cases, the water table drawdown initially produces a large volume of unsaturated zone residual oil (Fig. 6A), but oil drainage releases much of this oil by the end of pumping. The effect is greatest in the homogeneous soil, where pumping causes 67 m^3 of oil to be trapped initially in the unsaturated zone, followed by drainage of 35 m^3. In contrast, both heterogeneous soils trap approximately 52 m^3 initially, while 24 m^3 drain in case 1 and 22 m^3 drain in case 2. Thus, the variability of the soil properties limits the amount of vertical oil drainage, even though heterogeneities in the vertical direction were not considered explicitly. In reality, vertical heterogeneities may further limit oil drainage from low conductivity regions of the unsaturated zone (Catalan and Dullien 1995).

6 Conclusions

Results of these simulations revealed that the coefficients of variations (CVs) for H_o were lower than those for V_{of}, indicating that H_o is a more regular variable than V_{of}. Thus, for analysis of actual field data, interpolation between known values of H_o should provide a more reliable estimate of the areal distribution of oil than interpolation using V_{of}. The effects of soil heterogeneity were reduced when K_{sw} and α were uncorrelated; however, soil conditions such as fractures that reduce correlation are also likely to cause the input parameters to display unusually high variances. Therefore, caution should be exercised if soils with uncorrelated K_{sw} and α are modeled as homogeneous media with effective parameters. Similar caution should be observed in modeling spills where free oil area spans a limited number of correlation scales, and spills where the observed H_o are low. In these situations, the distribution and flow of oil are more sensitive to soil heterogeneities than in cases where the spill area and H_o values are relatively large. Nevertheless, these multiphase flow simulations suggest that the geometric mean soil properties provide a useful and accurate estimate of the potential for oil recovery from large spills that span a large number of correlation scales.

Acknowledgement. The work reported in this paper was performed as part of a project sponsored by the Binational Agricultural Research and Development Fund (BARD Project IS-1646-91).

References

Ababou R (1988) Three-dimensional flow in random porous media. PhD Thesis, Massachusetts Institute of Technology, Cambridge

Carsel RF, Parrish RS (1988) Developing joint probability distributions of soil water retention characteristics. Water Resour Res 24:755–769

Catalan LJJ, Dullien FAL (1995) Application of gravity drainage to the recovery of residual LNAPL in homogeneous and lensed sand packs. J Contam Hydrol 18:279–306

Chen Z, Govindaraju RS, Kavvas ML (1994) Spatial averaging of unsaturated flow equations under infiltration conditions over areally heterogeneous fields. 2. Numerical simulations. Water Resour Res 30:535–548

Corey A (1986) Mechanics of immiscible fluids in porous media. Water Resources Publications, Littleton, Colorado

Dagan G (1990) Transport in heterogeneous porous formations: spatial moments, ergodicity and effective dispersion. Water Resour Res 26:1281–1290

ES&T: Environmental Systems & Technologies Inc (1996) Areal multiphase organic simulator for free phase hydrocarbon migration and recovery: ARMOS Version 5.2 User's Guide. ES&T, Blacksburg, Virginia

Essaid HI, Herkelrath WN, Hess KM (1993) Simulation of fluid distributions observed at a crude-oil spill site incorporating hysteresis, oil entrapment, and spatial variability of hydraulic properties. Water Resour Res 29:1753–1770

Gelhar LW (1986) Stochastic subsurface hydrology from theory to applications. Water Resour Res 22:135S–145S

Jury WA, Russo D, Sposito G, Elabd H (1987) The spatial variability of water and solute spatial structure in unsaturated soil 1. Analysis of property variation and spatial structure with statistical models. Hilgardia 55:1–32

Kaluarachchi JJ (1996) Effect of subsurface heterogeneity on free-product recovery from unconfined aquifers. J Contam Hydrol 22:19–37

Kaluarachchi JJ, Parker JC (1989) An efficient finite element method for modeling multiphase flow in porous media. Water Resour Res 25:43–54

Kaluarachchi JJ, Parker JC, Lenhard RJ (1990) A numerical model for areal migration of water and light hydrocarbon in unconfined aquifers. Adv Water Resour 13:29–40

Lenhard RJ, Parker JC (1990) Estimation of free hydrocarbon volume from fluid levels in monitoring wells. Ground Water 28:57–67

Lumley JL, Panofsky A (1964) The structure of atmospheric turbulence. Wiley, New York

Mantoglou A, Wilson JL (1982) The turning bands method for simulation of random fields using line generation by a spectral method. Water Resour Res 18:1379–1394

Parker JC, Lenhard RJ (1989) Vertical integration of three-phase flow equations for analysis of light hydrocarbon plume movement. Transport Porous Media 5:187–206

Parker JC, Lenhard RJ, Kuppusamy T (1987) A parametric model for constitutive properties governing multiphase flow in porous media. Water Resour Res 23:618–624

Parker JC, Waddill DW, Johnson JA (1994) UST corrective action technologies: engineering design of free product recovery systems. US EPA Risk Reduction Engineering Laboratory, Edison, New Jersey

Russo D (1991) Stochastic analysis of simulated vadose zone solute transport in a vertical cross section of heterogeneous soil during nonsteady water flow. Water Resour Res 27:267–283

Russo D, Bressler E (1981) Soil hydraulic properties as stochastic processes 1. An analysis of field spatial variability. Soil Sci Soc Am J 45:682–687

Russo D, Zaidel J, Laufer A (1994) Stochastic analysis of solute transport in partially saturated heterogeneous soil 1, Numerical experiments. Water Resour Res 30:769–779

Sposito G, Jury WA, Gupta VK (1986) Fundamental problems in the convection-dispersion model of solute transport in aquifers and field soils. Water Resour Res 22:77–88

Tompson AF, Ababou A, Gelhar L (1987) Application and use of the three-dimensional turning bands random field generation: single realization problem. Research Report #313, Ralph M Parsons Laboratory, MIT, Cambridge, Massachusetts

van Genuchten MT (1980) A closed-form equation for predicting the hydraulic conductivity of unsaturated soils. Soil Sci Soc Am J 44:892–898

Viera SR, Nelson DR, Biggar JW (1981) Spatial variability of field measured infiltration rate. Soil Sci Soc Am J 45:1040–1048

Waddill DW, Parker JC (1996) Recovery of light, non-aqueous phase liquid from porous media: laboratory experiments and model validation. J Contam Hydrol (in press)

Wilson JL, Conrad SH, Mason WR, Peplinski W, Hagan E (1990) Laboratory investigation of residual liquid organics from spills, leaks, and the disposal of hazardous wastes in groundwater. EPA/600/6-90/004

Yeh T-C, Gelhar LW, Gutjahr AL (1985) Stochastic analysis of unsaturated flow in heterogeneous soils 2, Statistically anisotropic media with variable α. Water Resour Res 21:457–464

16 Characteristics of Entrapped LNAPL Dissolution and Transport in a Contaminated Fractured Permeable Medium

H. Rubin[1], K. Rathfelder[2], and L. M. Abriola[2]

Contents

1 Introduction

The impetus of the mathematical investigation described herein lies in the need to reclaim sandstone aquifers whose top layers are contaminated by entrapped oil, specifically LNAPLs (light non-aqueous-phase-liquids). According to classic descriptions of immiscible fluid flow in the subsurface (e.g. Schwille 1975), LNAPL travels vertically downward though the vadose zone with limited spreading until it arrives at the saturated zone. The LNAPL will then spread horizontally, forming a pancake shaped lens located with the capillary fringe zone, above the water table. Fluctuations of the groundwater table due to natural, as well as artificial, factors may lead to penetration of LNAPL into the saturated zone (Rubin and Mechrez 1989). Commercial equipment can be used to skim and pump the free floating LNAPL on top of the saturated zone. Pumping alone, however, cannot entirely remove all LNAPL; capillary entrapment acts to retain a small yet significant fraction of the LNAPL in the soil pores. Reclamation of these entrapped LNAPLs requires further remediation applications such as traditional pump-and-treat flushing (Powers et al. 1991, 1992), vapor extraction, or advanced methods employing surfactant additives to enhance LNAPL solubilization and/or mobilization.

[1]Department of Civil Engineering, Environmental & Water Resources Engineering, Technion-Israel Institute of Technology, Haifa 32000, Israel
[2]Department of Civil & Environmental Engineering, The University of Michigan, Ann Arbor, Michigan 48109-2125, USA

H. Rubin/N. Narkis/ J. Carberry
Soil and Aquifer Pollution

Sandstone aquifers can very often be classified as fractured permeable formations (Birkhölzer et al. 1993a,b). Such formations are characterized by their ability to transmit significant fluid flow through both the fracture network and the hosting permeable matrix. We consider the specific example of a fractured permeable sandstone aquifer along the coastal plain of Israel. Portions of this water supply aquifer have been contaminated by a large spill of jet fuel (Kanfi 1986; Rubin et al. 1994). The top layers of the aquifer are contaminated with large quantities of floating LNAPL some of which are entrapped in the saturated zone due to water table fluctuations. Following the skimming of the floating jet fuel, remediation measures are required to address the LNAPL entrapped in the permeable formation. This study presents a numerical investigation of pump-and-treat measures for the remediation of entrapped LNAPL in a fractured permeable formation.

Previous studies of the authors (Rubin et al. 1997a,b) refer to a fractured permeable formation in which all layers include entrapped NAPL. The present study addresses the problem of a fractured permeable formation in which it is assumed that only the top layers contain entrapped LNAPL within the permeable blocks. The application of pump-and-treat measures in such an aquifer results in flow through the permeable blocks, inducing the gradual dissolution of the entrapped LNAPL and enabling the subsequent transport of LNAPL solute to recovery wells. However, mixing between the permeable block flow and the fracture flow additionally causes the advection of LNAPL solute into down-gradient and transverse regions of the aquifer. Investigations described in this chapter focus on early behavior of pump-and-treat operations during which time initial LNAPL solute is transported to recovery wells and quasi-steady state LNAPL dissolution conditions are established. Investigation of processes during this period provides insight into the potential effectiveness and limitations of pump-and-treat operations in fractured permeable formations.

2 Conceptual Model and Basic Formulation

We adopt the simplified conceptual model shown in Fig. 1 which represents the fractured permeable formation. The x^* axis represents the groundwater table; the y^* axis is directed downwards and represents the vertical coordinate. The formation incorporates permeable blocks in which oblique fractures are embedded. The fracture network is made of two sets of parallel equidistant fractures whose aperture is constant. LNAPL ganglia are entrapped in a portion of the permeable blocks comprising several top layers in the domain. The horizontal extent of the contaminated region is also limited to a number of permeable blocks. Beneath the layers of contaminated blocks are layers of permeable blocks free of entrapped LNAPL.

At time $t^* \leq 0$, there is negligible flow in the domain. Beginning at time $t^* \geq 0$, pump-and-treat operations from an injection/pumping arrangement shown by Powers et al. (1991) are assumed to produce a steady flow of water in the horizontal

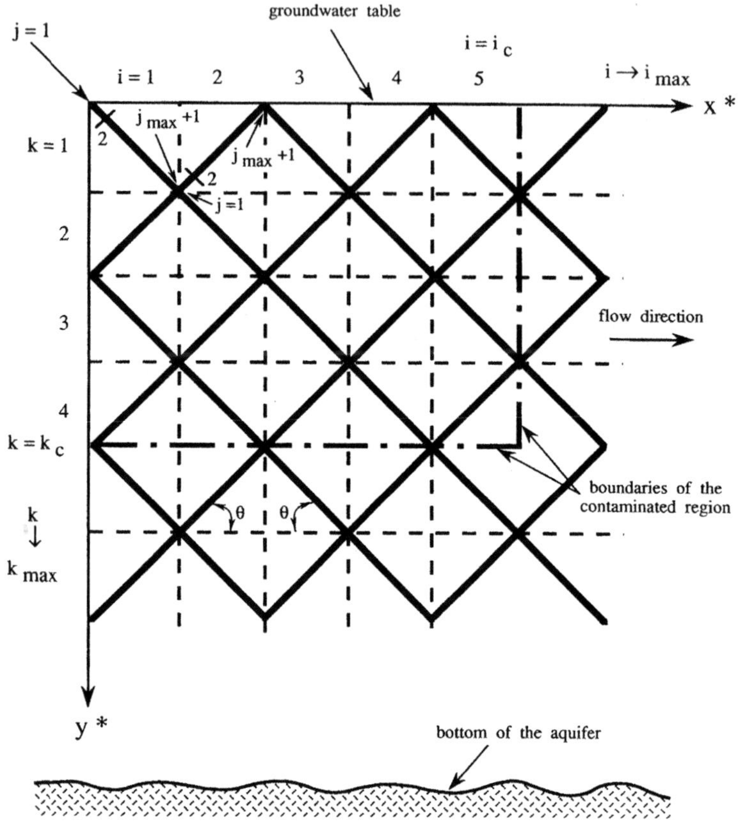

Fig. 1. Conceptual model of the fractured permeable formation

direction through the permeable blocks. Unsteady state behavior stemming from aquifer storage is assumed to be negligible due to the short time scale of compressibility effects. The specific discharge of the permeable block flow is denoted by q_b. This flow is minimally affected by the presence of the entrapped LNAPL (Rubin et al. 1997a,b).

The flow rate in each single fracture is Q_f. The distance between two adjacent fracture intersections measured along the fractures is B. Therefore, the specific discharge, q_f, stemming from the fracture presence is given by

$$q_f = \frac{Q_f}{B\sin\theta}, \tag{1}$$

where θ is the fracture orientation angle with regard to the permeable block flow direction.

The initial LNAPL solute concentration at $t^* \leq 0$ is assumed to be at the equilibrium concentration, C_s, in permeable blocks containing entrapped LNAPL. It is zero elsewhere. Following the start of pump-and-treat operations at $t^* = 0$, contaminant-free water arrives at $x^* = 0$. As flow advances into the formation, water particles are gradually contaminated due to their contact with the LNAPL ganglia entrapped in the permeable blocks. The solute concentration in the fracture flow is always zero at $x^* = 0$, but is subject to growth along the fracture sectors in which the contaminated permeable block flow mixes with the fracture flow. Because the fracture network is a conduit into transverse regions, mixing and transfer between the permeable block and fracture flow significantly influences the transverse migration of LNAPL solute into uncontaminated regions of the aquifer.

Mathematical representation of transport in the fracture flow is based on the mass balance of LNAPL solute, ignoring mass storage in the fracture and assuming that diffusion and dispersion are negligible in comparison to advection (Birkhölzer et al. 1993a):

$$Q_f \frac{\partial C^*}{\partial x'} dx' + q_b \left(C^* - C_b^*\right) dy^* = 0, \tag{2}$$

where C^* is the fracture flow concentration; C_b^* is the concentration of the permeable block flow which enters the fracture and mixes with the fracture flow; x' is a local coordinate representing the fracture centerline; and y^* is the vertical downward coordinate. The second term accounts for mixing between the fracture and permeable block.

The following relationships may be obtained from Fig. 1:

$$dx' = dx^*/\cos\theta; \quad dy^* = dx^* \tan\theta. \tag{3}$$

Introducing Eq. (3) into Eq. (2) we obtain

$$Q_f \frac{\partial C^*}{\partial x^*} = q_b \tan\theta \left(C^* - C_b^*\right) = 0. \tag{4}$$

Next, we define the following dimensionless space coordinates and dimensionless concentrations:

$$x = \frac{x^*}{B\cos\theta}; \quad y = \frac{y^*}{B\sin\theta}; \quad C = \frac{C^*}{C_s}; \quad C_b = \frac{C_b^*}{C_s}. \tag{5}$$

Introducing Eq. (5) into Eq. (4) we obtain

$$\frac{\partial C}{\partial x} + N_M C = N_M C_b. \tag{6}$$

Here, N_M is the mobility number defined as

$$N_M = \frac{q_b}{q_f} = \frac{K_b B \tan\theta}{M_f}, \tag{7}$$

where K_b is the hydraulic conductivity of the permeable blocks, and M_f is the mobility of the single fracture. The mobility of the single fracture represents the ratio between the fracture flow-rate and the hydraulic gradient imposed along the fracture. Its value can be estimated by the employment of the viscous flow equation, namely

$$M_f = gw^3/(12\nu),\tag{8}$$

where g is gravitational acceleration, w is the width of the fracture, and ν is the kinematic viscosity of the fluid. We ignore any effect of the dissolved LNAPL on the latter parameter.

The distribution of the dissolved LNAPL in the fracture flow is determined by the numerical solution of E_q. (6). The solution, however, requires the specification boundary conditions for the LNAPL solute concentration at the fracture junctions, C, and values of C_b, the solute concentration in the permeable blocks. These concentrations are time-dependent due to solute transport from the LNAPL-contaminated domain into regions that are initially uncontaminated, as well as the transfer and mixing of LNAPL solute between fractures and permeable blocks. Following this transient phase, there is an extended period of quasi-steady-state conditions during which the entrapped LNAPL saturation gradually decreases until complete reclamation of the aquifer is achieved. We ignore changes in the entrapped LNAPL saturation during the initial transient phase prior to the establishment of quasi-steady state conditions.

To develop a complete set of dimensionless expressions, we define the dimensionless time coordinate, t, as

$$t = t^* V_b/(B\cos\theta),\tag{9}$$

where V_b is the interstitial velocity of the water flowing through the permeable block.

Mathematical representation of dissolution and transport in the permeable blocks is similarly developed from a mass balance on the LNAPL solute, again ignoring dispersion effects (Powers et al. 1991):

$$\frac{\partial C_b^*}{\partial t^*} + V_b \frac{\partial C_b^*}{\partial x^*} = \frac{k_f a(C_s - C_b^*)}{\phi_b s_w},\tag{10}$$

where k_f is the effective mass transfer coefficient; a is the specific interfacial area; ϕ_b is the porosity of the permeable block; s_w is the water phase saturation. To simplify the solution we consider that

$$s_w \gg s_n;\ s_w \approx 1,\tag{11}$$

where s_n is the LNAPL saturation.

A general solution of E_q. (10) is obtained by the employment of dimensionless variables and a moving coordinate system, obtaining (Rubin et al. 1997a);

$$\frac{1 - C_b}{1 - C_o} = \exp[-K_f(x - x_o)],\tag{12}$$

where C_o is the value of C_b at a specified x_o; K_f is the dimensionless phase transfer coefficient. The latter parameter is given by

$$K_f = k_f \, a \, B \cos\theta / q_b . \tag{13}$$

Regarding mixing in fracture intersections, we assume a complete mixing in such locations, namely, solute concentrations in the outgoing discharges are equal to the flux average solute concentrations of the incoming fracture and permeable block flows.

3 The Numerical Model

The numerical model employed in the present study utilizes the basic features of the authors' model used in a previous study (Rubin et al. 1997a). However, simulations in this study cover a much larger number of fracture sectors in a wider region of the simulated domain. Other changes originate from mixing in fracture intersections and average contaminant concentration evaluation.

The domain shown in Fig. 1 is subdivided into blocks, defined by i_{max} longitudinal segments and k_{max} vertical levels. Each discrete sector (linear segment) of the fracture network is identified by two indices; i which represents the longitudinal position, and k which represents the vertical level in which the particular fracture sector is located. Within the blocks, each fracture sector is subdivided into j_{max} intervals bounded by $j_{max} + 1$ nodes. The length of a single interval measured along the x-axis is Δx. Values of C and C_b at each nodal point of the fracture network are represented by a four-dimensional matrix referring to the segment number (i), vertical level number (k), nodal point (j), and time step number (m).

The finite difference approximation of E_q. (6) is given as

$$\frac{C^m_{i,k,j+1} - C^m_{i,k,j}}{\Delta x} + N_M \frac{C^m_{i,k,j+1} + C_{i,k,j}}{2} = N_M \frac{C^m_{bi,k,j+1} + C^m_{bi,k,j}}{2} . \tag{14}$$

Birkhölzer et al. (1993b) showed that the use of identical values for Δx and Δt is applicable and convenient. We therefore apply, $\Delta t = \Delta x$. Rearranging Eq. (14) yields the following explicit numerical scheme:

$$C^m_{i,k,j+1}\left[1 + \frac{\Delta x}{2} N_M\right] = C^m_{i,k,j}\left[1 + \frac{\Delta x}{2} N_M\right] + \frac{\Delta x}{2} N_M \left[C^m_{bi,k,j+1} + C^m_{bi,k,j}\right] . \tag{16}$$

Fracture sectors incorporated in the contaminated domain are specified by the following i and k values:

$$1 \le i \le i_c$$
$$1 \le k \le k_c . \tag{17}$$

These fracture sectors are fed with LNAPL solute by influx from the permeable block flow through the LNAPL contaminated domain. The source of the LNAPL solute is from both the initial storage in the contaminated domain prior to the start of pump-and-treat remediation and the gradual dissolution of entrapped LNAPL following the start of pump-and-treat operations. The fracture sectors along the right edge of the contaminated domain are similary fed with LNAPL solute by influx from the permeable block flow, however, the entrapped LNAPL is located only in a portion of the permeable blocks. These fracture sectors are denoted by;

$$i = i_c + 1$$
$$1 \le k \le k_c. \tag{18}$$

Portions of the domain free of entrapped LNAPL are specified by

$$i_c + 1 \le i \le i_{max}$$
$$k_c + 1 \le k \le k_{max}. \tag{19}$$

In these locations, the dissolved LNAPL is subject only to advection by the permeable block and fracture flows as well as mixing between these two types of flow and mixing in fracture intersections.

In the following paragraphs we consider the various boundary conditions which should be applied for each set of fracture sectors in the domain of Fig. 1.

For fracture sectors located at the entrance of the contaminated region we apply

$$C^m_{1,k,1} = 0 \quad \text{at} \quad 1 \le k \le k_c. \tag{20}$$

For fracture sectors located at $k_c + 1 \le k \le k_{max}$ we apply

$$C^m_{k-k_c,k,1} = 0 \quad \text{at} \quad k_c < k \le k_{max}. \tag{21}$$

This equation implies that if $k > k_c$ and $1 \le i < k - k_c$ then C is zero for all nodal points of the fracture sector.

At interfaces between adjacent fracture sectors, namely fracture intersections, we apply

$$C^m_{2i+1,1,1} = C^m_{2i,1,j_{max}+1} \quad \text{at} \quad 2 \le 2i \le i_{max} - 1 \tag{22}$$

$$C^m_{2i,k_{max}-1} = C^m_{2i-1,k_{max},j_{max}+1} \quad \text{at} \quad 2 \le 2i \le i_{max} - 1 \tag{23}$$

$$C^m_{i+1,k,1} = C^m_{i+1,k+1,1} = 0.5\left(C^m_{i,k,j_{max}+1} + C^m_{i,k+1,j_{max}+1}\right)$$
$$\text{at} \quad 1 \le i < i_{max}; \quad 1 \le k < k_{max}. \tag{24}$$

The permeable block flow is the main mechanism for dissolution of the entrapped LNAPL as well as transport to the fracture network. Based upon Eq. (12), the following relationships are applicable with regard to C_b values:

$$C^m_{b1,k,j} = 1 \quad \text{at} \quad 1 \le k \le k_C \quad \text{if} \quad m < j-1 \tag{25a}$$

$$C^m_{b1,k,j} = 1 - \exp\left[-K_f(j-1)\Delta x\right] \quad \text{at} \quad 1 \le k \le k_c \quad \text{if} \quad m \ge j-1 \tag{25b}$$

$$C^m_{b1,k,j} = 0 \quad \text{at} \quad k_c < k \le k_{max} \tag{26}$$

$$C^m_{b_i,k,j} = 1 \quad \text{at} \quad \begin{Bmatrix} 1 \le k \le k_c \\ 1 < i \le i_c \end{Bmatrix} \quad \text{if} \quad m < 2(j-1) \tag{27a}$$

$$C^{m}_{b_i,k,j} = \left\{1 - C^{m-2j+2}_{i-1,k,j_{max}+1-j}\right\} \exp\left[-2K_f(j-1)\Delta x\right]$$
$$\text{at} \quad \left\{\begin{matrix} 1 \le k \le k_c \\ 1 < i \le i_c \end{matrix}\right\} \quad \text{if} \quad m \ge 2(j-1) \tag{27b}$$

$$C^{m}_{bi_c+1,k,j} = 0 \quad \text{if} \quad m < j-1 \tag{28a}$$

$$C^{m}_{bi_c+1,k,j} = 1 \quad \text{if} \quad j-1 \le m < 2(j-1) \tag{28b}$$

$$C^{m}_{bi_c+1,k,j} = 1 - \left\{1 - C^{m-2j-2}_{i_c,k,j_{max}+1-j}\right\} \exp\left[-2K_f(j-1)\Delta x\right] \quad \text{if} \quad m \ge 2(j-1) \tag{28c}$$

$$C^{m}_{bi,k,j} = 0] \quad \text{at} \quad i > i_c + 1 \quad \text{or} \quad k > k_c \quad \text{if} \quad m < 2(j-1) \tag{29a}$$

$$C^{m}_{bi,k,j} = C^{m-2j+2}_{i-1,k,j_{max}+1-j} \quad \text{at} \quad i > i_c + 1 \quad \text{or} \quad k > k_c \quad \text{if} \quad m \ge 2(j-1). \tag{29b}$$

To quantify the flushing and dissolution process, we calculate variations in a flux averaged dissolved LNAPL concentration in each block and fracture sector cross section incorporating the central point of the permeable block. We consider cross sections at the downstream side of the i-th segments at the k and k + 1 levels. For the calculation of C_b in such cross sections, we apply the following relationships:

$$\left[C^{m}_{b,k,j}\right]_{x=i} = 1 \quad \text{at} \quad m < j-1; \quad i \le i_c; \quad k \ge k_c \tag{30a}$$

$$\left[C^{m}_{b,k,j}\right]_{x=i} = 1 - \left\{1 - C^{m-j+1}_{i,k,j_{max}+1-j}\right\} \exp\left[-K_f(j-1)\Delta x\right] \quad \text{at} \quad m \ge j-1; \quad i \le i_c; \quad k \le k_c \tag{30b}$$

$$\left[C^{m}_{b,k,j}\right]_{x=i} = 0 \quad \text{at} \quad m < j-1; \quad i > i_c \quad \text{or} \quad k > k_c \tag{31a}$$

$$\left[C^{m}_{b,k,j}\right]_{x=i} = C^{m-j+1}_{i,k,j_{max}+1-j} \quad \text{at} \quad m \ge j-1; \quad i > i_c \quad \text{or} \quad k > k_c. \tag{31b}$$

Employing the trapezoidal integration procedure, the flux average concentration of the dissolved LNAPL is obtained as

$$\left[C^{m}_{a,v_k}\right]_{x=i} = \frac{0.5}{1+N_M}\Big\{C^{m}_{i,k,j_{max}+1} + C^{m}_{i,k+1,j_{max}+1} + \Delta N_M\Big[\frac{1}{2}\left(C^{m}_{i,k,j_{max}+1} + C^{m}_{i,k,1} + C^{m}_{i,k+1,j_{max}+1} + C^{m}_{i,k+1,1}\right) + \sum_{j=2}^{j_{max}}\left(C^{m}_{bk,j} + C^{m}_{bk+1,j}\right)_{x=i}\Big]\Big\}. \tag{32}$$

4 Simulation Results

In the framework of this study, numerous numerical simulations were performed with regard to unsteady state as well as apparent steady-state conditions. We use the term "apparent steady-state conditions" with regard to flow conditions established after the whole domain has been flushed of the initial LNAPL solute. Here, it was assumed that, during this flushing, only negligible quantities of entrapped LNAPL are subject to dissolution, and therefore the entrapped LNAPL saturation is not changed.

Simulation results presented here were obtained on a domain incorporating $i_{max} = 50$ segments, from which $i_c = 40$ segments are contaminated with entrapped LNAPL. The domain incorporates $k_{max} = 20$ levels, from which $k_c = 6$ levels are contaminated with entrapped LNAPL.

Figure 2 exemplifies the flushing processes in which the initial LNAPL solute is flushed from the aquifer and new quantities of solute are introduced due to the dissolution of the entrapped LNAPL. The dissolution of the entrapped LNAPL leads to growth of the dissolved LNAPL concentration with x. At the exit of the contaminated region. namely at $x = 40$, the dissolved LNAPL concentration is at the solubility limit ($C \approx 1$). However, between $x = 40$ and $x = 50$, the value of C_{av} at the top of the aquifer decreases due to the transverse dispersion stemming from the fracture presence in the domain. The dissolved LNAPL concentration at the aquifer surface, namely at $y \approx 0$, is smaller at $x = 50$ than at $x = 40$ since the region contaminated with entrapped LNAPL ends at $x = 40$. The LNAPL concentration in deep layers of the aquifer, however, namely at $6 \leq y \leq 20$, increases between $x = 40$ and $x = 50$ due to transverse dispersion.

Following the dimensionless time period $t = i_{max}$, apparent steady-state conditions are established in the domain. This time period represents the time interval needed for the movement of a fluid particle, moving with the permeable block flow, from $x = 0$ to the exit of the domain.

Figure 3 illustrates the apparent steady-state conditions typical of a high mobility number. In such cases the penetration of the dissolved LNAPL into deep layers of the aquifer is very limited, due to the small transverse dispersion effect attributed to the fracture flow (Rubin and Buddemeier 1996). When the mass transfer coefficient is small, such as for results shown in Fig. 3a ($K_f = 0.1$) a large number of LNAPL contaminated blocks (i.e., long travel distance through the contaminated domain) is required to obtain equilibrium solute concentrations in the effluent stream. For this example, even 40 contaminated blocks do not lead to $C_{av} = 1$. Notice, however, that at $x = 50$ the dissolved LNAPL concentration at the aquifer surface is quite similar to that existing at $x = 40$ due to the very limited transverse dispersion.

Results for the case of an intermediate mass transfer rate ($K_f = 1$) are presented in Fig. 3b. Here, equilibrium solute concentrations are established over a much shorter travel distance, approximately ten LNAPL-contaminated fracture sectors. Since the transverse dispersion effect is very minor, all profiles of dissolved LNAPL concentration are similar. However, there is a minor decrease in that concentra-

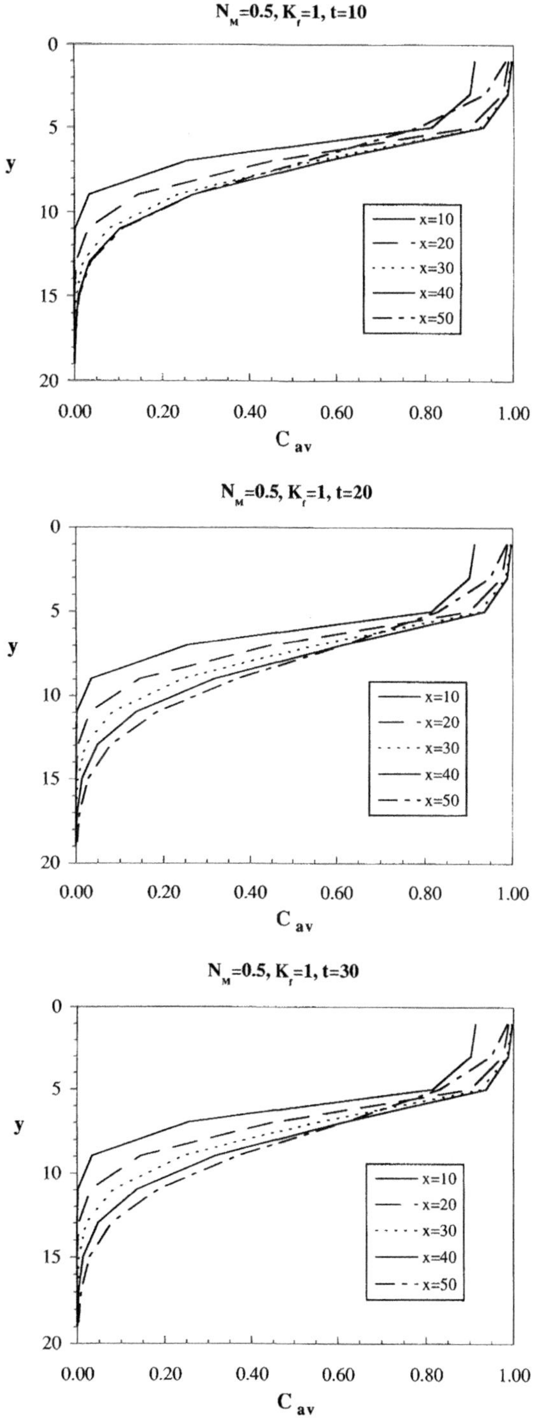

Fig. 2. Development of concentration profiles in the domain ($N_M = 0.5$, $K_f = 1$). **a** t = 10; **b** t = 20; **c** t = 30; **d** t = 40

Fig. 2. *Continued*

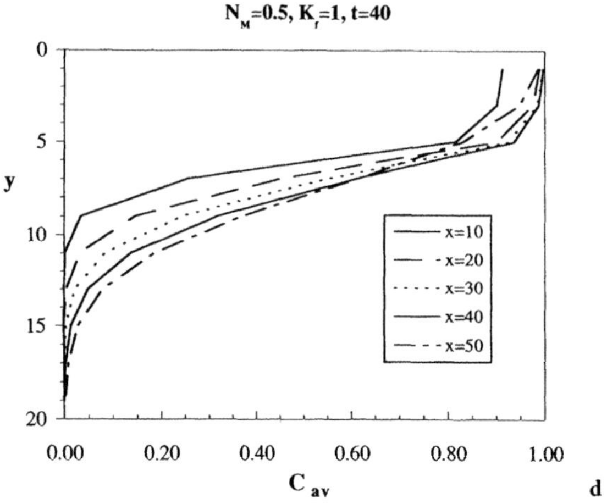

tion at the aquifer surface between x = 40 and x = 50. Similar results are obtained for the case of a large mass transfer rate ($K_f = 10$) shown in Fig. 3c. Here, differences between Fig. 3b and c are apparent only in several upstream blocks of the domain.

The effect of the mobility number on dissolution processes is examined in results shown in Figs. 4 and 5. Figure 4 presents apparents steady-state conditions typical of small values of the mobility number, namely $N_M = 0.5$ (i.e., flow through the fracture network is twice the flow through the permeable blocks). In such cases the effect of transverse dispersion stemming from the fracture flow is significant (Rubin and Buddemeier 1996), and the dissolved LNAPL penetrates to large depths of the aquifer.

Figure 4a refers to a very small value of $K_f = 0.1$. The 40 contaminated blocks are not sufficient to load the groundwater with its full capacity LNAPL solute. Simulations using higher mass transfer rates (Fig. 4b–c) show that equilibrium solute concentrations are established within a travel distance of less than 20 fracture sectors. Downstream of that location, changes in profiles of dissolved LNAPL are attributed only to transverse dispersion which causes deep vertical penetration of solute into the aquifer.

Figure 5 concerns apparent steady-state conditions typical of a very small mobility number, $N_M = 0.1$. In these cases, very large numbers of contaminated blocks are needed in order to achieve equilibrium solute concentrations in the flowing groundwater. Moreover, the effect of transverse dispersion due to the fracture flow is very significant, and the dissolved LNAPL penetrates to large depths in the aquifer.

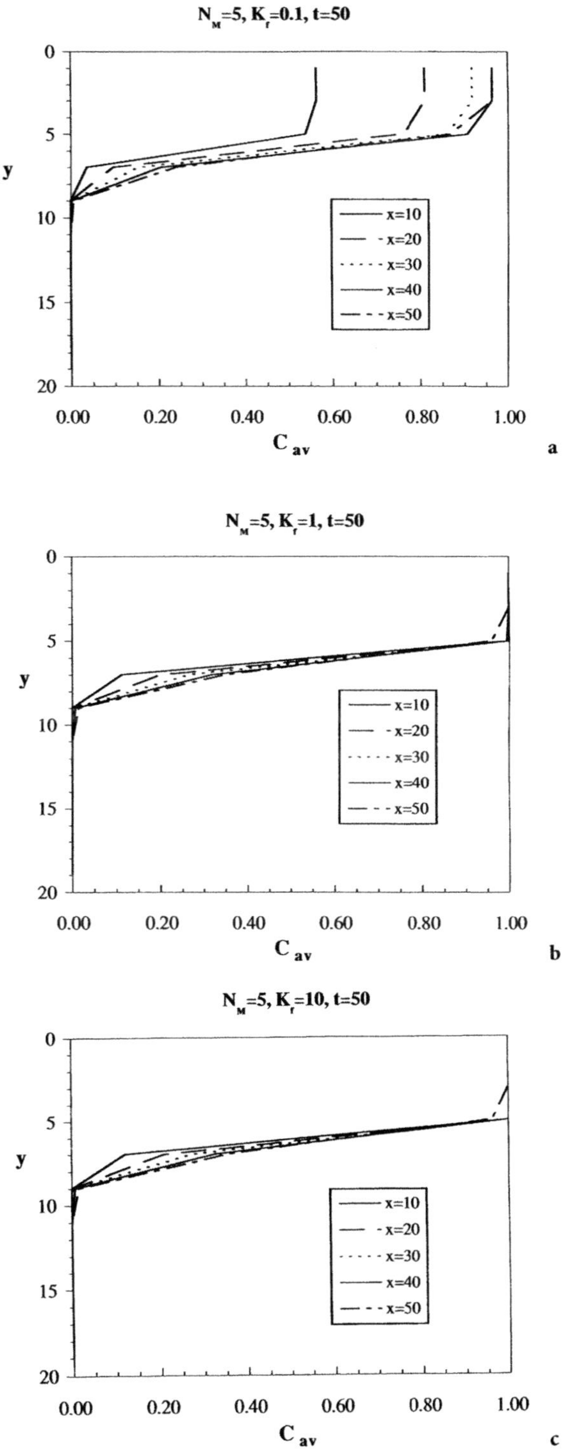

Fig. 3. Apparent steady-state concentration profiles typical of a high mobility number ($N_M = 5$). **a** $K_f = 0.1$; **b** $K_f = 1$; **c** $K_f = 10$

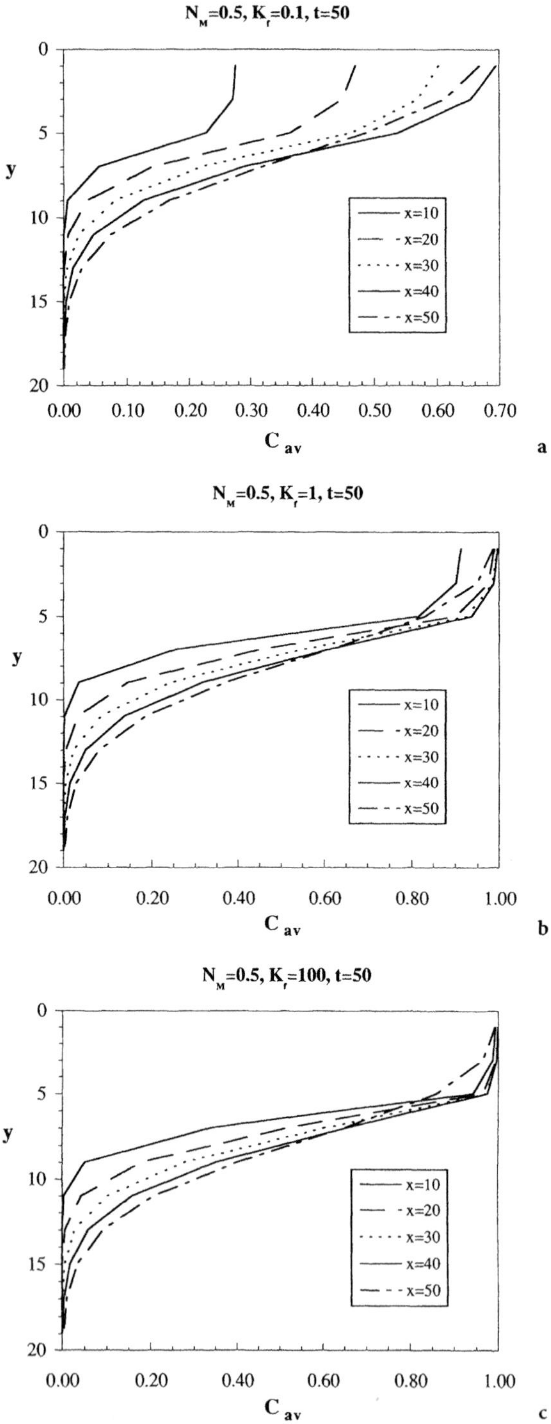

Fig. 4. Apparent steady-state concentration profiles typical of low a mobility number ($N_M = 0.5$). **a** $K_f = 0.1$; **b** $K_f = 1$; **c** $K_f = 100$

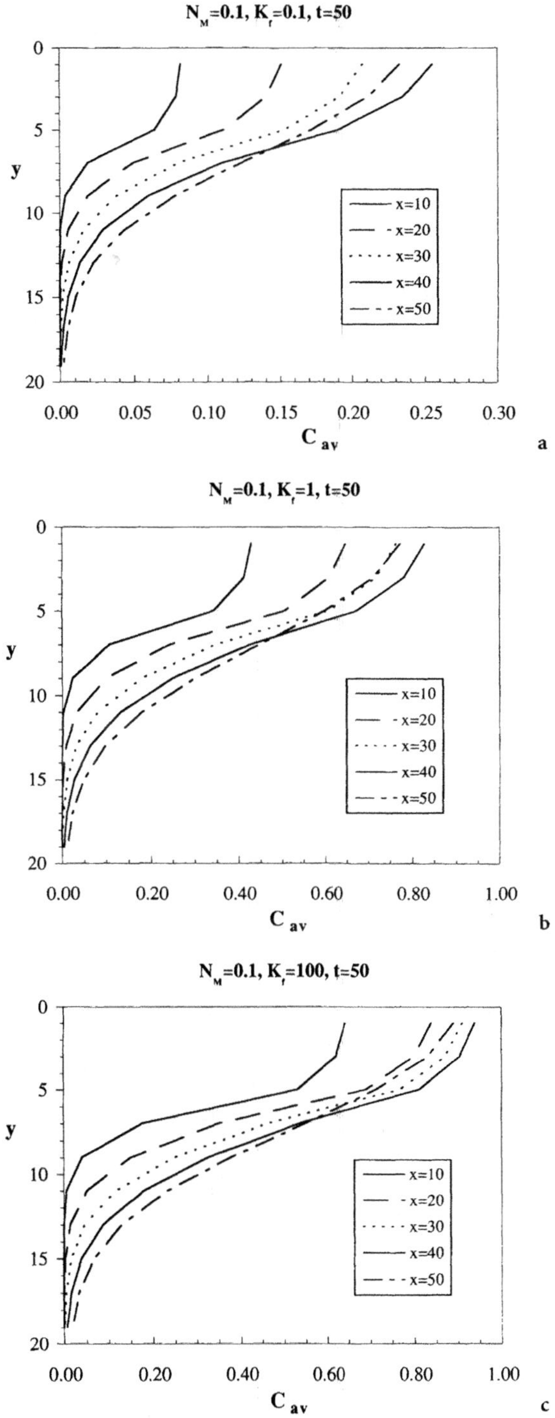

Fig. 5. Apparent steady-state concentration profiles typical of a very low mobility number ($N_M = 0.1$). **a** $K_f = 0.1$; **b** $K_f = 1$; **c** $K_f = 100$

5 Calculation of Continuum Parameters

The applicability of continuum parameters to represent transport phenomenon in fractured permeable formations was analyzed by Birkhölzer et al. (1993a,b). They determined the size of the REV (representative elementary volume) needed for the continuum representation. If the considered domain is smaller than the REV, then transport phenomena are dominated by non-Fickian processes. Birkhölzer et al. (1993a,b) claimed that the continuum approach can be useful for the calculation of longitudinal dispersion in the fractured permeable domain provided that the domain incorporates at least four permeable blocks and its length is larger than the loss of identity length. The latter is defined in dimensionless terms by

$$L = 4.6/N_M \,. \tag{33}$$

For domains larger than L, Birkhölzer et al. (1993a,b) could identify the equivalent porosity and the longitudinal dispersivity of the representative continuum domain. In a later study, Rubin and Ruddemeier (1996) determined the transverse dispersivity of the continuum representing the fractured permeable formation.

In previous studies (Rubin et al. 1997a,b), the phase transfer coefficient of the continuum representing the fractured permeable formation have been determined. In the present study we intend to apply the results obtained in all studies mentioned above and consider the possible application of the continuum approach to the determination of phenomena associated with dissolution, advection and dispersion of LNAPL in the aquifer.

In a continuum permeable domain contaminated by entrapped LNAPL, the longitudinal advection of water, dissolution of entrapped LNAPL, and dispersion of solute are represented by

$$\frac{\partial c*}{\partial t*} + V\frac{\partial c*}{\partial x*} = a_L^* V\frac{\partial^2 c*}{\partial x*^2} + a_T^* V\frac{\partial^2 c*}{\partial y*^2} + \frac{k_{con} a(C_s - c)*}{\phi s_w}, \tag{34}$$

where c* is the dissolved NAPL concentration in the continuum; t*, x* and y* are the time and space coordinates; V is the effective interstitial flow velocity; k_{con} is the effective mass transfer coefficient for the continuum; a is the specific interfacial area: ϕ is the equivalent porosity of the continuum; s_w is the water saturation; a_L^* and a_T^* are the longitudinal and transverse dispersivities.

We apply Eqs. (5) and (9) to nondimensionalize Eq. (34), obtaining

$$\frac{\partial c}{\partial t} + v\frac{\partial c}{\partial x}x = a_L v\frac{\partial^2 c}{\partial x^2} + av\frac{\partial^2 c}{\partial y^2} + K_{con}(1 + c), \tag{35}$$

where

$$c = \frac{c*}{c_s};\ a_L = \frac{a_L^*}{B\cos\theta};\ a_T = \frac{a_T^*\cos\theta}{B\sin^2\theta};\ k_{con} = \frac{k_{con} aB\cos\theta}{\phi V_b s_w}. \tag{36}$$

Birkhölzer et al. (1993a,b) showed that

$$\phi = \phi_b; \; v = 1 + 1/N_M. \tag{37}$$

It was also found that the mobility number can be used to determine values of longitudinal dispersivity (Birkhölzer et al. 1993a,b) and transverse dispersivity (Rubin and Buddemeier 1996). Other studies (Rubin et al. 1997a,b), have shown that the value of K_{con} is fully determined by the values of K_f and N_M. For a constant value of N_M, there is a nonlinear relationship between K_f and K_{con} as

$$\begin{aligned} &K = K_f \quad \text{at} \quad K_f < K_{fc} \\ &\text{and} \quad K_{con} = A_1 K_f^{\alpha} / \left(A_2 K_f^{\alpha}\right) \quad \text{at} \quad K_f > K_{fc}, \end{aligned} \tag{38}$$

where K_{fc} is a certain critical value of K_f; A_1, A_2 and α are constant coefficients. The value of K_{fc} and the coefficients depend on N_M. Their values are given by Rubin et al. (1997a).

Under steady-state conditions Eq. (35) yields

$$\frac{\partial c}{\partial x} = a_L \frac{\partial^2 c}{\partial x^2} + a_T \frac{\partial^2 c}{\partial y^2} + K_{eff}(1 - c), \tag{39}$$

where

$$K_{eff} = K_{con}/v. \tag{40}$$

Consider a simulated domain discretized with a finite difference grid of 51 × 21 nodes. The upper LNAPL contaminated region of the domain contains 41 × 7 nodes. The finite difference approximation of Eq. (39) is given by

$$\begin{aligned} c_{i,k}\left[1 + k_{eff}\Delta x + \frac{2a_L}{\Delta x} + \frac{2a_T \Delta x}{(\Delta y)^2}\right] &= c_{i-1,k}\left[1 + \frac{a_L}{\Delta x}\right] + \frac{a_L}{\Delta x} c_{i+1,k} \\ &+ \frac{a_T \Delta x}{(\Delta y)^2}\left(c_{i,k-1} + c_{i,k+1}\right) + K_{eff}\Delta x, \end{aligned} \tag{41}$$

where i is the node number in the longitudinal direction, x; k is the node number in the vertical direction, y; Δx is the longitudinal interval; Δy is the vertical interval. We consider the following boundary conditions of the continuum domain:

$$\begin{aligned} &c = 0 \quad \text{at} \quad x = 0 \\ &\frac{\partial c}{\partial y} = 0 \quad \text{at} \quad y = y_{max} \\ &\left(\frac{\partial c}{\partial x}\right)_{x_{max}} = \left(\frac{\partial c}{\partial x}\right)_{x_{max} - \Delta x}, \end{aligned} \tag{42}$$

where x_{max} is the length of the simulated domain, and y_{max} is the depth of the domain.

The solution of Eq. (41), subject to the boundary conditions of Eq. (42), is obtained by using the SOR (successive over-relaxation) scheme

$$c_{i,k}^{(n+1)} = (1-\omega)c_{i,k}^{(n)} + \omega\left[\alpha_1 c_{i-1,k}^{(n+1)} + \alpha_2 c_{i+1,k}^{(n)} + \alpha_3\left(c_{i,k-1}^{(n+1)} + c_{i,k+1}^{(n)}\right) + \alpha_4\right], \tag{43}$$

where

$$\alpha_0 = 1 + k_{eff}\Delta x + \frac{2a_L}{\Delta x} + \frac{2a_T\Delta x}{(\Delta y)^2}$$

$$\alpha_1 = \left(1 + \frac{2a_L}{\Delta x}\right)\Big/\alpha_0;\ \alpha_2 = \frac{2a_L}{\alpha_0\Delta x};\ \alpha_3 = \frac{a_T\Delta x}{\alpha_0(\Delta y)^2};\ \alpha_4 = K_{eff}\frac{\Delta x}{\alpha_0}. \tag{44}$$

In Eq. (43), n is the iteration number and ω is the over-relaxation coefficient.

Figure 6 illustrates the reproduction of the results given in Fig. 4a by the employment of the continuum approach. In Fig. 7 we compare two representative concentration profiles of Fig. 6 with their relevant profiles of Fig. 4a. The agreement between the profiles obtained by the discrete and continuum approaches is very good.

The calculations performed in this section indicate that the dissolution and transport of the LNAPL in the fractured permeable formation can be represented by considering appropriate continuum parameters. Such parameters of the continuum are nonlinearly related to those of the fractured permeable formation.

6 Summary and Conclusions

In this chapter, results were presented from a numerical investigation of pump-and-treat remediation of entrapped LNAPL in a fractured permeable formation. It

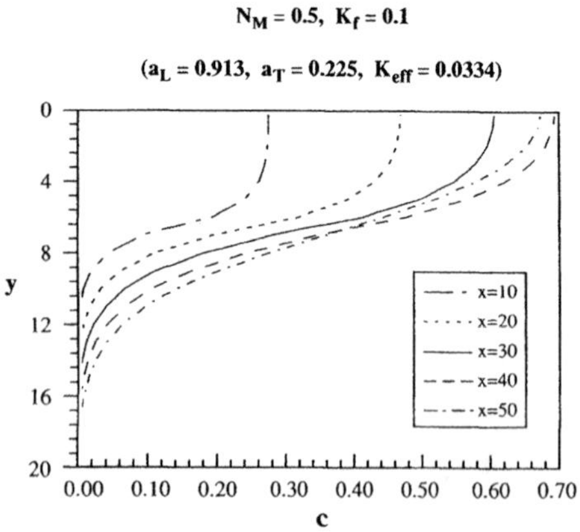

Fig. 6. Apparent steady-state concentration profiles obtained by the continuum approach ($N_M = 0.5$, $K_f = 0.1$)

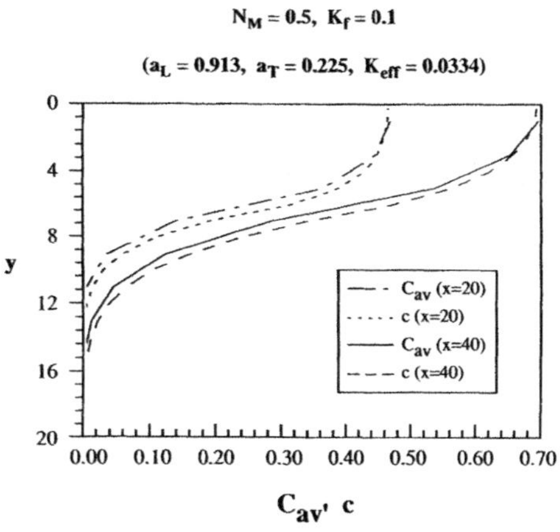

Fig. 7. Comparison between apparent steady-state profiles obtained by reference to fractured permeable formation as well as reference to a representative continuum-permeable medium

was assumed that the entrapped LNAPL is located in the permeable blocks which form the top layers of the aquifer. Reclamation of that portion of the aquifer can be accomplished by generating flow through the contaminated region using some arrangement of pumping and injection wells. The flow system is used to gradually solubilize the entrapped LNAPL and to transport the solute to recovery points. Pump-and-treat remediation was simulated in this study by assuming a steady flow through the contaminated domain.

The fractured permeable medium was represented by a conceptual model incorporating permeable blocks in which two sets of parallel, oblique, equidistant fractures are embedded. LNAPL storage in the fracture network was considered negligible and permeable block flow was assumed to control the dissolution of the entrapped LNAPL. Due to mixing between the permeable block flow and the fracture flow, the LNAPL solute is advected by both types of flow. Additionally, the fracture network can function as a conduit for conveying LNAPL solute into regions in which there is no entrapped LNAPL. This work focused on the simulation of the early behavior of pump-and-treat processes. During this period, the initial LNAPL solute present in the domain is advected to recovery points and apparent steady-state conditions are established in the domain. Within this initial stage the saturation of the entrapped LNAPL is subject to negligible changes.

Finite difference simulation results for a representative continuum scenario indicated that the characteristics of the aquifer reclamation process depend on the geometry of the contaminated region (its length and thickness), the mobility number, and the interphase mass transfer coefficient. If the mobility number is small, then the injection-pumpage arrangement leads to penetration of dissolved LNAPL into deep layers of the aquifer. The depth of the penetration of the dissolved LNAPL, however, was found to be independent of the phase transfer coefficient. This process represents a phenomenon of transverse dispersion that is fully determined by the mobility number.

If the contaminated region incorporates a large number of contaminated segments, namely blocks in the horizontal direction, then the characteristics of the reclamation process can be analyzed by the use of continuum parameters. Such parameters include longitudinal and transverse dispersivities as well as continuum phase transfer coefficients. The longitudinal and transverse dispersivities are fully determined by the mobility number. The continuum phase transfer coefficient K_{con} is determined by the mobility number N_M, and the phase transfer coefficient K_f of the fractured permeable formation. The dependence of K_{con} upon N_M and K_f is nonlinear. When K_f is very small, then K_{con} and K_f have identical values. On the other hand, for very large values of K_f, the value of K_{con} may be smaller than K_f by some orders of magnitude. Therefore, pump-and-treat procedures may need much more time and effort within fractured permeable formations than with regular porous media. Due to the presence of fractures, the application of pump-and-treat remediation may lead to penetration of LNAPL solute into deep layers of the aquifer which would otherwise remain free of contamination.

In the framework of this study, we have performed numerical experiments concerning the ability of the continuum approach to reproduce physical phenomena associated with aquifer reclamation. For a comparatively large region, contaminated with entrapped LNAPL, an adequate continuum description of the dissolved LNAPL distribution has been determined.

Notation

a	interfacial area per unit volume of porous media, L
$l_{ongitudinal}$	dimensionless longitudinal dispersivity
a_L^*	longitudinal dispersivity, L
a_T	dimensionless transverse dispersivity
a_T^*	transverse dispersivity, L
A_1,A_2	coefficient defined in Eq. (32)
B	characteristic length between adjacent fracture intersections, L
c	dimensionless continuum concentration
c^*	continuum concentration, ML^{-3}
C	dimensionless fracture flow concentration
C^*	fracture flow concentration, ML^{-3}
C_o	dimensionless concentration of reference
C_{av}	dimensionless average concentration
C_b	dimensionless concentration of the permeable block flow
a_b^*	concentration of the permeable block flow, ML^{-3}
C_s	concentration of equilibrium, ML^{-3}
g	gravitational acceleration, LT^{-2}
i	number of segment or longitudinal nodal point
i_c	number of segment contaminated with entrapped fuel
i_{max}	number of segments incorporated in the simulated domain
j	number of nodal point in the fracture sector

j_{max}	number of intervals in the fracture sector
k	number of level or vertical nodal point
k_c	number of levels contaminated with entrapped fuel
k_{max}	number of levels incorporated in the simulated domain
k_f	coefficient of mass transfer between the fuel and water phase, LT^{-1}
k_{con}	continuum coefficient of mass transfer between the fuel and water phase, LT^{-1}
K_f	dimensionless coefficient of mass transfer between the fuel and water phase
K_{fc}	critical value of K_f
K_b	hydraulic conductivity, LT^{-1}
K_{con}	continuum dimensionless coefficient of mass transfer between the fuel and water phase
K_{eff}	continuum effective dimensionless coefficient of mass transfer between the fuel and water phase
L	loss of identity length
m	number of time step
M_f	mobility of the fracture, L^2T^{-1}
n	number of iteration
N_M	mobility number
q_b	specific discharge of the permeable block flow, LT^{-1}
q_f	specific discharge of the fracture flow, LT^{-1}
Q_f	discharge of a single fracture, L^2T^{-1}
S_n	saturation of entrapped LNAPL
S_w	saturation of the water phase
t	dimensionless time
t*	time, T
v	dimensionless interstitial flow velocity in the continuum permeable medium
V	interstitial flow velocity in the continuum permeable medium, LT^{-1}
V_b	interstitial flow velocity in the permeable block, LT^{-1}
w	width of the fracture, L
x	dimensionless longitudinal coordinate
x*	longitudinal coordinate, L
x_o	reference value of x
x′	local coordinate extended along the fracture centerline, L
x_{max}	dimensionless length of the domain
y	dimensionless vertical coordinate
y*	vertical coordinate, L
y_{max}	dimensionless depth of the domain
α	coefficient defined in (38)
$\alpha_0, \alpha_1, \alpha_2, \alpha_3, \alpha_4$	coefficients defined in (48)
Δt	dimensionless time step

Δx	dimensionless longitudinal interval
Δy	dimensionless vertical interval
θ	orientation of the fractures with regard to the flow direction
ν	kinematic viscosity, L2T-1
ϕ	equivalent porosity
ϕ_b	porosity of the permeable block
ω	coefficient of over-relaxation

References

Birkhölzer J, Rubin H, Daniels H, Rouè¯ G (1993a) Contaminant advection and spreading in a fractured permeable formation 1. Parametric evaluation and analytical solution. J Hydrol 144:1–33

Birkhölzer J, Rubin H, Daniels H, Rouè¯ G (1993b) Contaminant advection and spreading in a fractured permeable formation 2. Numerical simulation. J Hydrol 14:33–52

Kanfi Y (1986) Groundwater contamination by oil in the coastal plain aquifer of Israel, Ministry of Agriculture, Water Commission, Tel Aviv, Israel

Powers SE, Loureiro CO, Abriola LM, Weber WJ Jr (1991) Theoretical study of the significance of nonequilibrium dissolution of non-aqueous phase liquids in subsurface system. Water Resour Res 27:463–477

Powers SE, Abriola LM, Weber WJ Jr (1992) An experimental investigation of NAPL dissolution in saturated subsurface systems: steady-state mass transfer rate. Water Resour Res 28:2691–2705

Rubin H, Buddemeier RW (1996) Transverse dispersion of contaminants in fractured permeable formations. J Hydrol 176:153–180

Rubin H, Mechrez E (1989) Transport of organic pollutants in a multiphase system. In: Gerstl Z, Chen Y, Minglegrin U, Yaron B (eds) Toxic organic chemicals in porous media. Springer, Berlin Heidelbery New York, pp 231–250

Rubin H, Braxein A, Daniels H, Rouv¯ G (1994) Migration of oil and petroleum pollutants in soils and groundwater. In: Zoller U (ed) Groundwater contamination and control. Marcel Dekker, New York pp 355–375

Rubin H, Rathfelder K, Abriola LM (1997a) Modeling quasisteady NAPL dissolution in fractured permeable media. ASCE J Environ Eng 123:205–216

Rubin H, Rathfelder K, Abriola LM (1997b) Simulation of NAPL dissolution and pumpage from a contaminated fractured permeable formation. (submitted)

Schwille F (1975) Groundwater pollution by mineral oil products. IAHS-AISH Publ 103:226–240

17 Three-Dimensional Modeling of Multiphase Flow and Mass Transfer in Porous Media

V. Lagendijk[1], C. Forkel[1], J. Köngeter[1], and A. Braxein[2]

Contents

1 Introduction

The number of documented aquifers contaminated by organic pollutants has rapidly increased within recent years, and it has become evident that they are one of the major threats to groundwater quality. Due to their hydrophobic nature, organic contaminants in aqueous systems primarily exist as a non-aqueous phase. This organic phase usually enters the subsurface in the unsaturated zone, where it forms a three-phase system with soil-water and soil-air. Although immiscible with water, most organic pollutants are soluble in water to an extent far beyond the threshold of drinking water standards. Since these dissolved fractions of the contaminant are much more mobile, their movement will often have the major impact

[1]Institute of Hydraulic Engineering and Water Resources Management Aachen, University of Technology, Mies-van-der-Rohe Str. 1, 52056 Aachen, Germany
[2]iwu, Ingenieurgesellschaft für Wasser und Umwelt mbH, Technologiezentrum, Jülicher Str. 336, 52070 Aachen, Germany

H. Rubin/N. Narkis/ J. Carberry
Soil and Aquifer Pollution

on environmental and water resources concern. Once the pollutant is in the subsurface, one of three different pathways may dominate further migration: movement as a separate organic phase, dissolution and subsequent transport of dissolved fractions in water, or volatilization together with gaseous organics in soil-air. Depending on soil and fluid parameters, each of these migration paths might be important in natural aquifer systems. In most cases the different paths are strongly coupled and it is not obvious which path actually dominates. Thus, only a thorough investigation of the complex interactions between all of these mechanisms can give full understanding of the aquifer contamination that derives from these NAPLs (non-aqueous phase liquids). This is essential for the selection of appropriate remediation measures and the evaluation of their efficiency.

Particularly helpful laboratory experiments on the problem of organic contaminations in porous media were performed by Schiegg (1979, 1980) and Schwille (1988). In their experimental work they investigated a wide range of conditions, in order to gain physical understanding of multiphase flow and transport behavior. For example, they studied the distribution of an organic phase in layered media and in the capillary fringe, the role of residual saturations, microscopic effects, and the role of solubilization and vaporization.

Within the last decade, several numerical models have been developed in order to study the movement of organic pollutants in aquifer systems. However, most of them reduce the physical three-phase, three-dimensional problem by some simplifying assumptions to two- or even only one-phase and two or one dimension, respectively, before solving it. Most common is the assumption of a constant gas pressure equal to the atmospheric pressure, which has been used by Abriola (1984), Abriola and Pinder (1985a,b), Kaluarachchi and Parker (1989), Faust et al. (1989), Dorgarten (1989a,b), and others. This assumption limits the applicability of these models to cases where volatilization and gas phase transport are of minor importance.

Other models have been developed especially to simulate volatile organics, but only at the cost of other restrictions. Sleep and Sykes (1989) assume the organic phase to be immobile, and simulate a two-phase system of water and gas. In contrast, Falta et al. (1989) assume that water is at residual saturation, and reduce the problem to a two-phase system of organics and air.

Only recently have models simulating real three-phase movement of organic pollutants been presented (Dorgarten and Tsang 1990; Mayer et al. 1988; Dorgarten et al. 1991). Additionally, concerning the spatial influence, Sleep and Sykes (1993) and Braxein (1994) developed three-dimensional models being capable of simulating real three-phase processes.

Another important limitation of most of these models is that the interphase mass transfer is either neglected completely or described by equilibrium approaches. It has become evident that reaction rates may have an important role in phase transfer, especially when flow rates are artificially increased by remedial measures such as pump-and-treat techniques or gas extraction. This effect has only recently been taken into account by Dorgarten (1989a,b), Sleep and Sykes

(1989, 1993), Dorgarten and Tsang (1990), Dorgarten et al. (1991) and Braxein (1994).

The three-dimensional finite element model presented here is able to simulate simultaneous flow of all three phases. Furthermore, the interphase mass transfer is taken into account by means of kinetic formulations for dissolution, volatilization and air-water partitioning. Additionally, transformation of pollutants due to adsorption and biodegradation in the solid phase is included in the model.

2 Governing Flow Equations

The problem under consideration consists of the coupled processes of multiphase flow, transport, interphase mass transfer, and reactions. Here, only a brief equation summary is given. The complete mathematical formulation of the presented model is given by Rouve and Braxein (1994) and Braxein (1994).

In a general formulation of the present multiphase problem, three separate phases are involved. These phases are a water or aqueous phase (denoted by w), an organic, oil or non-aqueous phase (denoted by o), and an air or gas phase (denoted by g). Flow of each phase in this system can be described by combining a generalized form of Darcy's law with the mass continuity equation (e.g. Bear 1970; Aziz and Settari 1979). This results in three coupled, non-linear differential equations;

$$\frac{\partial}{\partial t}(\rho_w \phi S_w) - \frac{\partial}{\partial x_i}\left[\frac{\rho_w}{\mu_w} k_{rw} k_{ij}\left(\frac{\partial p_w}{\partial x_j} + \rho_w g \frac{\partial z}{\partial x_j}\right)\right] + \rho_w Q_w + \Gamma_w = 0 \tag{1}$$

for the water phase,

$$\frac{\partial}{\partial t}(\rho_o \phi S_o) - \frac{\partial}{\partial x_i}\left[\frac{\rho_o}{\mu_o} k_{ro} k_{ij}\left(\frac{\partial p_o}{\partial x_j} + \rho_o g \frac{\partial z}{\partial x_j}\right)\right] + \rho_o Q_o + \Gamma_o = 0 \tag{2}$$

for the non-aqueous phase and

$$\frac{\partial}{\partial t}(\rho_g \phi S_g) - \frac{\partial}{\partial x_i}\left[\frac{\rho_g}{\mu_g} k_{rg} k_{ij}\left(\frac{\partial p_g}{\partial x_j} + \rho_g g \frac{\partial z}{\partial x_j}\right)\right] + \rho_g Q_g + \Gamma_g = 0 \tag{3}$$

for the gas phase. In this formulation, ρ is the density (kg/m^3), ϕ is the porosity of the porous medium (–), S is the phase saturation (–), μ is the dynamic viscosity (kg/m·s), k_r is the relative permeability factor (–), k_{ij} is the intrinsic permeability tensor of the porous medium (m^2), p is the phase pressure (kg/m·s^2), g is the gravitational constant (m/s^2), and z is the vertical spatial coordinate (m). $\rho \cdot Q$ represents external sinks and sources (kg/m^3·s), while Γ (kg/m^3·s) represents internal sinks and sources due to reactions and interphase mass transfer. Although this final term may be negligible in most cases, it will be significant, e.g. for highly volatile organics, or when residual amounts of the organic phase are diminished by dissolution or volatilization.

Equations (1) to (3) are non-linear and strongly coupled due to dependencies of saturations on capillary pressures and of the relative permeabilities on the saturations. These relationships have been proposed in the literature in several formulations. Here, the three-phase parametric model of Parker et al. (1987) is used.

3 Transport and Reactions

3.1 Water and Gas Phase Transport

Considering ecological and water resource concerns, the major issue of subsurface contamination by immiscible fluids is pollution of groundwater and soil-air by dissolved and volatilized fractions of the organic phase, respectively. Thus, transport of those fractions within the other phases is an essential part of investigation and simulating organic contamination in the subsurface.

Transport is described by conventional convection-dispersion equations, which may be formulated for partially saturated media as

$$\frac{\partial}{\partial t}\left(\phi S_w C_w\right)+\frac{\partial}{\partial x_i}\left(q_{wi} C_w\right)-\frac{\partial}{\partial x_i}\left(\phi S_w D_{wij}\frac{\partial C_w}{\partial x_j}\right)+Q_w\overline{C}_w+\Gamma_w=0 \tag{4}$$

for the water phase, and

$$\frac{\partial}{\partial t}\left(\phi S_g C_g\right)+\frac{\partial}{\partial x_i}\left(q_{gi} C_g\right)-\frac{\partial}{\partial x_i}\left(\phi S_g D_{gij}\frac{\partial C_g}{\partial x_j}\right)+Q_g\overline{C}_g+\Gamma_g=0 \tag{5}$$

for the gas phase. Here, C is the concentration (kg/m^3), q_i is the Darcian velocity (m/s), $D_{\alpha ij}$ (α = w or g, respectively) is the tensor of hydrodynamic dispersion (m^2/s), $\overline{C}$ (kg/m^3) is the concentration of the sink and source flux Q and all other parameters have been defined previously.

The tensor $D_{\alpha ij}$ covers mechanical dispersion as well as molecular diffusion. Mechanical dispersion is represented by the conventional approach as defined by Scheidegger (1961), and molecular diffusion is represented by the model of Millington and Quirk (1961). The tensor of hydrodynamic dispersion is given by

$$\phi S_\alpha D_{\alpha ij}=\delta_{ij} a_t\left|v_\alpha\right|+\left(a_l-a_t\right)\frac{v_{\alpha i}v_{\alpha j}}{\left|v_\alpha\right|}+\delta_{ij}\phi^{4/3}S_\alpha^{10/3}D_{0\alpha}, \tag{6}$$

where a_l and a_t are longitudinal and transversal dispersivities (m), D_0 is the molecular diffusivity (m^2/s) in the pure phase (water or air), and δ_{ij} is the Kronecker delta.

3.2 Interphase Mass Transfer

For low-solubility organic contaminants, a proper description of water and gas phase transport requires the consideration of interphase mass transfer. Dissolution and volatilization of the organic phase are responsible for the input of the pollutants into groundwater and soil-air. Air-water partitioning can significantly affect transport within these phases.

In the case of highly volatile organics or when residual amounts of the organic phase are diminished by dissolution or volatilization, interphase mass transfer is also relevant for the mass balance of the fluid flow equations. Such conditions are very likely when remedial measures are studied, such as gas extraction or pump-and-treat procedures.

Usually, if interphase mass transfer is not totally ignored, it is described by means of equilibrium approaches. With the presence of high flow velocities, this assumption is no longer valid. This enforces the implementation of kinetic formulations, as they have been used by Dorgarten (1989a,b) and Sleep and Sykes (1989). In the presented model, the same formulations are used as in Dorgarten and Tsang (1990), Dorgarten et al. (1991), Braxein (1994) and Rouve and Braxein (1994). The dissolution rate of the organic phase in water is given by

$$\frac{\partial C_w}{\partial t} = K_{ow}\left(\tilde{C}_w - C_w\right), \tag{7}$$

where K is a rate coefficient (s^{-1}). $\tilde{C}_w$ is the equilibrium concentration in the aqueous phase, which is usually given by the contaminant's solubility. Similarly, volatilization of the organic phase into the gas phase is described by

$$\frac{\partial C_g}{\partial t} = K_{og}\left(\tilde{C}_g - C_g\right), \tag{8}$$

where $\tilde{C}_g$ is the saturated vapor concentration. Transfer of organics between water and gas is formulated as

$$\frac{\partial C_g}{\partial t} = K_{wg}\left(HC_w - C_g\right). \tag{9}$$

Henry's law constant H (–) is the relation between gas and water phase concentrations at equilibrium, and can be derived from $H = \tilde{C}_g/\tilde{C}_w$.

3.3 Reactions

Although other reactions may take place, adsorption and biodegradation seem to be the most important ones for the problem under consideration. The adsorption introduces a fourth, non-moving phase, into the problem. This phase is the solid phase (denoted by s). For low-solubility organic contaminants, it is widely accepted that adsorption can be represented by a linear relationship between solid and water phase concentrations, which results in the kinetic formulation

$$\frac{\partial C_s}{\partial t} = K_{ws}\left(K_d C_w - C_s\right), \tag{10}$$

where C_s is the mass-related concentration in the solid phase [kg/kg], and K_d is the distribution coefficient (m^3/kg).

Karickhoff et al. (1979), Schwarzenbach and Westall (1981, 1984) and Karickhoff (1984) pointed out that the distribution coefficient K_d can be calculated as the product of the fraction of the organic carbon content f_{oc} (–) and the organic carbon partition coefficient K_{oc} (m^3/kg), where K_{oc} may be estimated from the octanol/water partition coefficient of the contaminant.

The presence of micro-organisms below the root zone and under saturated conditions has become evident during recent years. Thus, there is a potential for biodegradation, depending on the presence of nutrients, substrate, toxicants and other factors (Lee et al. 1988). Biodegradation Kinetics are mostly represented by first order decay equations for the water and the solid phase

$$\frac{\partial C_w}{\partial t} = -\lambda_w C_w \tag{11}$$

and

$$\frac{\partial C_s}{\partial t} = -\lambda_s C_s, \tag{12}$$

where λ is a rate coefficient (s^{-1}) depending on the microbial concentration, its utilization rate and many other factors.

3.4 Equation Summary of Interphase Mass Transfer and Reactions

In order to take the reactions and transfer processes of Eqs. (7) to (12) into account within the multiphase flow and transport problem, they have to be inserted into the Γ terms of Eqs. (1) to (3), and Eqs. (4) and (5), respectively. Therefore, the terms of the different processes are summed for each phase, and related to the bulk volume of the porous medium. For the water phase, e.g., summation of the dissolution and volatilization expressions yields

$$\begin{aligned} \Gamma_w = & -\phi S_w K_{ow}\left(\tilde{C}_w - C_w\right) + \phi S_g K_{wg}\left(HC_w - C_g\right) \\ & + (1-\phi)\rho_s K_{ws}\left(K_d C_w - C_s\right) + \phi S_w \lambda_w C_w . \end{aligned} \tag{13}$$

A complete formulation of all Γ terms and a discussion of the transfer and reaction parameters are given by Dorgarten and Tsang (1990) and Braxein (1994).

4 Numerical Model

The coupled multiphase flow and transport problem as defined in the previous sections is solved by the Galerkin finite element method, using linear three-dimensional tetraeder elements. In order to decrease the CPU time, the global problem is split into two sub-systems by decoupling flow and transport. The flow equations are solved primarily, with the influence of the concentrations lagged by one time step.

4.1 Three-Phase Flow

The three-phase flow problem is solved simultaneously for the fluid pressures. The strong non-linearities are treated with a Newton-Raphson iteration scheme with relaxation. By choosing fluid pressures instead of saturations, the flow equations can be solved simultaneously for the saturated and unsaturated zone without switching variables. In general, the numerical formulation follows the standard finite element formulation, and therefore only the special features are described here. For more details of the numerical code, see Braxein (1994).

4.2 Zero Saturations

Normally, only the water phase is continuously present in a multiphase system, while the NAPL phase and the gas phase exist only in certain regions. Because of this, a multiphase formulation must be found that is capable of handling up to four different combinations of phases at one time. These combinations are pure water, water-NAPL, water-gas and the complete three-phase system water-NAPL-gas.

To solve this problem, several methods can be found in the literature. Pinder and Abriola (1986) suggest working with a minimum saturation for a phase that is absent. However, this method will fail in cases of reduction of the residual saturation by dissolution or volatilization. Other methods are the reduction of the solution domain for a phase if it is absent, or switching between different primary variables, but both procedures require huge computational effort and disturb the vectorization of the code. Here another approach, developed by Dorgarten et al. (1991), is applied. The simplicity of using minimum saturations is combined with the correctness of reduced solution domains. In order to cover the whole solution domain without switching variables, all three flow equations are solved for phase pressures. If a phase is absent, its phase pressure will decrease until it reaches a certain threshold. Below this threshold the phase is not of interest and the respective nodal equation vanishes.

The problem now is to assign suitable pressures to such nodes that disturb neither mass balance nor stability, and additionally allow the phase to reenter the system. These criteria are satisfied best by a pressure just below the threshold where the phase would reappear. If the same pressure is described as a Dirichlet-type boundary condition, when the nodal equation has vanished, the linear equations can be solved without numerical problems.

4.3 Mass Conservation

An essential part of the numerical solution of differential equations with non-linear accumulation terms, is to develop a fully mass-conservative scheme. For the unsaturated flow problem Celia et al. (1990) showed that the standard h-based (or p-based) form of the Richard's equation may lead to serious mass balance errors, while by selecting a mixed form of the equation an accurate solution can be achieved.

The same consideration can be applied to the multiphase flow problem. In a p-based formulation, the time derivatives would be written as

$$\frac{\partial S_\alpha}{\partial t} = \sum_\nu \frac{\partial S_\alpha}{\partial p_\nu} \cdot \frac{p_\nu^{t+\Delta t} - p_\nu^t}{\Delta t} \quad (\nu = w, o, g), \tag{14}$$

where $\partial S_\alpha/\partial p_\nu$ are the fluid capacities (s^2/kg·m), which are used in a similar way in most multiphase flow models (e.g. Abriola and Pinder 1985a; Kaluarachchi and Parker 1989). This formulation is equivalent to a linearization of the S–p relations within each timestep Δt, which results in numerical "loss" or "production" of mass. In the mixed form, in contrast, the time derivative is written in terms of saturations

$$\frac{\partial S_\alpha}{\partial t} = \frac{S_\alpha^{t+\Delta t} - S_\alpha^t}{\Delta t}, \tag{15}$$

which means that differentiation of the S-p relations and introduction of the fluid capacities is avoided. Now a linearization is not required between the old and new time levels, but only between old and new iteration levels. Thus, the mass balance error becomes zero when the iteration scheme approaches its solution.

4.4 Convergence and Stability

Again the results of Celia et al. (1990) from unsaturated flow simulations are extended to the multiphase flow problem. They found that, in addition to using the mixed form of Richard's equation, diagonalized (lumped) mass matrices help to improve the numerical solution. Simulations with the present multiphase code indicated the same behavior. Consistent (distributed) mass matrices led to large numbers of iterations and, in addition, resulted in serious undershooting. With mass lumping in contrast, convergence was achieved much faster, and no undershooting occurred.

4.5 Transport, Mass Transfer and Reactions

The two transport Eqs. (4) and (5) for the water and gas phases, respectively, are solved simultaneously, with mass transfer and reaction terms taken into account. The solid phase equation is treated by an efficient algorithm described by Dorgarten (1989b) that allows its solution without additional matrix equations.

5 Model Verification and Application

The model was verified against several analytical and semi-analytical solutions for flow and transport. Out of these many verifications, a comparison is presented here of the numerical code against the two-dimensional semi-analytical solution for two-phase flow and against the analytical solution for nonconservative transport.

The model has also been applied to some two-phase and three-phase studies. Here, two different studies have been chosen to demonstrate the applicability of the three-dimensional numerical model which has been developed.

5.1 Verification Examples

One flow verification example is the transient displacement of oil by water in a horizontal one-dimensional water-oil system. With the comparison of the semi-analytical solution of Kueper and McWorther (1987) and the numerical results, the following features of the code can be verified:

1. Multiphase flow.
2. Influence of capillary pressure and pressure-saturation relationship.
3. Extreme saturations (totally oil or water saturated).

Figure 1 shows the structure of the verification example and the discretization used. The chosen parameters can be taken from Table 1.

Figure 2 shows the comparison of the results of Kueper and McWorther (1987) and the numerical results with a boundary condition on the left side for water saturation equal to unity. In general, the agreement is quite good, considering that the numerical solution is not capable of simulating zero saturations for the water phase. This can be attributed to the fact that the pressure saturation relationship

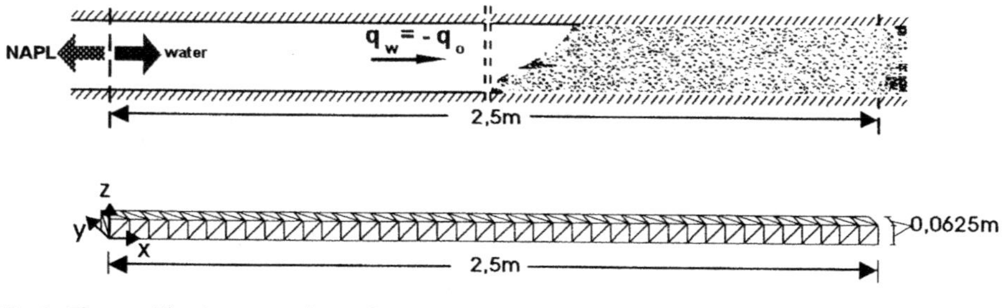

Fig. 1. Flow verification example configuration

Table 1. Parameters used for the flow verification sample

Parameter	Value	Unit
Permeability k	10^{-10}	m^2
Porosity n	0.4	–
Viscosity water μ_w	10^{-3}	kg m^{-1} s^{-1}
Viscosity oil μ_o	10^{-3}	kg m^{-1} s^{-1}

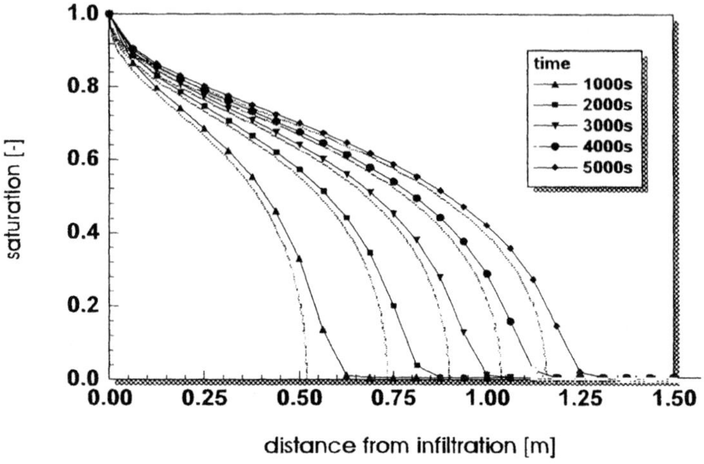

Fig. 2. Comparison of the quasi-analytical solution (*dashed lines*) with the numerical results for the multiphase flow verification simulation (*symbols*)

Table 2. Parameters used for the transport verification sample

Parameter	Value	Unit
Porosity n	0.3	–
Effective porosity n_{eff}	0.3	–
Darcian velocity q_f	3.94×10^{-5}	m/s
Longitudinal dispersion coefficient α_l	0.05	m
Degradation constant λ_w	0.05×10^{-3}	1/s

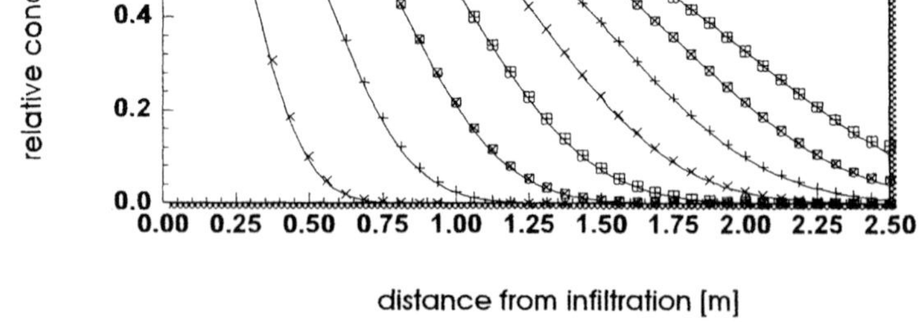

Fig. 3. Comparison of the analytical solution (*solid lines*) with the numerical results for a transport simulation (*symbols*)

has an asymptotic behavior in this region. This is the reason for the faster movement of the whole water front within the numerical results as well as for the problems at the sharp edge of the front.

Several additional flow verification examples have been simulated and in all cases a good agreement between quasi-analytical and numerical solutions was achieved. The results showed neither stability nor convergence problems and even extreme pressure gradients could easily be simulated with the new code.

The mass transport and transfer has been verified by comparing the numerical solution of a contamination within the same geometry as in the flow example (Fig. 1), infiltrating from the left to the right. The chosen analytical solution was evaluated by Ogata and Banks (1961) with the modified approach of van Genuchten (1981). This one-dimensional solution includes advection, dispersion, biodegradation and adsorption. The parameters used are as shown in Table 2.

Figure 3 illustrates the perfect agreement between the numerical and analytical solution, which could also be achieved in computations with other material parameters as well as in the gas phase.

5.2 Transport of TCE in the Water and Gas Phase

In this first application, the movement of trichlorethylene (TCE) in an unsaturated soil has been simulated. A system drawing of the studied domain is given in Fig. 4, together with the mesh geometry. The discretization is based on tetrahedral elements with 4662 nodes and 13200 elements. The described problem is of two-dimensional character.

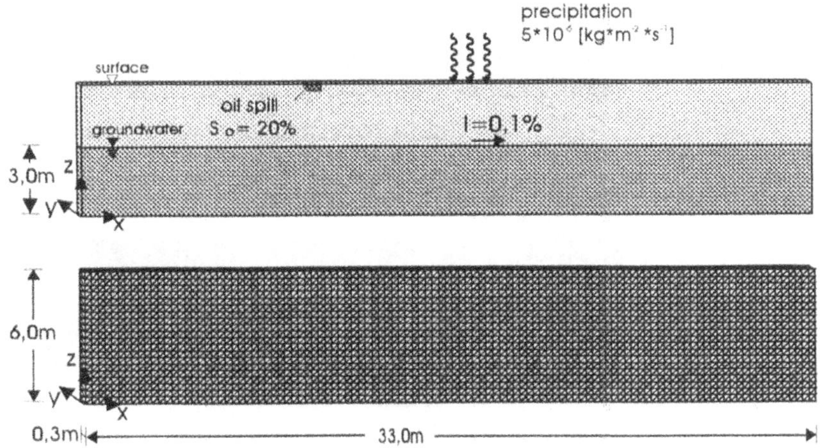

Fig. 4. Model area and boundary conditions for TCE simulations

The data used is documented in Braxein (1994), based on the study of Dorgarten and Tsang (1990). As far as possible the data reflect real values taken from numerous references. Soil properties are chosen to represent a coarse sandy aquifer.

The application of the presented model has been focused on the study of gas phase phenomena. Thus, TCE input is not simulated as inflow of a separate organic phase, but by dissolution and volatilization of a non-moving organic phase. The initial conditions for TCE were set to its residual saturation, assumed to be in this case at the value of 0.2 within an area of 0.5 m^2 at the ground surface, resulting in a total TCE mass of 58.4 kg.

Figure 5 illustrates the results for the reference simulation. In this simulation all transfer-processes in between the phases are possible. Water and gas phase concentrations are shown after a simulation period of 1 year. The distribution of the contaminant appears mainly in the unsaturated zone.

To represent the different distribution mechanisms, an additional simulation has been computed. In this simulation the TCE distribution in water and gas is decoupled. When K_{wg} equals zero, mass transfer between the water and gas phases is eliminated. The results for this simulation of the concentration in the water and gas phase after 1 year can be taken from Fig. 6.

In comparison to the reference simulation, the distribution in the gas phase has been faster. This effect can be attributed to the fact that the contaminant in the gas phase is mainly influenced by diffusion and convective transport caused by density driven flow. In contrast to this behavior, the distribution velocity within the seepage water decreased. The alignment between both concentrations in the reference simulation can be explained by the TCE transfer from gas to water.

Introducing a further simulation, no. 2, without transfer into the gas phase, i.e. neglecting transport within the gas phase at all, the results are as illustrated in Fig.

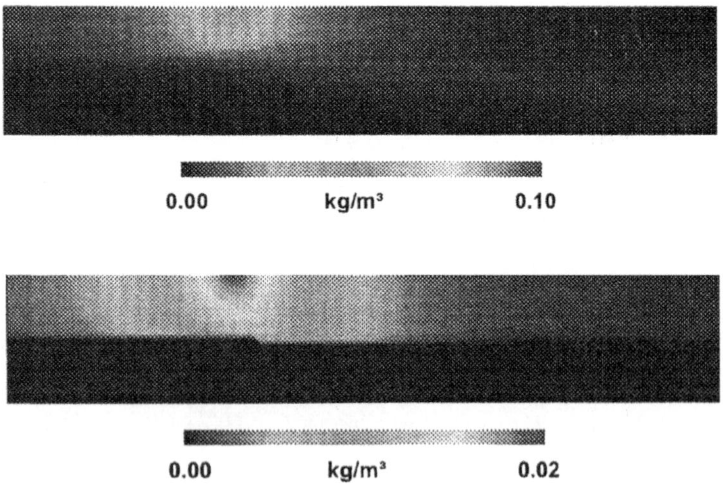

Fig. 5. TCE concentrations [kg/m^3] in water (*top*) and gas (*bottom*) after 1 year (reference simulation)

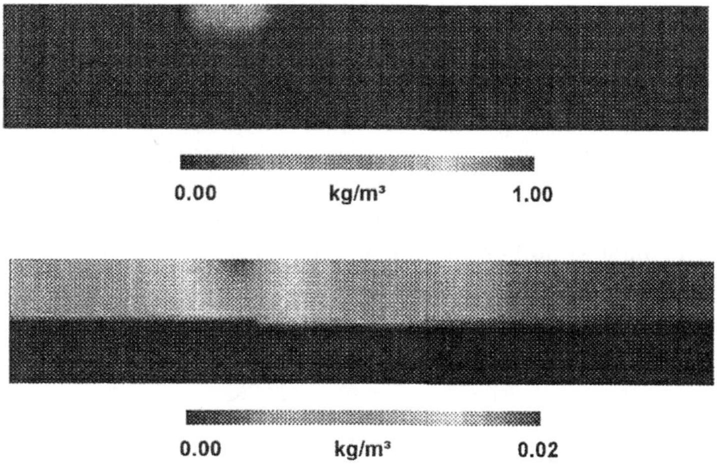

Fig. 6. TCE concentrations [kg/m^3] in water (*top*) and gas (*bottom*) after 1 year (simulation no. 1)

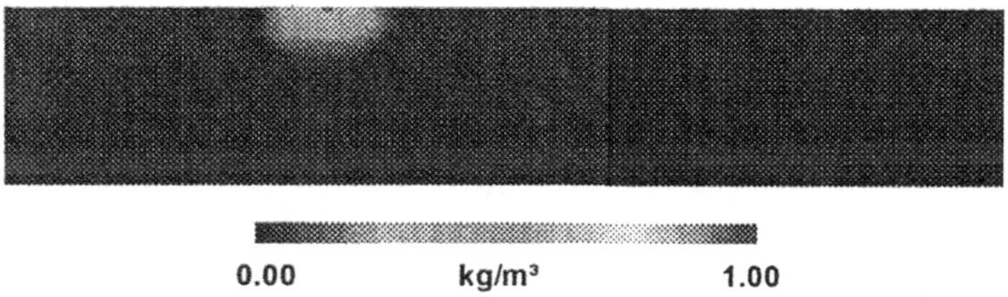

Fig. 7. TCE concentrations [kg/m^3] in water after 1 year (simulation no. 2)

7. The concentrations of TCE in the water phase are similar to those of simulation no.1 with slightly higher local concentrations. These higher concentrations result from the fact that the TCE can only dissolve in the water phase.

The previous simulations and additional simulations showed that the pathway of TCE from the infiltration site into the groundwater consists of the following steps:

1. Flow as a separate organic phase.
2. Volatilization of the organic phase.
3. Diffusive transport into the gas phase and density-induced flow in the gas phase.
4. Phase transfer from the gas phase to the groundwater within the capillary zone.
5. Reactive transport in the groundwater.

This complex mechanism demonstrates the need for the coupled simulation of multiphase flow, water and gas phase transport, interphase mass transfer and reactions.

5.3 NAPL Movement in a Realistic Aquifer

Finally, the model was applied to study the movement of two different NAPLs, one heavier and one lighter than water, in a hypothetical, yet realistic aquifer. Figure 8 shows the area examined and the boundary conditions for both simulations. The mesh geometry consists of 3751 nodes and 18000 tetrahedral elements.

In the beginning of the simulation, the whole area is free from contamination. The initial conditions for the different phase pressures have been calculated with

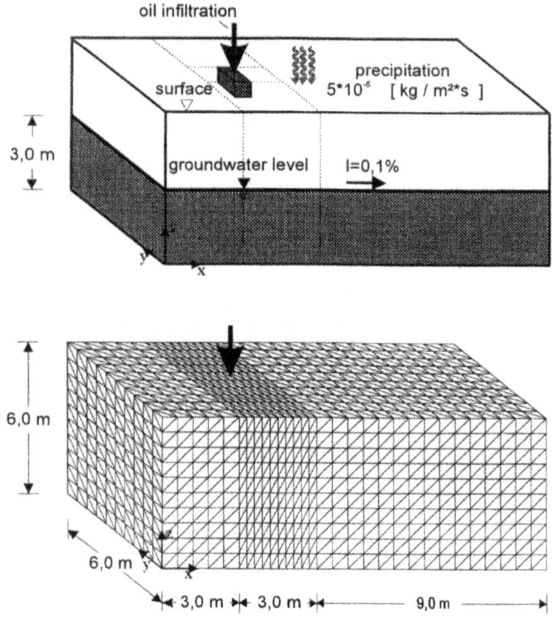

Fig. 8. Model area and boundary conditions

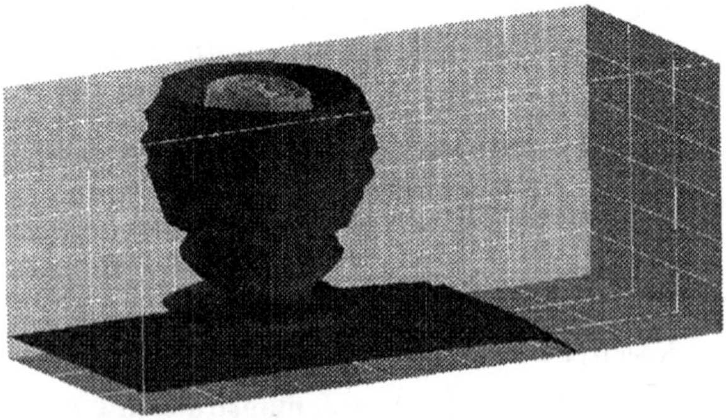

Fig. 9. Isosurfaces of NAPL saturation after t = 200 000 s, values from the inside to the outside: 0.45; 0.35; 0.25; 0.15; 0.05

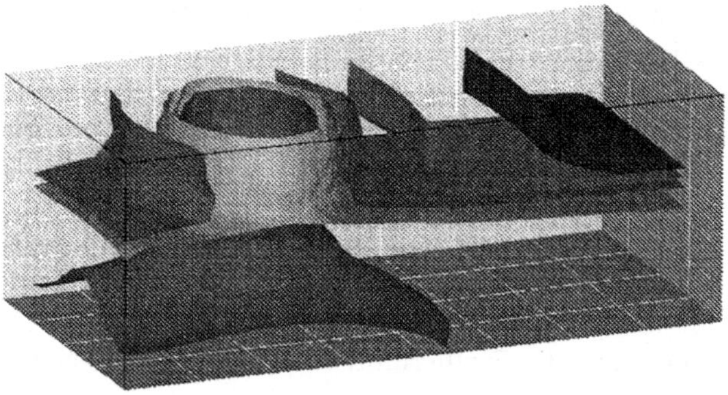

Fig. 10. Isosurfaces of the NAPL concentration in the water phase [kg/m^3] after t = 100 days, values from the inside to the outside: 1.0; 0.8; 0.6; 0.4; 0.2

the restriction of a residual saturation for water of $S_{rw} = 0.2$. In the first part of our study we focused on the spreading of the NAPL phase itself. No mass transfer between the phases was considered. A continuous disposal of NAPL was assumed at the top of the model area. Figure 9 shows saturation isosurfaces for a NAPL that is heavier than water ($\rho_0 = 1460\,\mathrm{kg\,m^{-3}}$), after 200 000 s.

The spreading of the NAPL is strongly three-dimensional and the NAPL accumulates at the bottom of the aquifer.

In a second step, the mass transfer between the phases was taken into account, and the transport of contaminants within the gas and water phases was examined. The simulation time was increased up to 100 days, due to the fact that mass transfer processes are much slower than flow processes. The initial conditions for the saturation have been taken from the results of the first simulation step.

Figure 10 shows the contaminant concentration solved in the water phase. The contaminated area is much larger than the extension of the NAPL phase itself. Furthermore, the spreading of the contaminant differs between the saturated and nonsaturated zone. In the saturated zone, the concentrations are much higher, sometimes up to the equilibrium concentration. This can be put down to the fact that the flow of water is faster in the saturated zone.

This study shows that for both NAPLs, the contamination of an aquifer is mainly dominated by the interphase mass transfer and the transport of the dissolved and volatilized fractions in the water and gas phases. The contamination by the NAPL itself is limited to a small fraction of the aquifer adjacent to the contamination source.

6 Summary

A three-dimensional finite element model is presented for the simulation of coupled three-phase flow and transport problems. After a detailed description of the physical processes and their mathematical formulation, the special numerical features of the model for mass conservation, convergence, stability and the simulation of zero saturations are presented. An extract of the verification is shown. Different applications of the model show that the mass transfer kinetics are of significant importance, especially for the correct prediction of mass transport within the gas phase.

The results show that the contamination of an aquifer is mainly dominated by the interphase mass transfer phenomena and the transport of dissolved and volatilized fractions in the water and gas phases. The developed finite element model is an excellent tool to predict the contamination of an aquifer caused by organic contaminants. Furthermore, it is possible to simulate and optimize different countermeasures for reclamation.

Acknowledgments. A major part of this work was made possible by the Deutsche Forschungsgemeinschaft (DFG), Bonn, Federal Republic of Germany under contract No. Ro 365/41-2.

References

Abriola LM (1984) Multiphase migration of organic compounds in a porous medium, a mathematical model. Springer, Berlin Heidelberg New York

Abriola LM, Pinder GF (1985a) A Multiphase approach to the modeling of porous media contamination by organic compounds, 1. Equation development. Water Resour Res 21:11–18

Abriola LM, Pinder GF (1985b) A multiphase approach to the modeling of porous media contamination by organic compounds, 2. Numerical simulation. Water Resour Res 21:19–26

Aziz K, Settari A (1979) Petroleum reservoir simulation. Applied Science Publishers, London

Bear J (1970) Two-liquid flows in porous media. In: Chow VT (ed) Advances in hydroscience, vol 6. Academic Press, New York, pp 141–252

Braxein A (1994) Dreidimensionale Modellierung von Mehrphasenströmungs- und Stofftransportvorgängen in porösen Medien. Mitteilung des Instituts für Wasserbau und Wasserwirtschaft, RWTH Aachen, Heft 95

Celia MA, Bouloutas ET, Zabra RL (1990) A general mass-conservative numerical solution for the unsaturated flow equation. Water Resour Res 26:1483–1496

Dorgarten H-W (1989a) Das Verhalten hydrophober Stoffe in Boden und Grundwasser. Mitteilung des Instituts für Wasserbau und Wasserwirtschaft, RWTH Aachen, Heft 73

Dorgarten H-W (1989b) Finite element simulation of immiscible slightly soluble pollutants in soil and groundwater. In: Kobus HE, Kinzelbach W (eds) Proc Int Symp on contaminant transport in groundwater, Stuttgart. AA Balkema, Rotterdam/Brookfield, pp 389–396

Dorgarten H-W, Tsang C-F (1990) Three-phase simulation of organic contaminations in aquifer systems. In: Weyer KU (ed) Proc Conf on Subsurface contamination by immiscible fluids, Calgary, Alberta, Canada

Dorgarten H-W, Braxein A, Rouve G (1991) Multiphase flow and transport modeling of surface contamination by immiscible fluids. In: Ben Sari D, Brebbid CA, Duazdr D (eds) Computer Methods in Water Research II, vol 1. Groundwater and pressure flow, Proc 2nd Int Conf Marrakesh, Marocco, 1991. Springer, Berlin Heidelberg New York, pp 129–140

Falta RW, Javandel I, Pruess K, Witherspoon PA (1989) Density-driven flow of gas in the unsaturated zone due to the evaporation of volatile organic compounds. Water Resour Res 25:2159–2169

Faust CR, Guswa JH, Mercer JW (1989) Simulation of three-dimensional flow of immiscible fluids within and below the unsaturated zone. Water Resour Res 25:2449–2464

Karickhoff SW (1984) Organic pollutant Sorption in aquatic systems. J Hydraul Eng 110:707–735

Kaluarachchi JJ, Parker JC (1989) An efficient finite element method for modeling multiphase flow. Water Resour Res 25:43–54
Karickhoff SW, Brown DS, Scott TA (1979) Sorption of hydrophobic pollutants on natural sediments. Water Res 13:241–248
Kueper BH, McWorther DB (1987) Mechanics and mathematics of immiscible fluids in porous media. Univ of Waterloo, Waterloo, Ontario (preprint)
Lee MD, Thomas JM, Borden RC, Bedient PB, Wilson CH, Ward CH (1988) Biorestoration of aquifers contaminated with organic compounds. CRC Crit Rev Environ Control 18:29–89
Mayer AS, Miller CT, McNeill MM (1988) Mass transfer from non-aqueous phase liquids to the aqueous phase in groundwater systems. Eos 69:1189
Milllington RJ, Quirk JP (1961) Permeability of porous solids. Dept of Agronomy and Agricultural Chemistry. Trans Faraday Soc 57:1200–1207
Ogata A, Banks RB (1961) A solution of the differential equation of longitudinal dispersion in porous media. US Geol Survey, Prof Pap 411-A
Parker JC, Lenhard RJ, Kuppusamy T (1987) A parametric model for constitutive properties governing multiphase flow in porous media. Water Resour Res 23: 618–624
Pinder GF, Abriola LM (1986) On the simulation of nonaqueous phase organic compounds in the subsurface. Water Resour Res 22:109–119
Rouve G, Braxein A (1994) Simulation der Strömung in Dreiphasensystemen unter Berücksichtigung veränderlicher Durchlässigkeiten. Unpublished Research Report, Institut für Wasserbau und Wasserwirtschaft, RWTH Aachen, Ro 365/41-2
Scheidegger AE (1961) General theory of dispersion in porous media. J Geophys Res 66:3273–3278
Schiegg HO (1979) Verdrängungssimulation dreier nicht mischbarer Fluide in poröser Matrix. Mitteilung der Versuchsanstalt für Wasserbau, Hydrologie und Glaziologie, ETH Zürich, Switzerland, No 40
Schiegg HO (1980) Grundlagen, Aufbau und Resultate von Laborexperimenten zur Erforschung von Ölausbreitungen in Grundwasserträgern. Mitteilung der Versuchsanstalt für Wasserbau, Hydrologie und Glaziologie, ETH Zürich, Switzerland, No 43
Schwarzenbach RP, Westall J (1981) Transport of nonpolar organic compounds from surface water to groundwater. Laboratory sorption studies. Environ Sci Technol 15(11):1360–1367
Schwarzenbach RP, Westall J (1984) Sorption of hydrophobic trace organics in groundwater systems. LAWPRC Seminar on Degradation, retention and dispersion of pollutants in groundwater, Kopenhagen, Denmark
Schwille F (1988) Dense chlorinated solvents in porous and fractured media. Lewis, Chelsea, Michigan, 146 pp
Sleep BE, Sykes JF (1989) Modeling the transport of volatile organics in variably saturated media. Water Resour Res 25:81–92
Sleep BE, Sykes JF (1993) Compositional simulation of groundwater contamination by organic compounds, 2. Model applications. Water Resour Res 29(6):1709–1718
Van Genuchten M-TH (1981) Analytical solutions for chemical transport with simultaneous adsorption, zero-order production and first-order decay. J Hydrol 49:213–233
Zilliox L, Muntzer P, Schwille F (1974) Untersuchungen über den Stoffaustausch zwischen Mineralöl und Wasser in porösen Medien. Dtsch Gewasserkd Mitt 18(2):35–37

18 Problems of Flow Through Fractured Rock Formations Related to Contamination of Aquifers

D. G. Berger[1] and C. Braester[2]

Contents

1 Introduction

The problem investigated here is that of pollution due to oil spilled on the ground surface, which flows through the unsaturated zone and into the aquifer in fractured rock formations.

Fractured rocks are pervious or impervious formations cut by a fracture network. Geologists refer to fractured rock formations with impervious blocks as hard rocks or crystalline rock formations; fractured rocks with pervious blocks are called fractured rocks of double porosity (Barenblatt et al. 1960). Permeability of porous blocks is negligible in comparison to that of the fracture network, so that in both types of formations, the permeability of the formation is represented by that of the fracture network. The overall porosity of a double porosity rock formation is represented by that of the porous blocks, since the fracture network has a negligible porosity in comparison to that of the blocks.

Fractured aquifers are not potential water reservoirs for water supply; however, pollution of such aquifers should be avoided when there is contact with reservoir aquifers, to which pollutants may be conveyed (Fig. 1).

One of the main governing factors in multiphase flow is the wettability of the solid by the couples of fluids. In natural rock formations, oil is the non-wetting liquid in the presence of water, and the wetting liquid in the presence of air.

In the unsaturated zone of a deep water table, the flow is two-phase oil-air; in the case of a shallow water table, the flow is three-phase oil-air-water.

[1] Mekoroth Watershed Unit, Jordan District, Nazareth-Elith 17105, P.O. Box 610, Israel
[2] Technion – Israel Institute of Technology, Faculty of Civil Engineering Haifa 32000, Israel

H. Rubin/N. Narkis/ J. Carberry
Soil and Aquifer Pollution

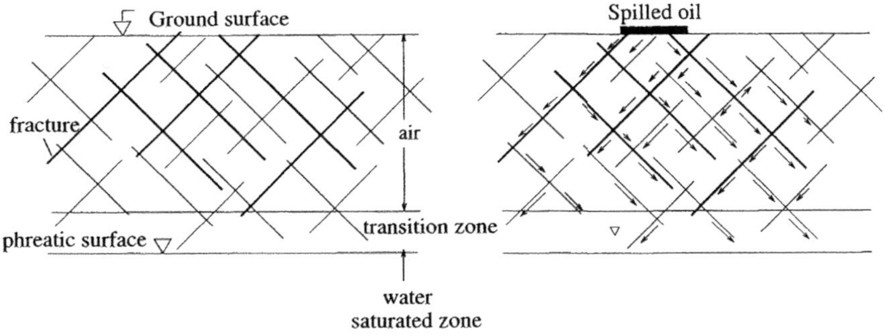

Fig. 1. Flow of oil spilled on the ground surface of a fractured rock formation

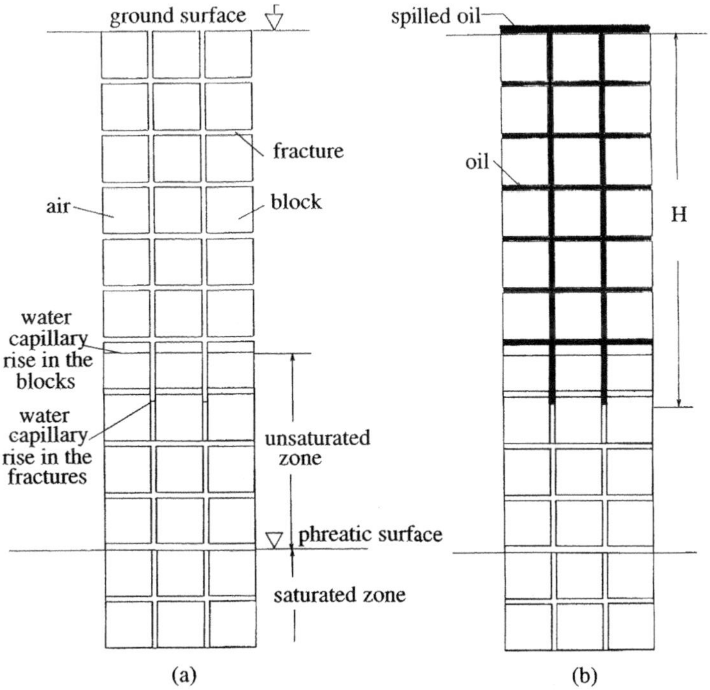

Fig. 2. Schematic representation of water capillary rise in fractures and porous blocks (**a**), and the depth *H* of oil penetration (**b**)

After oil reaches the water-saturated aquifer, the flow is two-phase, oil-water displacement.

In the air-saturated region, from the ground surface down to the transition air-water zone, in both fractures and porous blocks, oil displaces air by imbibition. Due to the low permeability of the blocks, the rate of oil penetration into the blocks will be low, and the oil will rapidly reach the water table of the saturated aquifer through the high permeability fracture network.

In the water-saturated aquifer, oil-water displacement is a drainage process. As is well known, displacement of a wetting fluid by a non-wetting one requires a threshold pressure or an *entry value*. The entry value is related to the permeability of the medium; the lower the permeability, the higher the entry value. Thus, it is expected that the blocks of a fractured rock of double porosity, and of a very low permeability, have a very high entry value. The maximum pressure that the spilled oil may reach is related to the oil column H (as shown in Fig. 2), the length between the ground surface and the water table of the saturated aquifer provided that the oil column is continuous. This pressure may be approximated by the hydrostatic pressure value $\rho_o gH$, where ρ_o is oil density, and g is the acceleration of gravity. If this oil pressure is lower than the required entry value for displacing the water from the blocks, the water-saturated blocks will behave like impervious blocks, and there will be no difference between the behavior of a hard rock and that of a double porosity rock formation.

2 Continuum vs. Discrete Approach

Two basic approaches may be considered when solving problems of flow through fractured rock formations: (1) the discrete approach and (2) the continuum approach.

Treatment of flow problems through the discrete fractures requires a detailed description of the fracture network geometry, and of the geometrical and hydraulic properties of the individual fractures. Obviously, such a detailed description, similar to the description of an ordinary porous medium at the scale of the pores seems practically impossible.

It would seem that the continuum approach, which circumvents the need for the discrete fracture representation, is the appropriate method for investigation of flow problems in such media. The continuum approach is the classic means of investigating complex systems for which details on the constituents are not accessible. The fictitious continuum is a macroscale representation using average properties, and as such there is an associated loss of information.

The continuum approach requires the possibility of defining average media properties over a representative elementary volume (REV). In most cases an REV of a fractured rock flow domain cannot be defined; even when it can be defined, it is expected that the resulting continuum will exhibit a large degree of heterogeneity and anisotropy, which implies a large number of experiments over the investigated domain. Although formulation of the governing equations of flow and transport through the continuum approach is possible, there is no practical way of determining the continuum properties through field and laboratory experiments.

As will be shown, except for one-phase mass flow through large flow domains, flow and transport through fractured rock formations exhibit different behavior from that seen in granular porous formations. One example is the shape of a breakthrough curve C/C_o vs. time in a hard rock formation, i.e., a fractured rock

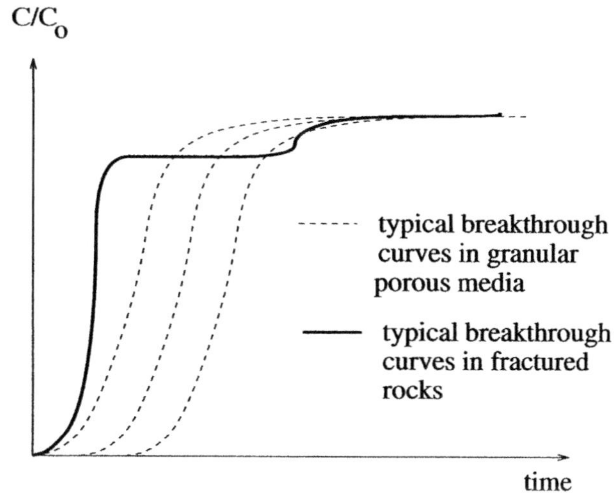

Fig. 3. Typical breakthrough curves

with impervious blocks. Experiments carried out on granite show typical breakthrough curves as presented in Fig. 3. These breakthrough curves are steeper than the ones exhibited by porous media, present one or several jumps, and have a longer tail. Nowadays the concept of channelling flow replaces the concept of flow through fracture planes. According to this concept, contact surface between the blocks creates channels of variable width, which may be considered an interconnected network of tubes and pockets. The channelling concept may explain the behavior of the breakthrough curve in fractured rock formations. The steepest shape may be attributed to the relatively fast breakthrough through fractures of larger permeability; the different jumps may be attributed to the subsequent breakthrough through different fractures of lower permeability, and the long tail may be explained by the existence of pockets, low velocity regions, which may supply a pollutant during a relatively long period of time. Such a shape of a breakthrough curve cannot be fitted by the typical shapes exhibited by granular porous media shown in Fig. 3; however, they may be fitted by a model of channels with pockets (Berkowitz and Braester 1991). The same shape of the breakthrough curve is expected to be obtained in a fracture formation of double porosity, where the effect of the pockets will be enhanced by the pervious blocks.

A second example is the shape of the drawdown curve when pumping at a constant rate of flow from a fractured rock formation with porous blocks. Fractured rock formations of double porosity are characterized by a permeability represented by the medium of the fractures, and a porosity represented by the medium of the blocks. Thus, the fluid stored mainly in the blocks flows through the medium of the fractures. One-phase flow of a slightly compressible fluid through fractured rocks of double porosity was formulated through the continuum

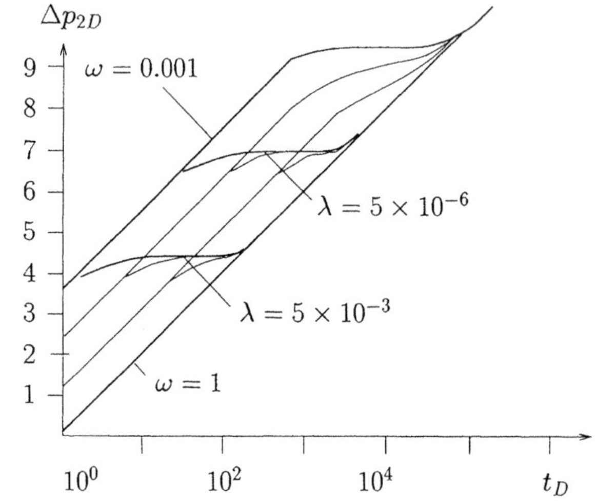

Fig. 4. Graphic representation of the Warren and Root solution

approach by Barenblatt et al. (1960). The medium of the fractures and the medium of the blocks are regarded as two overlapping continua, with a source term in the equations of conservation of mass, representing the transfer of fluids between the two media, fractures and blocks.

Warren and Root (1963) present a solution of these equations for pumping at constant rate of flow from an infinite homogeneous and isotropic aquifer. The solution obtained, presented graphically in Fig. 4, contains, in addition to the parameters of granular aquifer formations, two characteristic parameters of the fractured formation of double porosity, λ and ω.

The parameter $\lambda = sk_1/k_2$ in Fig. 4 characterizes the transition time from the first straight line to the second, and it is related to the permeabilities of the blocks (k_1), to the permeability of the fractures (k_2), and to the specific surface of the blocks through the coefficient s. The expression $\omega = c_1\phi_1/(c_1\phi_1 + c_2\phi_2)$ characterizes the displacement between the two straight lines and is related to the porosities (ϕ) and to the compressibilities (c) of the two media (subscript 1 for the blocks and 2 for the fractures).

Braester (1984) presented results of numerical solutions for two media of double porosity in which the second one is obtained from the first one by dividing the blocks into two equal parts, and diminishing the distance between the fracture by half. This way, one obtains two double porosity media of equal fracture permeability and different specific surfaces of the blocks. The problem was solved under identical initial and boundary conditions as considered by Warren and Root (1963), using the formation and fluid properties considered by Kazemi (1969). The results shown in Fig. 5 indicate that for this set of parameter values, the solution is insensitive to the coefficient s related to the specific surface of the block.

Braester and Zeitoun (1993) solved Barenblatt's equations in which the permeability of the fractures was considered a stochastic parameter. Solving the problem

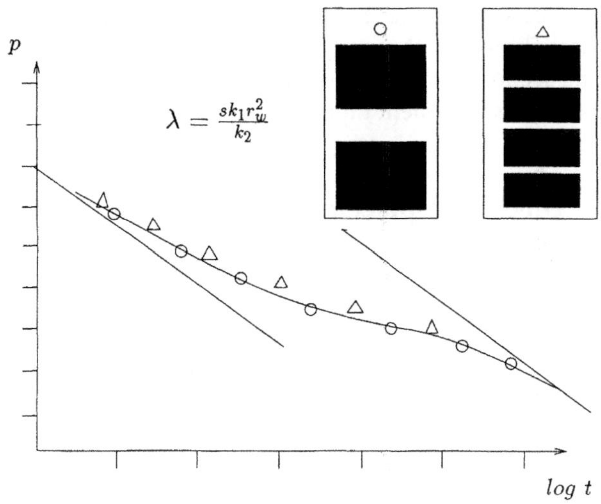

Fig. 5. Insensitivity of the Warren and Root solution to the block size

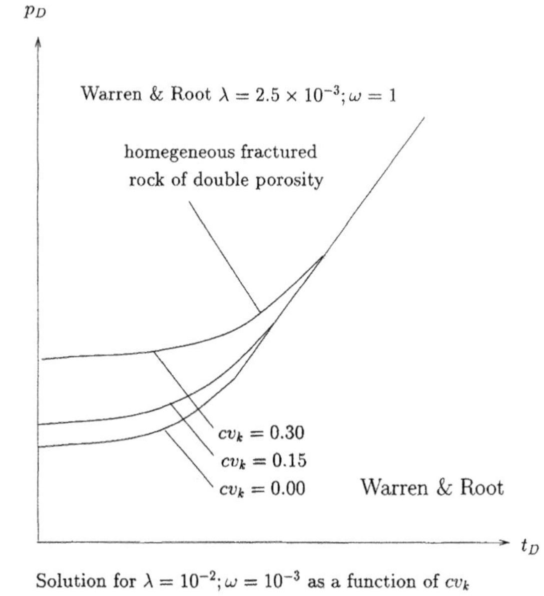

Fig. 6. Identical solutions for inhomogeneous and homogeneous fractured formations

of flow to a well, under identical initial and boundary conditions as considered by Warren and Root, they obtained a similar behavior as for a homogeneous medium, as shown in Fig. 6. Using the Warren and Root solution, one obtains λ and ω values different from those corresponding to the actual formation.

These examples show that the continuum approach may not be satisfactory in describing flow through fractured rock formations. As will be shown in the following section, the continuum approach is especially inadequate when studying two-phase displacement liquid-gas in the fracture network of a fractured rock.

3 Oil-Air Displacement in Fractured Rocks

Figures 7 and 8 are typical junctions of fracture networks shown in Fig. 1. Figure 7 shows two fractures of different apertures and equal inclination, converging into a single one. Due to the different apertures, oil spilled on the ground surface will advance faster through the fracture of a larger aperture, through which, according to Poiseuille's Law, the velocity is higher. For the same reason, unequal advance will take place in the junction shown in Fig. 8a. When the displacement takes place in a plane different from the horizontal, as shown in Fig. 8b, the gravity force component may lead to unequal advance of the liquid even in fractures of equal aperture but of different inclination.

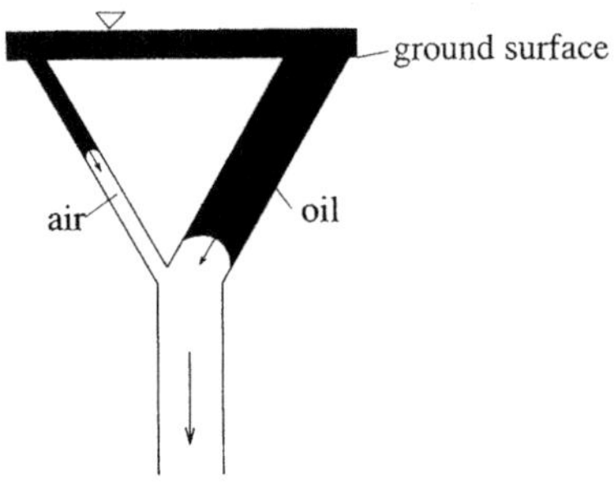

Fig. 7. Unequal advance of oil in fractures of different width

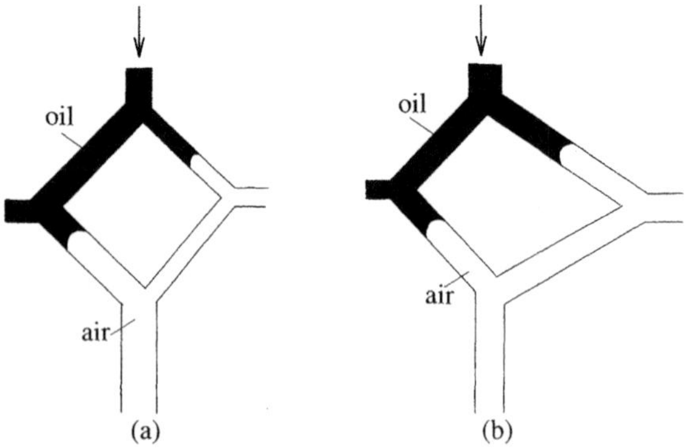

Fig. 8. Unequal oil advance: **(a)** in fractures of the same inclination and different width, and **(b)** in fractures of the same width but different inclinations

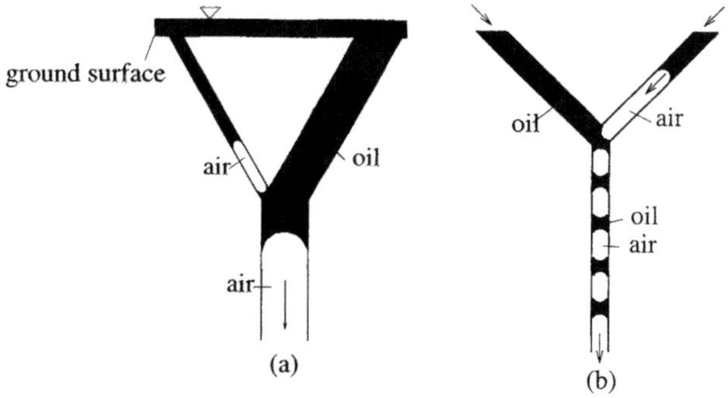

Fig. 9. Trapped air in the thinner fracture (**a**), and oil and air reaching a fracture junction simultaneously (**b**)

Because of the unequal advance of the oil in the different fractures, air may be trapped in one of the fractures (Fig. 9). In a fracture of constant aperture, the trapped air will move upwards and finally escape at the ground surface; however, in a fracture of variable aperture, as occurs in reality, the capillary forces may prevent the upward movement of the air, and thus the air will remain trapped.

When air and oil reach a junction simultaneously, as shown in Fig. 9b, they will intermittently enter the fracture following the junction.

To study the oil-air displacement in fracture networks, one should formulate the mathematical flow model, and solve the equations for the actual fracture pattern.

4 Model Equations

Following field and laboratory experiments, the concept of channeling flow is now accepted; in this study the network of interconnected channels of variable sections are conceptualized as a network of capillary tubes, each tube of different length and different circular cross section. In the capillary tube network, the flow is three-dimensional, while in the individual capillary tube the flow is one-dimensional.

A sharp interface is assumed to separate the two phases in a capillary tube. Thus at every cross section only one fluid phase is present.

Averaged expressions of the momentum and of the continuity equations for flow of a compressible fluid through a conduit of circular cross section are:

The Momentum Equation:

$$\frac{\partial u}{\partial t}+u\frac{\partial u}{\partial s}+\frac{1}{\rho}\frac{\partial p}{\partial s}-g\frac{dz}{ds}+\frac{fu^2}{2d}=0, \tag{1}$$

where z is the vertical coordinate positive downward, s is a coordinate along the capillary tube, d is the tube diameter and u and p are average velocity and pressure respectively, in a cross-section. Considering that $\frac{dz}{ds} = \cos\theta$, with θ, the angle between the fracture and the vertical axis z, we obtain:

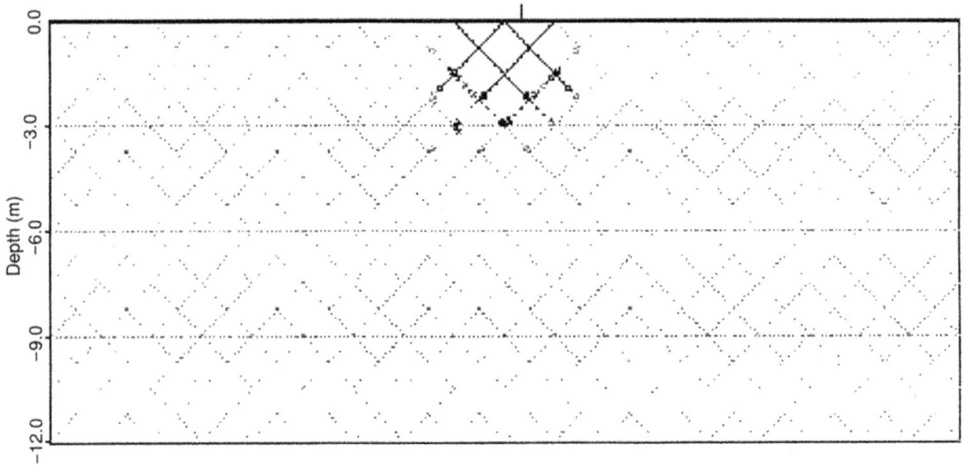

Fig. 10. The spilled oil advance through the fracture network at time = 0.41 h from the beginning of infiltration

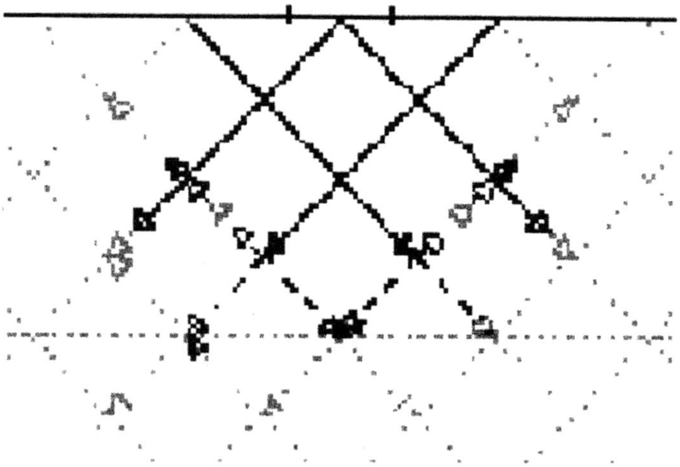

Fig. 11. The spilled oil advance through the fracture network at time = 0.41 h from the beginning of infiltration – a window through the upper region of Fig. 10

$$\frac{\partial u}{\partial t}+u\frac{\partial u}{\partial s}+\frac{1}{\rho}\frac{\partial p}{\partial s}-g\cos\theta+\frac{fu^2}{2d}=0. \tag{2}$$

The Continuity Equation:

$$\frac{\partial \rho}{\partial t}+u\frac{\partial \rho}{\partial s}+\rho\frac{\partial u}{\partial s}=0. \tag{3}$$

Equations of State:

For the gas (superscript g)

$$\rho=\rho^g=\frac{pM}{ZRT}, \tag{4}$$

where ρ is density, M is the molecular weight of the gas, R is the universal gas constant, Z is the deviation factor for real gases, and T is temperature.

For the liquid (superscript l)

$$\rho=\rho^l=\rho_0^l\left[1+\beta(p-p_0)\right], \tag{5}$$

where ρ_0^l is reference density, p_0 is the reference pressure, and β is the compressibility of the liquid.

Gas Phase Equations:

From Eq. (4) we obtain

$$\frac{\partial \rho}{\partial t}=\frac{\partial p}{\partial t}\frac{M}{ZRT} \tag{6}$$

$$\frac{\partial \rho}{\partial s}=\frac{\partial p}{\partial s}\frac{M}{ZRT}. \tag{7}$$

Substituting Eqs. (4), (6), and (7) in Eq. (3), we obtain:

$$\frac{\partial p}{\partial t}+u\frac{\partial p}{\partial s}+p\frac{\partial u}{\partial s}=0. \tag{8}$$

Combining the momentum Eq. (2) and the equation of state for the gas we obtain the following expression:

$$p\frac{M}{ZRT}\left(\frac{\partial u}{\partial t}+u\frac{\partial u}{\partial s}-g\cos\theta\right)+\frac{\partial p}{\partial s}+\frac{fu^2\rho}{2d}=0, \tag{9}$$

Since the friction factor f depends on the type of flow (laminar, critical, or turbulent) and the roughness of the tube walls, different expressions are obtained depending on the flow regime and the nature of the walls with which the fluid is in contact. For smooth walls the friction factor is a function of the *Reynolds* number, only. Blasius formula was used for turbulent flow; for critical flow interpolated functions were used.

Liquid Phase Equations:

From Eq. (5) we obtain:

$$\frac{\partial \rho}{\partial t} = \rho_0^l \beta \frac{\partial p}{\partial t} \tag{10}$$

$$\frac{\partial \rho}{\partial s} = \rho_0^l \beta \frac{\partial p}{\partial s}. \tag{11}$$

Substitution of Eqs. (5), (10), and (11) in Eq. (3) yields

$$\begin{gathered} \rho_0^l \beta \left(\frac{\partial p}{\partial t} + u \frac{\partial p}{\partial s} \right) + \rho_0^l \left[1 + \beta (p - p0) \right] \frac{\partial u}{\partial s} = 0 \\ \beta \left(\frac{\partial p}{\partial t} + u \frac{\partial p}{\partial s} \right) + \left[1 + \beta (p - p0) \right] \frac{\partial u}{\partial s} = 0. \end{gathered} \tag{12}$$

Combining the state equation for water and the momentum Eq. (2) we obtain

$$\rho_0^l \left[1 + \beta (p - p0) \right] \left(\frac{\partial u}{\partial t} + u \frac{\partial u}{\partial s} - g \cos\theta \right) + \frac{\partial p}{\partial s} + \frac{f u^2 \rho}{2d} = 0. \tag{13}$$

Boundary Condition at the Gas-Liquid Interface:

$$u_{fr}^l = u_{fr}^g, \tag{14}$$

where subscript *fr* denotes the front. There is a pressure jump at the interface that equals the capillary pressure p_c. Since the gas is the non-wetting phase and the liquid is the wetting one, the capillary pressure is

$$p_{fr}^g - p_{fr}^l = p_c. \tag{15}$$

5 A Numerical Simulation

In the following example, we investigate the infiltration of oil spilled on the ground surface, which travels through the unsaturated zone of a hard rock formation with a developed fracture network. The fracture network was synthetically generated. The fractures were considered to be of equal length and inclinations of 45° and 135° with respect to the horizontal, so that the two fracture systems intersect each other orthogonally, as illustrated in Figs. 10 through 13. The conduit diameters, generated randomly, were assumed to follow a lognormal distribution with a mean value $m_d = 10^{-4}$ m and a standard deviation $s_d = 10^{-5}$ m.

Oil was supposed to be spilled on the ground surface at the location of one fracture. The results presented graphically in Figs. 10 through 13 show the advance of the spilled oil through the fracture network after 0.41 h and after 0.55 h from the beginning of infiltration.

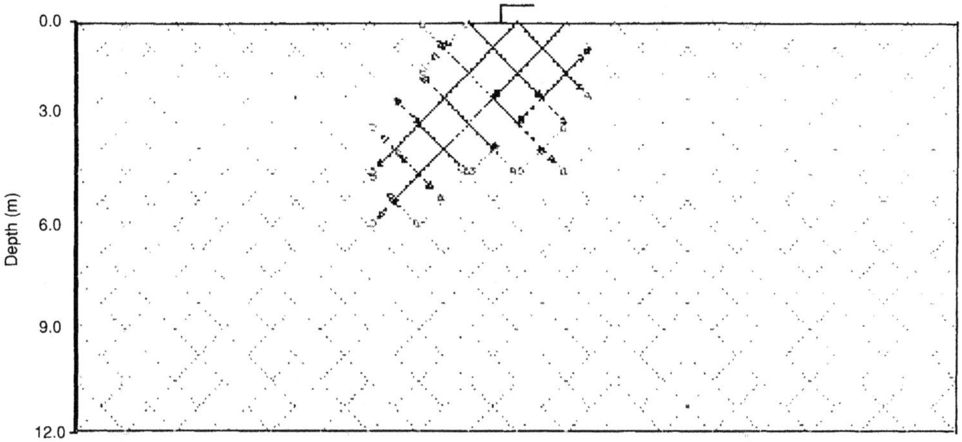

Fig. 12. The spilled oil advance through the fracture network at time = 0.55 h from the beginning of infiltration

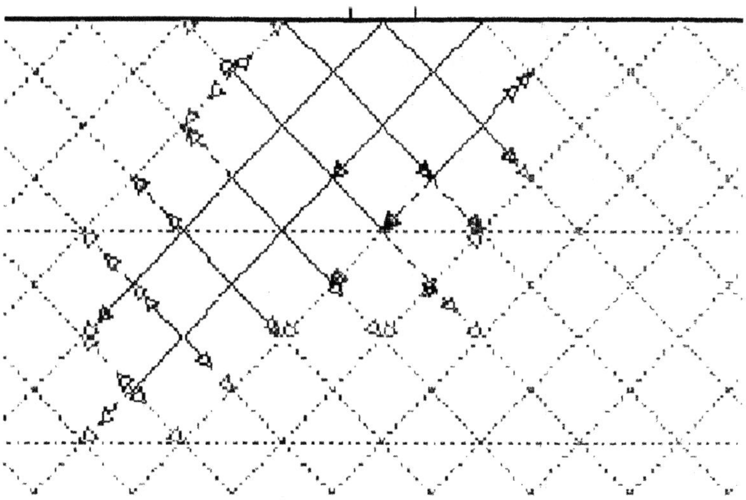

Fig. 13. The spilled oil advance through the fracture network at time = 0.55 h from the beginning of infiltration - a window through the upper region of Fig. 12

6 Discussion and Conclusions

The investigation of the oil-air displacement in the unsaturated zone was carried out with a synthetically generated fracture network.

Although we have the mathematical tools for solving the flow problem, it seems that determination of the properties of real fracture networks is impossible, and studies on fractured rock formations are limited to theoretical investigations.

Some characteristic features of the oil-air displacement follow from the results of the present simulation: (1) although the oil spill extends to one fracture as shown in Fig. 11, the pressure exerted by the oil at the ground surface causes an upward oil flow in the adjacent lateral fracture, close to the ground surface; (2) in some fractures, one may observe regions of slug flow of oil and air; and (3) the oil advances through *preferred* flow paths, those with an equivalent higher permeability; in the present simulation, this is illustrated by flow development in the left part of the investigated region.

Because of the relatively high hydraulic conductivity of the individual fractures and the low porosity of the rock mass, even small quantities of oil spilled on the ground surface will reach the saturated aquifer within a time interval in the order of days.

In hard rocks, flow takes place through the fractures, so that the flow paths are dictated by the fracture pattern, and the oil will first reach the aquifer through a

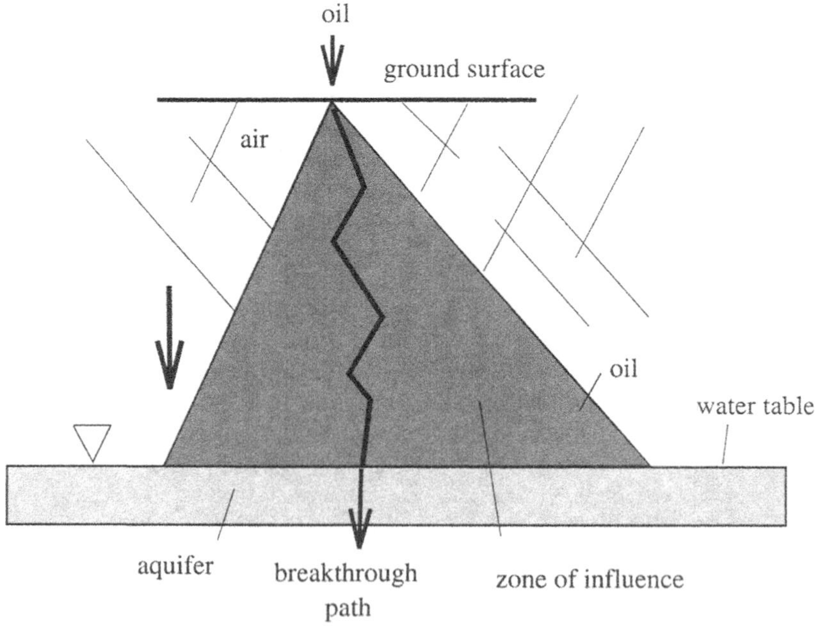

Fig. 14. The oil-polluted region between the ground surface and the water table

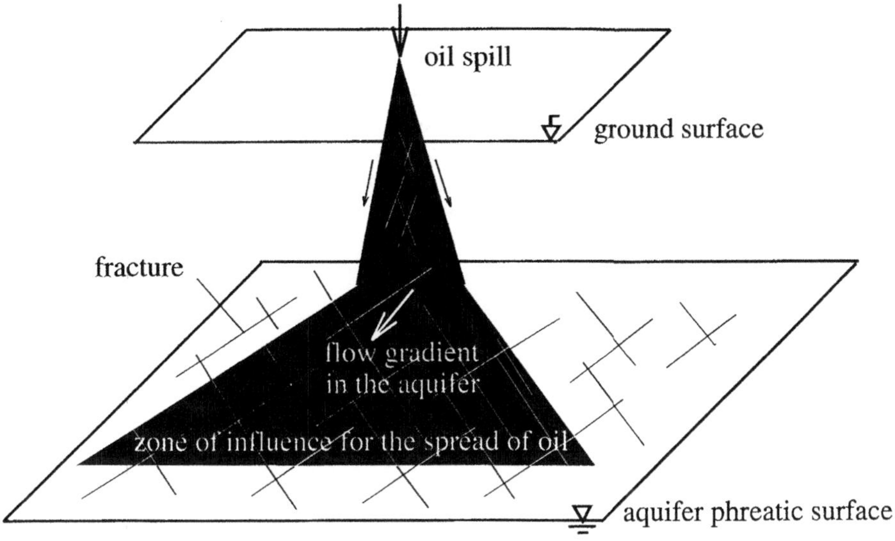

Fig. 15. Oil pollution in a fractured rock formation

preferential path (*breakthrough path*) and then through the system of fractures through which the oil may flow (*zone of influence*). Thus, in addition to the flow gradient, the flow paths are dictated by the orientation of fractures. These characteristics are illustrated in Fig. 14.

In a fractured rock formation with pervious blocks, in the unsaturated zone, part of the oil will penetrate the blocks. However, due to the low block permeability, except for the first stages of infiltration during which the process is dominated by the capillary forces, the rate of oil infiltration into the blocks will be relatively low. Depending on the amount of oil spilled on the ground surface and the depth of the saturated aquifer, the oil may or may not reach the saturated aquifer.

Also, as shown in Fig. 15, flow in a saturated aquifer will be controlled by the fracture network geometry, spreading in the direction of the flow gradient through the paths imposed by the direction of the fractures. In fractured rock formations of double porosity, it is expected that oil floating on the surface of water will not reach the *entry value* for displacing water from the blocks, and with respect to the oil, the blocks will behave imperviously. Thus, in the saturated aquifer, in both hard rocks and double porosity fractured rock formations, the oil will flow only through the fractures, displacing water.

References

Barenblatt GI, Zheltov Yu P, Kochina IN (1960) Basic concepts in the theory of seepage of homogeneous liquids in fissured rocks. PMM – Sov Appl Math Mech: 852–864

Berkowitz B, Braester C (1991) Solute transport in fracture channel and parallel plate models. Geophys Res Lett 18:(2)227–230

Braester C (1984) Influene of block size on the transition curve for a drawdown test in a naturally fractured reservoir. Soc Petrol Eng J 24(5):498–504

Braester C, Zeitoun D (1993) Pressure transient response of stochastically heterogeneous fractured reservoirs. TIPM – Transport in Porous Media 11:263–280

Kazemi H (1969) Pressure transient analysis of naturally fractured reservoirs. Soc Petrol Eng J 451–462

Warren JE, Root PJ (1963) The behavior of naturally fractured reservoirs. Soc Petrol Eng J 245–255

19 Residual Nonaqueous Phase Liquid Dissolution in Micromodels

M. Yavuz Corapcioglu, S. Chowdhury, and Sharon E. Roosevelt

Contents

1 Introduction

Nonaqueous phase liquids (NAPLs) such as gasoline and halogenated solvents enter the subsurface after a spill, or from leaking underground storage tanks. NAPLs trapped by capillary forces in subsurface environments pose a long-term threat to drinking water supplies. The main challenge in site remediation lies in dealing with the dissolving NAPL source (Corapcioglu and Baehr 1987). An understanding and quantification of the NAPL dissolution process at pore scale are of utmost importance in estimating the source of existing dissolved compounds and in evaluating the efficiency of in situ remediation of NAPL-contaminated aquifers.

The use of small-scale artificial models of porous media known as micromodels is a novel approach to simulating flow and transport through porous media. Micromodels are artificial two-dimensional network models of interconnecting pores and throats that can simulate natural porous media up to a certain degree. Micromodels give insight to the pore-scale interplay of various aspects of transport phenomena. In recent years, micromodels have proven to be an easy and economical tool for conducting small-scale experiments that provide the opportunity to discover yet unrecognized processes and enhance the understanding of existing theories and assumptions (Wilson 1994).

Micromodels are commonly fabricated by etching the desired pore network pattern on two plates of mirror glass which are then fused together. Using this

Department of Civil Engineering, Texas A & M University, College Station, Texas 77843-3136, USA

H. Rubin/N. Narkis/ J. Carberry
Soil and Aquifer Pollution

method, highly intricate and detailed patterns can be etched with the dimensions of pores and throats as low as a few microns. The models are provided with an inlet and an outlet at the two ends. Details of the model production procedure are given by McKeller and Wardlaw (1982), Dawe et al. (1987), Buckley (1992), and Wilson (1994).

The types of micromodel studies reported in the literature vary widely in their methods and applications. Buckley (1992) provides a comprehensive overview of evolution of micromodels as research tools and summarizes all major micromodel studies of multiphase flow reported in the literature. Residual trapping and blob formation at pore scale have been investigated by various researchers. Wardlaw (1982) investigated the trapping mechanism during displacement of one fluid by another. Relation of pore structure and pore network to residual saturation was studied by Chatzis et al. (1983). Conrad et al. (1992) described flow visualization experiments which examined migration of organic fluids during two-phase flow and the distribution of residual organic fluid in saturated micromodels. Micromodels were initially water-saturated, then they were flooded by an immiscible organic liquid to simulate migration in the saturated zone. After the fluid stabilized, the medium was left with a residual immiscible phase saturation. The observed formation and distribution of trapped residual organics gave insight into dissolution of organic liquid into water. The effect of wettability changes on oil displacement was studied by Morrow et al. (1986) by direct flow visualization in micromodels. Soll et al. (1993) investigated the physical processes involved in a three-fluid system. They obtained the three-fluid (water, oil, and air) capillary pressure-saturation relationships at pore scale. Digital image analysis was used to measure the relative saturation of the three phases in the micromodel.

In this study, we investigated the dissolution kinetics of residual NAPLs in a porous medium by conducting micromodel experiments. Dissolution of residual NAPLs was quantified by direct observation of the dissolution mechanism in a micromodel. We processed video images with an image analyzer to measure the temporal change of irregularly shaped NAPL blobs in a heterogeneous porous medium in order to calculate the mass transfer coefficient as well as the mass transfer rate coefficient. Relationships between the mass transfer rate coefficient, mass transfer coefficient, residual NAPL saturation, and flow velocity were examined.

The dissolution process was also investigated by separating the mass transfer into two zones of mobile and immobile water. This type of bicontinuum approximation of the fluid velocity field assumes that convective-dispersive transport occurs in the mobile region, while the water remains stagnant in the immobile region which is connected to the mobile region. Although this concept is not new, the estimation of the interfacial area and the fraction of area in the mobile region have not been contemplated due to the inherent difficulty in measuring these parameters. We introduced a methodology to evaluate NAPL dissolution in the mobile and immobile water region by measuring the fraction of total residual NAPL in contact with mobile water, immobile water, and solid surfaces.

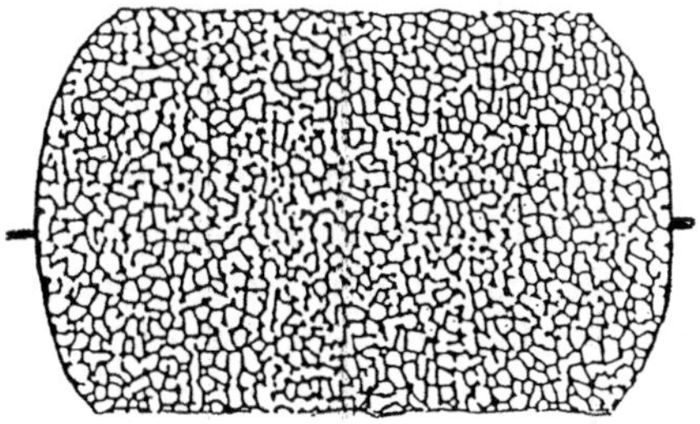

Fig. 1. Micromodel pattern

The micromodel used in this study was obtained from the New Mexico Petroleum Recovery Research Center at the New Mexico Institute of Mining and Technology (Fig. 1). The micromodel has a heterogeneous pattern with a porosity of 0.5 and hydraulic conductivity of 0.3 mm/s (Corapcioglu et al. 1997). The dissolution of TCE was observed in the micromodel and mass transfer was quantified by analyzing the images of the dissolving TCE blobs.

2 Dissolution Kinetics of Residual DNAPLs

After an initial spill or leak of a dense nonaqueous phase liquid, the free product migrates downward through the subsurface as a separate phase, leaving behind a residual phase trapped in the aquifer material by capillary forces. The mechanisms of residual entrapment in a soil matrix have been studied in detail by Chatzis et al. (1983) and Conrad et al. (1992). Hunt et al. (1988) examined the physics governing the emplacement and movement of a separate phase in porous media, the role of sorption, and the conditions necessary to mobilize a separate phase. Residual blobs are held in pores by capillary forces which cannot be overcome under normal hydraulic gradients. Therefore, mobilization of the blobs is not a viable alternative for remediation in most cases. These immobile residual hydrocarbons create very long-term sources of contamination in groundwater aquifers. Therefore, an understanding of mass transfer rates for the dissolution of residual hydrocarbons in a groundwater aquifer is critical for the prediction of groundwater quality and remediation schemes.

The rate-limited mass transfer between a hydrocarbon phase and the water phase can be expressed by

$$\rho_N \theta \frac{\partial S}{\partial t} = -k^*\left(C_{SAT} - C\right). \tag{1}$$

The relationship between the mass transfer rate coefficient k^* (T^{-1}) and the mass transfer coefficient k (L/T) is given by

$$k^* = k a_{na}. \tag{2}$$

After substituting Eq. (2) into (1), the mass transfer coefficient k is expressed as,

$$k=-\frac{\partial V_r}{\partial t}\frac{1}{A_{na}}\frac{\rho_N}{\left(C_{SAT}-C\right)}. \tag{3}$$

Brusseau (1992) and Hatfield and Stauffer (1993) proposed the modeling of liquid-liquid mass transfer processes by separating the mass transfer into two zones of mobile and immobile water. This type of bicontinuum approximation of the fluid velocity field assumes that convective-dispersive solute transport occurs in the mobile region, while the water remains stagnant in the immobile region (Hatfield and Stauffer 1993). The regions are assumed connected. Miller et al. (1990) also proposed the inclusion of a factor f in the assessment of total interfacial area of NAPL to account for the portion not in contact with the flowing water. However, the determination of the actual interfacial area and the fraction of area in the mobile zone have not been contemplated due to the inherent difficulty in measuring these parameters (Miller et al. 1990; Powers et al. 1991, 1992).

Hatfield and Stauffer (1993) proposed a one-dimensional analytical model to describe solute transport in heterogeneous media with a fixed NAPL residual. Hatfield and Stauffer (1993) took into account both instantaneous and rate-limited dissolution of the residual source. Powers et al. (1994) proposed a model for transient dissolution rates in terms of porous medium properties, Reynolds number, and volumetric fraction of NAPL. Miller et al. (1990) calculated the steady-state mass transfer rate coefficients of toluene in glass-bead columns. Imhoff et al. (1994) measured change in residual saturation of TCE with time by gamma attenuation in short sand columns. They did not find any variation in blob size and shape distribution along the column length. The k* value increased with increasing Darcy velocity and also decreased with decreasing NAPL saturation, as expected. With decrease in saturation, less interfacial area is available for dissolution. They also measured effluent concentration and calculated the mass balance for each experiment. Imhoff et al. (1994) did not calculate the mass transfer coefficient, and thus avoided the issue of NAPL blob size.

Kennedy and Lennox (1997) studied the dissolution kinetics of individual TCE blobs in a glass-bead micromodel. They measured the temporal change of the volume and surface area of spherical and ellipsoidal blobs using an image analyzer. They reported the microscale mass transfer coefficients of the individual blobs by taking into account the temporal change of specific interfacial area.

In the present study, we employed image analysis techniques to measure the temporal change of irregularly shaped blobs in a heterogeneous porous medium, enabling us to measure the transient mass transfer coefficient (L/T) as well as the mass transfer rate coefficient (1/T) which is a function of the porous media properties.

3 Experimental Methods

The experimental setup is shown in Fig. 2. The experimental apparatus consisted of a micromodel, syringe, syringe pump, Teflon valves, flasks, tubing, an Olympus

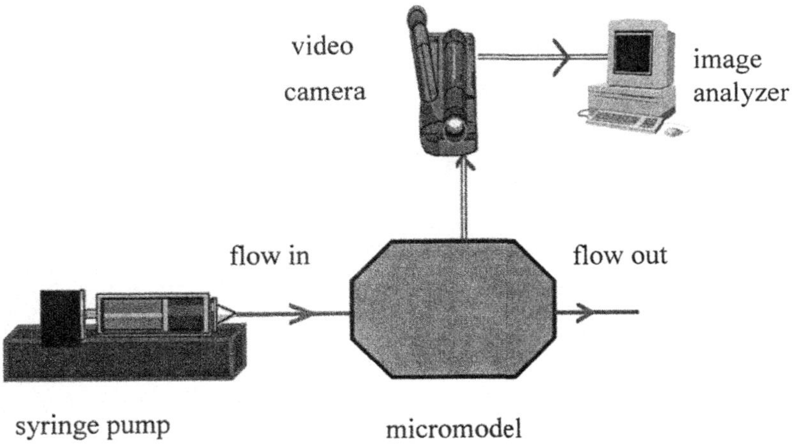

Fig. 2. Experimental setup

SZ-PT disectoscope, and a Sony-Iris CCD video camera and a computer with an image analyzer. The hydrocarbons used in the experiments were dyed with oil blue-N (Fisher Chemicals) at a concentration of 0.5 g/l to aid visualization.

In each experiment, the horizontally mounted micromodel was initially water-saturated and then flooded with the hydrocarbon. After the hydrocarbon saturation stabilized, a syringe pump (Sage Instruments, model no. 562) was used to flush the micromodel with distilled deionized water at a flow rate of 8 ml/h until only residual DNAPL was left in the pore spaces. The residual saturation was obtained in such a way that every run had a similar initial residual saturation with a similar size and shape distribution of blobs. These experiments represent a case in which a hydrocarbon infiltrates into the groundwater aquifer and then is displaced by ambient groundwater flow, and eventually only the residual hydrocarbon remains.

After the residual saturation stabilized, water was injected with the syringe pump through the micromodel at a constant flow rate. This flow rate was significantly less than the initial flushing flow rate and, hence, there was no further displacement of the DNAPL. Images of residual DNAPL saturation in the micromodel were recorded at certain times and analyzed with the Image-Pro analyzer (Media Cybernetics 1992).

4 Quantification of Dissolution Rates

A typical sample run of the dissolution of TCE (Aldrich Chemicals) in a heterogeneous micromodel with distilled deionized water is discussed. Figure 3 shows the progressive dissolution of TCE blobs in the micromodel for a water flow rate of 0.6 ml/h. The total area for each image in Fig. 3 is 61.8 mm^2. This is also the total

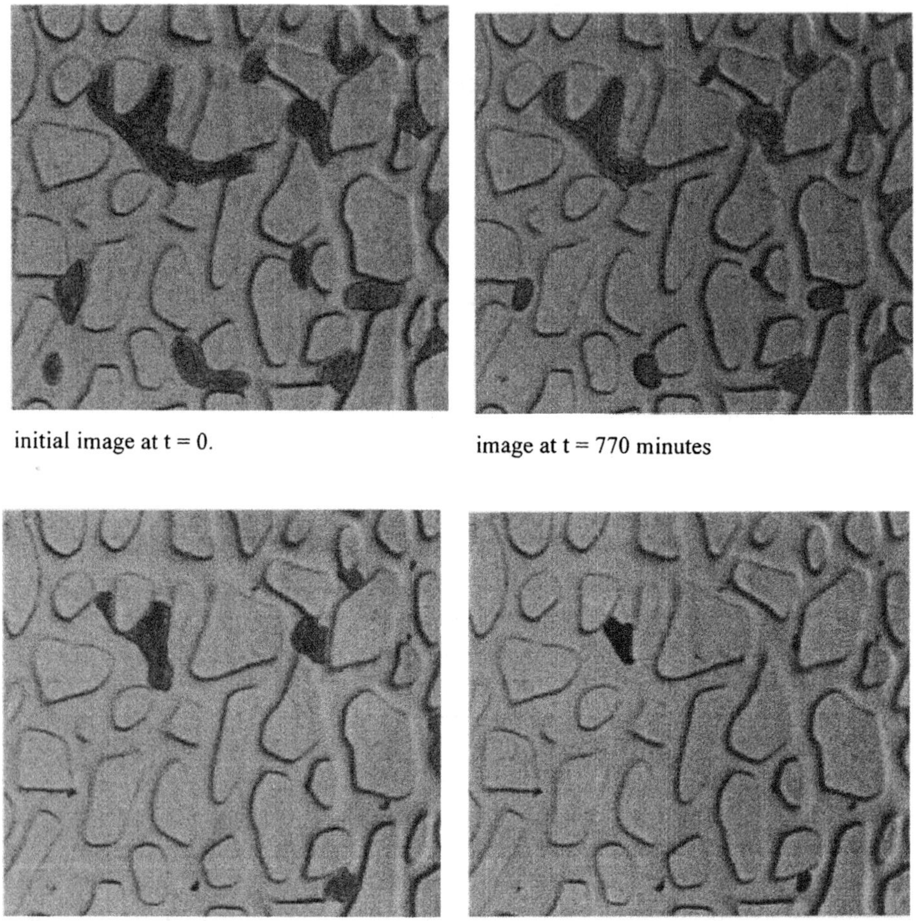

initial image at t = 0.

image at t = 770 minutes

image at t = 1700 minutes

final image at t = 2500 minutes

Fig. 3. Images of dissolving TCE blobs in a micromodel. The direction of water flow is from *left* to *right*

area used in all analyses. The two-dimensional area and perimeter of all blobs at different time intervals have been measured with the image analyzer. The removal of residual TCE saturation in the region of analysis is plotted in Fig. 4.

For a porous matrix with a uniform depth, Eq. (1) reduces to a two-dimensional form as

$$k = -\frac{\partial A_r}{\partial t}\frac{1}{P_{NA}}\frac{\rho_N}{C_{SAT}}. \tag{4}$$

The ambient concentration C is assumed negligible as compared to solubility C_{SAT}. This assumption is valid if dissolution is fast enough and velocity is high enough (Powers et al. 1994; Kennedy and Lennox 1997). The time derivatives of the

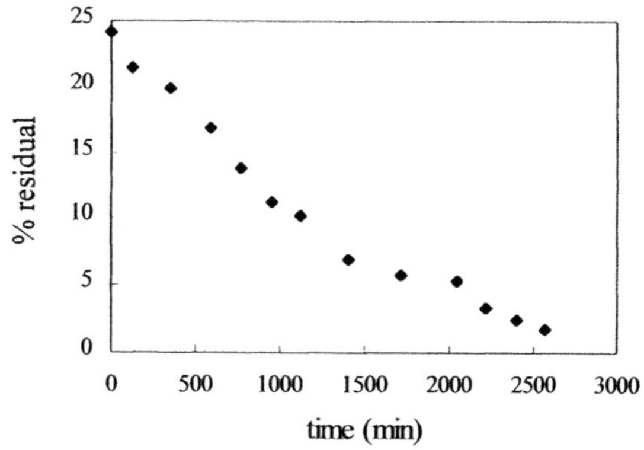

Fig. 4. The removal of residual TCE saturation in the micromodel shown as in Fig. 3

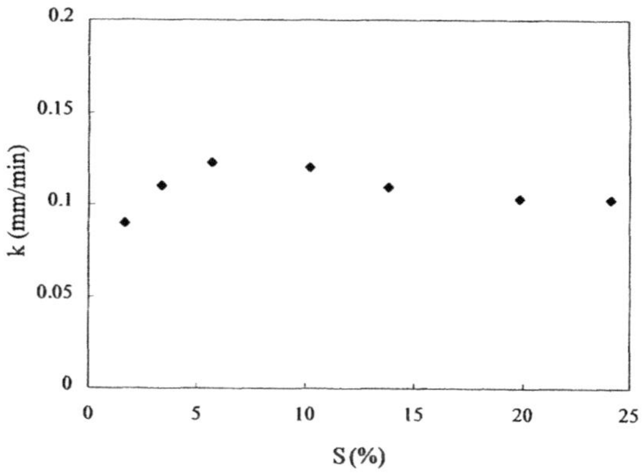

Fig. 5. Variation of mass transfer coefficient k (mm/min) of TCE with residual saturation at a water flow rate of 0.6 ml/h

area are determined from a second degree polynomial fit of the area vs. time graph. The mass transfer coefficient is then determined from Eq. (4), and its variation with residual saturation is shown in Fig. 5. For TCE, the density is 1.47 g/cm^3 and the solubility is 1270 mg/l. Finally, the mass transfer rate coefficient is obtained from Eq. (2), where a_{na} has been replaced by total interfacial perimeter per unit area of micromodel (Fig. 6).

The mass transfer coefficient increases with time as residual saturation decreases. The mass transfer rate coefficient (k*) decreased with dissolution, a trend also reported by Imhoff et al. (1994). The decrease in a_{na} with time offset the increase in k in Eq. (2). Most experimental investigations in the literature report a constant mass transfer rate coefficient at a certain velocity (Miller et al. 1990).

Imhoff et al. (1994) observed in their experiments that the dissolution front increased with time as it moved through the porous medium. This would indicate

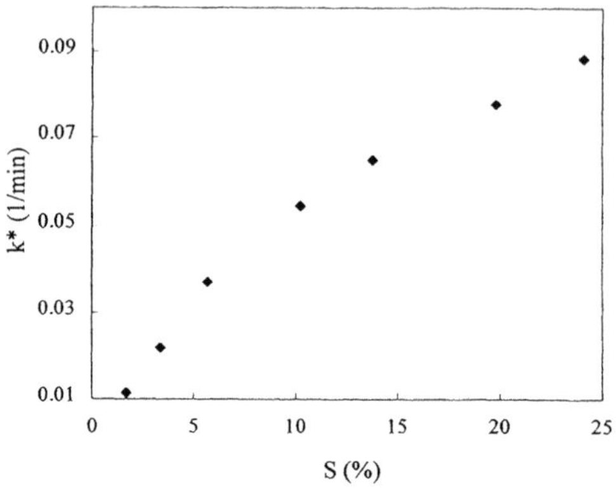

Fig. 6. Variation of mass transfer rate coefficient k* (l/min) of TCE with residual saturation at a water flow rate of 0.6 ml/h

that, in our micromodel, dissolution will be faster near the inlet and slower near the outlet.

5 Quantification of DNAPL in Mobile and Immobile Regions

As is evident from the sequence of photographs depicting the dissolution process (Fig. 3), none of the DNAPL blobs are exposed to mobile water. Based on the flow pattern, images of the dissolving blobs have been analyzed to estimate portions of interfacial perimeter in contact with mobile water, immobile water and solid surface (Fig. 7). Figure 8 shows the change in total perimeter and its three constituents. The fractions of the blob perimeter in contact with mobile, immobile water and solid are expressed as:

f_m = perimeter in contact with mobile water/total perimeter;
f_{im} = perimeter in contact with immobile water/total perimeter;
f_s = perimeter in contact with solid surfaces/total perimeter.

Figure 9 shows the change of the three fractions with time. It is evident from Fig. 9 that the fraction of perimeter in contact with mobile water increases as dissolution progresses, increasing the amount of DNAPL in contact with mobile water.

6 Conclusions

Mass transfer coefficients of dissolving DNAPL blobs in micromodels have been successfully quantified. Micromodel visualization enhances our understanding of the processes of dissolution of residuals and aids in more accurate model formulation and parameter estimation. In turn, this will lead to better prediction

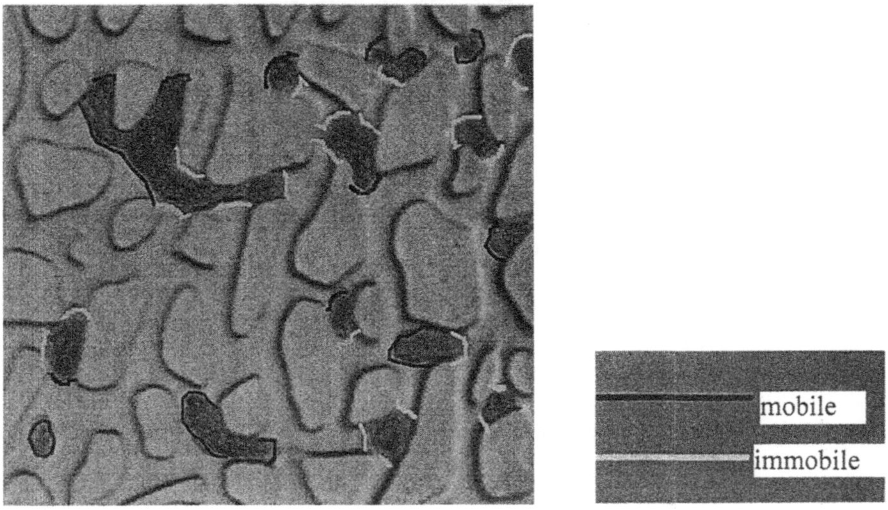

Fig. 7. Surfaces of DNAPL blobs in contact with regions of mobile and immobile water at initial time. The direction of flow is from *left* to *right*

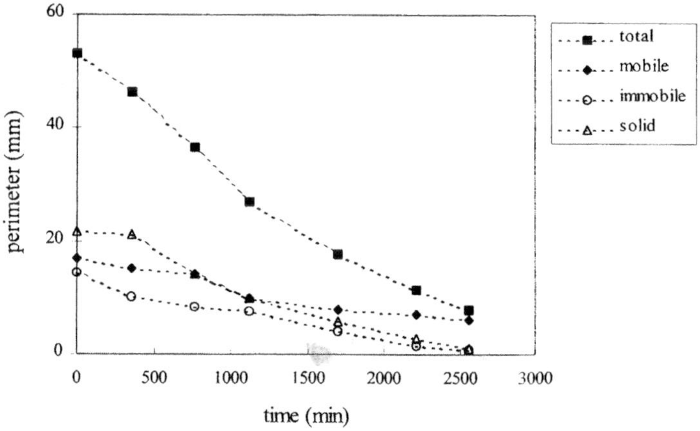

Fig. 8. Change in blob perimeters with time

of dissolution by ambient groundwater flow and during pump-and-treat remediation.

The information gathered in this study aids in the development of a more comprehensive understanding of the mechanisms of the dissolution, and subsequent contamination, of groundwater due to the presence of a DNAPL contaminant. Due to the difficulty of characterizing the overall nature of a DNAPL release in the subsurface, however, these results can only be relied on to provide a rough estimate of what would actually be observed in the field.

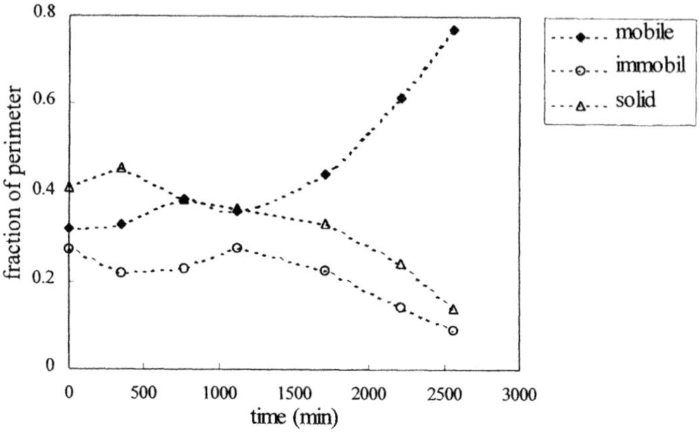

Fig. 9. Change in the TCE blob perimeter fractions with time

Acknowledgment. This research has been supported by the US Air Force under an SBIR subcontract.

Nomenclature

θ porosity of porous medium
A_{na} interfacial area between the NAPL and the aqueous phase
a_{na} specific interfacial area between the NAPL and the aqueous phase
A_r surface area (two-dimensional) of residual NAPL blob or ganglia
C aqueous concentration of NAPL
C_{SAT} saturation concentration of NAPL
k mass transfer coefficient
k* mass transfer rate coefficient
P_{NA} total (two-dimensional) interfacial perimeter of residual NAPL
ρ_N density of NAPL
S degree of saturation of NAPL
t time
V_r volume of residual NAPL blob or ganglia

References

Brusseau ML (1992) Rate-limited mass transfer and transport of organic solutes in porous media that contain immobile immiscible organic liquids. Water Resour Res 28:33–45

Buckley J (1992) Multiphase displacement in micromodels. In: Morrow N (ed) Interfacial phenomena in petroleum technology. Marcel Decker, New York, pp 157–189

Chatzis I, Morrow N, Lim HT (1983) Magnitude and detailed structure of residual oil saturation. Soc Petrol Eng J 23:311–325

Conrad SH, Wilson JL, Mason WR, Peplinski WJ (1992) Visualization of residual organic liquid trapped in aquifers. Water Resour Res 28:467–478
Corapcioglu MY, Baehr AL (1987) A compositional multiphase model for groundwater contamination by petroleum products. Water Resour Res 23:191–200
Corapcioglu MY, Chowdhury S, Roosevelt SE (1997) Micromodel visualization and quantification of solute transport in porous media, Water Resour Res: 33(11)
Dawe RA, Mahers EG, Williams JK (1987) Pore-scale physical modeling of transport phenomena in porous media. In: Bear J, Corapcioglu MY (eds) Advances in transport phenomena in porous media. Martinus Nijhoff, Dordrecht, pp 49–76
Hatfield K, Stauffer TB (1993) Transport in porous media containing residual hydrocarbon. I. Model. J Environ Eng 119:540–558
Hunt JR, Sitar N, Udel KS (1988) Nonaqueous phase liquid transport and cleanup, experimental studies. Water Resour Res 24:1259–1269
Imhoff PT, Jaffe PR, Pinder GF (1994) An experimental study of complete dissolution of a nonaqueous phase liquid in saturated porous media. Water Resour Res 30:307–320
Kennedy CA, Lennox WC (1997) A pore-scale investigation of mass transport from dissolving DNAPL ganglia. J Contam Hydrol, 24:221–246
McKeller M, Wardlaw N (1982) A method of making two-dimensional glass micromodels of pore systems. J Can Petrol Technol 21:39–41
Media Cybernetics (1992) Image-Pro Plus™ image processing system version 2.0. Silver Springs, Maryland
Miller CT, Poirier-McNeill MM, Mayer AS (1990) Dissolution of trapped nonaqueous phase liquids: mass transfer characteristics. Water Resour Res 26:2783–2796
Morrow NR, Lim HT, Ward JS (1986) Effect of crude-oil induced wettability changes on oil recovery. SPE Formation Evaluation 1:89–103
Powers SE, Loureiro CO, Abriola LM, Weber WJ Jr (1991) Theoretical study of the significance of nonequilibrium dissolution of nonaqueous phase liquids in subsurface systems. Water Resour Res 27:463–477
Powers SE, Abriola LM, Weber WJ Jr (1992) An experimental investigation of nonaqueous phase liquid dissolution in saturated subsurface systems. Steady state mass transfer rates. Water Resour Res 28:2691–2705
Powers SE, Abriola LM, Weber WJ Jr (1994) An experimental investigation of nonaqueous phase liquid dissolution in saturated subsurface systems. Transient mass transfer rates. Water Resour Res 30:321–332
Soll WE, Celia MA, Wilson JL (1993) Micromodel studies of three-fluid porous media systems: pore-scale processes relating to capillary pressure-saturation relationships. Water Resour Res 29:2963–2974
Wardlaw N (1982) The effects of geometry, wettability, viscosity and interfacial tension on trapping in single pore-throat pairs. J Can Petrol Technol 21:21–27
Wilson JL (1994) Visualization of flow and transport at the pore level. In: Dracos TH, Stauffer F (eds) Transport and reactive processes in aquifers. Balkema, Rotterdam, pp 19–36

20 On the Movement of NAPLs Above and Within a Phreatic Aquifer

J. Bear[1] and V. Ryzhik[2]

Contents

1 Introduction: Non-Aqueous Phase Liquid Contaminants and Their Specific Properties

Contamination of groundwater by NAPLs is a worldwide phenomenon, which is especially important in the industrialized countries. Potential sources of contamination are abundant: oil production facilities, petrochemical plants, gasoline distribution networks, to mention but a few.

The propagation of NAPL through the unsaturated and saturated zones constitutes a problem of multiphase flow through porous media – two-phase flow within the saturated zone and three-phase flow within the unsaturated one.

The issue of three-phase (oil-gas-water) is the bread and butter of reservoir engineers. It has been investigated and reported in numerous publications and books. Hydrological investigations of NAPL propagation within the unsaturated zone and in groundwater, as the two- and three-phase flow problem, probably began with van Dam (1967). A comprehensive review of the state of art in this field may be found in a paper by Schiegg and Schwille (1991). An experimental basis for the hydrodynamic analysis of NAPL equilibrium and flow in aquifers was proposed by Lenhard and Parker (Lenhard and Parker 1987a,b, 1988; Parker and

[1] Albert and Anne Mansfield Chair in Water Resources, Faculty of Civil Engineering, Technion – Israel Institute of Technology, Haifa, Israel
[2] Technion – Israel Institute of Technology, Haifa, Israel

H. Rubin/N. Narkis/ J. Carberry
Soil and Aquifer Pollution

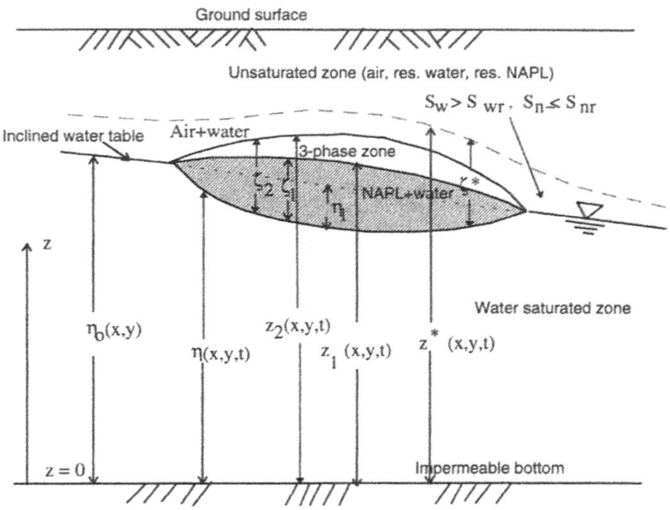

Fig. 1. The schematic disposition of an LNAPL lens over the water table

Lenhard 1987, 1990). The physical basis for three-phase vertical equilibrium above the water table was discussed by Blunt et al. (1995).

Beginning with the work of Parker and Lenhard (1990), the hypothesis of vertical equilibrium of NAPL in the unsaturated zone has been used to obtain simplified numerical solutions for problems of flow in this zone. A similar study was also conducted recently by Yu-Shu Wu et al. (1994). In this chapter, we shall also make use of the "vertical equilibrium" hypothesis to obtain simplified equations describing the movement of LNAPL in the form of a thin lens. However, our approach is limited to cases in which the thickness of the NAPL layer is small relative to that of the underlying saturated zone. This leads to a single mass balance equation, in terms of NAPL content, for the domain above the water table. Some exact analytical solutions of this (approximate) equation are obtained and presented. These solutions may be useful for predicting the propagation of NAPL in phreatic aquifers. They may also be useful for the verification of numerical schemes.

We shall limit the discussion to LNAPL flow problems in which the total amount of NAPL infiltrating through ground surface is relatively low, but sufficient for the formation of a NAPL lens, flowing along the water table (Fig. 1). The objective of this work is to discuss the movement of such a lens. The same approach is also applied to the case of a DNAPL lens flowing above the base of an aquifer that serves as an impermeable boundary to downward migration of DNAPL.

2 Static Vertical Saturation Distribution

We consider a situation in which a finite quantity of NAPL has penetrated into the unsaturated zone of an aquifer during some finite period of time, and a static vertical phase distribution has been reached. For the vertical static distribution of NAPL, air and water in the subsurface, we use a one-dimensional model with a homogeneous, isotropic, strongly water-wet soil, neglecting hysteresis phenomena. Within the unsaturated zone, NAPL is considered as the intermediate wetting phase.

The schematic static vertical distribution of NAPL, water and air in the unsaturated zone in a homogeneous soil above a water table is shown in Fig. 2. The distribution is governed by the interplay between gravitational and capillary forces.

When the total volume of spill is sufficiently large, some bulk (continuous) volume of LNAPL will accumulate above the water table at saturations in excess of residual S_{nr} (Fig. 2). The volume of soil in which $S_n \geq S_{nr}$ defines the "LNAPL lens", or simply the lens.

With p_w, p_n, p_a, denoting the pressures in water, NAPL, and air, respectively at a (macroscopic) point, we have two *capillary pressures*:

$$P_{cnw} = p_n - p_w; \quad P_{can} = p_a - p_n. \tag{1}$$

As analysis by Blunt et al. (1995) shows that, in the presence of LNAPL, there is no direct contact between water and air phases in the unsaturated zone.

Hence, as a good approximation, P_{can} may be considered as a function of the total liquid saturation, $S_l = S_w + S_n$, while P_{cnw} is a function of water saturation, S_w, only. These assumptions are supported by direct measurements performed by Lenhard and Parker (1987a).

Following Lenhard and Parker (1988), the following scaling rules can be used:

$$P_{caw}\left(\bar{S}_w\right) = \frac{1}{\beta_{aw}} P_c^*\left(\bar{S}_w\right); \quad P_{can}\left(\bar{S}_l\right) = \frac{1}{\beta_{an}} P_c^*\left(\bar{S}_l\right), \tag{2}$$

where S_{rw} is the residual water saturation, and

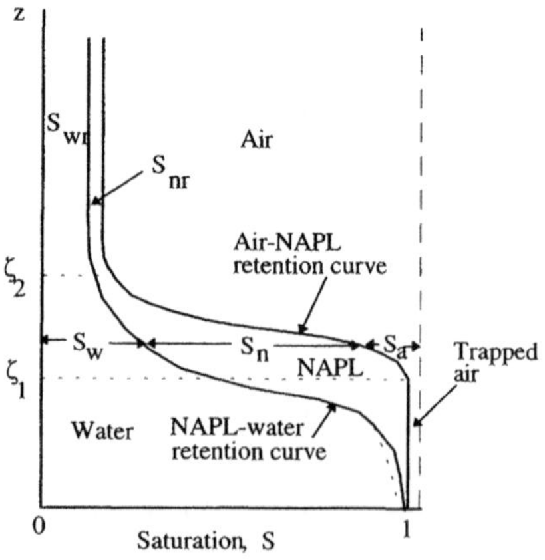

Fig. 2. Static saturation distribution. The case of a medium NAPL spill

$$\bar{S}_w = \frac{S_w - S_{wr}}{1 - S_{rw}}, \quad \bar{S}_l = \frac{S_l - S_{wr} - S_{nr}}{1 - S_l}.$$

The "reduced capillary pressure function", $P_c^\star(S)$, is a unique function for a given porous medium (soil), and β_{aw}, β_{an} are two constants. As an approximation, the scaling multipliers, β_{ij}, may be considered to be proportional to interfacial tensions between the corresponding phases.

To actually calculate the saturations' distributions, we shall employ the widely used analytical form of the capillary pressure-saturation dependence, $P_c^\star(\bar{S})$, introduced by van Genuchten (1980):

$$P_c^\star(\bar{S}) = P_{c0}\left(\bar{S}^{-\frac{\nu}{\nu-1}} - 1\right)^{\frac{1}{\nu}}, \quad S = \bar{S}_w, \bar{S}_l, \tag{3}$$

where P_{c0} (a characteristic capillary pressure) and ν (dimensionless) are empirical constants. Often the constant

$$\alpha = \frac{\rho_w g}{P_{c0}},$$

having the dimensions of reciprocal length, is also used, with $1/\alpha$ used as a characteristic height of the capillary fringe.

In the following, we shall neglect residual NAPL and water saturations, except in Section 4.4.

As LNAPL accumulates to form a lens, it displaces all the air in the void space, except for the entrapped air, up to some elevation ζ_1. This is shown in Figs. 1 and 2. Above $z = \zeta_1$, all three phases are present as mobile ones. The pressure difference between air and NAPL is described by the air-NAPL capillary pressure function. The air-liquid saturation distribution in the range $\eta + \zeta_1 < z < \eta + \zeta_2$, neglecting the density ρ_a (as $\rho_a << \rho_w, \rho_n$), is then described by:

$$z = \frac{P_{can}(S_l)}{g\rho_n} + \zeta_1 + \eta, \quad p_a = p_{atm}, \tag{4}$$

where g is acceleration due to gravity.

The capillary pressure between NAPL and water at $z = 0$ may be taken as equal to zero (assuming that there is no threshold capillary pressure). Then the NAPL-water saturation distribution for $\eta < z < \eta + \zeta_1$, is described by:

$$z = \frac{P_{cnw}(S_w)}{g\Delta\rho} + \eta, \tag{5}$$

where $\Delta\rho = \rho_w - \rho_n$.

At the upper boundary of the NAPL lens, $z = \zeta_2$, $S_n = 0$, and $S_l = S_w$. Using this boundary condition, the values of water saturation at ζ_2 and ζ_1 may be derived, and the entire water saturation distribution is then obtained.

The total volume, $\mathcal{U}$, of NAPL in the unsaturated zone, per unit of area the water table, is related to the thickness, ζ_1, by:

$$\mathcal{U} = \phi\int_0^{\zeta_2} S_n(z)dz = \phi\int_0^{\zeta_1}\left[1 - S_w(z)\right]dz + \phi\int_{\zeta_1}^{\zeta_2}\left[S_l(z) - S_w(z)\right]dz. \tag{6}$$

The accumulation of mobile NAPL in the form of a lens may take place not only over the water table. A DNAPL may penetrate through the water table, percolate through the saturated zone, and form a lens over an impermeable bottom. This situation is considered in the Section 5.

Both LNAPL and DNAPL may also accumulate over and within low permeability layers above the water table. For example, a mobile NAPL may form a lens in the unsaturated zone above a layer of water-saturated clayey soil, which is almost impermeable to the oil-phase. In a coarse-sand soil, if a low-permeability layer of fine sand exists above the water table, part of the NAPL may be retained within it, (the same as water), due to the capillary-barrier effect ("end-effect").

3 Flow of LANPL Along a Sloping Water Table

We consider the movement of an LNAPL lens along a slightly sloping, or a horizontal water table. We shall assume (1) that the thickness of the NAPL lens is much smaller than its lateral dimensions, (2) that it is also much smaller than the thickness of the (water) saturated zone, $\eta(x,y,t)$, and (3) that the lower boundary of the LNAPL layer is gently sloping (Fig. 1), i.e.

$$\left|\frac{\partial\eta}{\partial x}\right| << 1, \quad \left|\frac{\partial\eta}{\partial y}\right| << 1. \tag{7}$$

Taking into account the small thickness of the NAPL lens, we may apply the common hypothesis of "vertical capillary-gravity equilibrium" (≡ VE hypothesis). This hypothesis is based on the assumption that the flow of both NAPL and water is essentially horizontal. This is equivalent to stating that the vertical pressure distribution within each phase is hydrostatic, and may be described by (4) and (5).

The VE hypothesis implies that the saturations S_w and S_n at a point may be considered as known functions of two variables: the location of the point within the lens, above the water table, $\zeta = z - \eta$, and of $\mathcal{U}$, where $\mathcal{U}$ denotes the volume of mobile LNAPL within the lens, per unit horizontal area.

The VE hypothesis permits the use of a 2-D rather than a 3-D description of the NAPL behavior (see Bear 1979). To obtain a 2-D formulation of the flow model, we have to integrate the 3-D mass balance and flow equations for water and NAPL in the z-direction, from the impervious bottom ($z = 0$) to the elevation $z = z^*$, taking into account hydrostatic pressure and saturation distribution, and the boundary conditions along the surfaces at $z = z^*$ and $z = z_2 = \eta + \zeta_2$. After integration, and deletion of certain non-dominant terms, we obtain the balance equations for both phases:

$$\frac{\partial\mathcal{U}_w}{\partial t} - \nabla_2\cdot\frac{k}{\mu_w}\left(\eta + T_w^*\right)\nabla_2\Phi_w = 0, \quad \frac{\partial\mathcal{U}}{\partial t} - \nabla_2\cdot\frac{k}{\mu_w}T\nabla_2\Phi_n = 0, \tag{8}$$

where ∇_2 is the 2-D ∇-operator in the horizontal (x,y) plane, and Φ_w $(= p_w + \rho_w gz)$ and Φ_n $(= P_n + \rho_n gz)$ are water and NAPL potentials. The values T_w^* and T describe integrated relative permeabilities of the liquid phases within the lens:

$$T_w = \int_\eta^{z^*} k_{rw}(S_w, S_n)dz, \quad T_n = \int_\eta^{z_2} k_{rn}(S_n, S_w)dz. \tag{9}$$

From the previous section, it follows that the saturation distribution is totally defined by the value of $\mathcal{U}$. Hence T_w and T_n, as defined by (9), may also be considered as unique functions of $\mathcal{U}$.

In accordance with the VE hypothesis, the following approximate expressions for Φ_w and Φ_n may be used

$$\Phi_w = \rho_w g\left(\frac{\rho_n}{\rho_w}\zeta_1 + \eta\right), \quad \Phi_n = \rho_n g(\zeta_1 + \eta). \tag{10}$$

Since the thickness of the lens is assumed to be much smaller than the depth of the saturated zone in the aquifer, η_0, and consequently, the water mobility $(= (\eta + T_w)/\mu_0)$ is much greater than that of the NAPL, the pressure distribution throughout the water phase may be considered as hydrostatic.

Defining $\eta_1(x,y,t)$ (Fig. 1)

$$\eta(x,y,t) = \eta_0(x,y,t) - \eta_1(x,y,t), \tag{11}$$

where $\eta_0(x,y)$ is the elevation of the water table prior to the development of an LNAPL lens, then, under the conditions stated above, the following relation holds as a good approximation:

$$\eta_1 = \frac{\rho_n}{\rho_w} b_1. \tag{12}$$

Condition (12), equivalent to that of hydrostatic pressure distribution in the water phase, replaces the water flow equation.

Then, neglecting ζ_1 in comparison with η_0, we obtain from Eqs. (8) and (10) the final approximate equation for $u = \mathcal{U}\alpha/\phi$, in the scalar form:

$$\frac{\partial u}{\partial t} - C\frac{\partial}{\partial x}\left[\tau(u)\left(\chi\frac{\partial\varphi(u)}{\partial x} + i_{0x}\right)\right] - C\frac{\partial}{\partial y}\left[\tau(u)\left(\chi\frac{\partial\varphi(u)}{\partial y} + i_{0y}\right)\right] = 0, \tag{13}$$

where the following dimensionless variables have been introduced:

$$u = \frac{\mathcal{U}\alpha}{\phi}, \quad \tau(u) = T_n + (\mathcal{U})\alpha, \quad \varphi(u) = \frac{\zeta_1(\mathcal{U})\rho_n\alpha}{\Delta\rho},$$

with α denoting the van Genuchten's constant for the NAPL-water capillary pressure curve, and

$$C = \frac{k\rho_n g}{\phi\mu_n} = \frac{\rho_n}{\rho_w}\frac{\mu_w}{\mu_n}\frac{K}{\phi}, \quad \chi = \frac{\Delta\rho}{\alpha\rho_n}, \tag{14}$$

where $K = k\rho_w g/\mu_w$ is the water saturated hydraulic conductivity of the soil.

The functions $\varphi(u)$ and $\tau(u)$ may be obtained by calculating the integrals in Eqs. (6) and (9), using, as was stated above, the static saturation distributions.

To calculate the dimensionless functions $\tau(u)$ and $\varphi(u)$, we used Eqs. (4)–(6), Eq. (9), with van Genuchten's expression in Eq. (3) for the capillary pressure, and the simple power-law dependence of the relative permeability on NAPL saturation, S_n, in the form

$$k_{nr} = \left(S_n - S_{nr}\right)^m. \tag{15}$$

More sophisticated relationships may be used instead of Eq. (15).

The functions $\tau(u)$ and $\varphi(u)$ were calculated using the parameter values in Eq. (3): $\nu = 2$, 3 and 4, $\Delta\rho/\rho_w = 0.3$, or $\rho_n = 700\,\mathrm{kg/m^3}$. The LP scaling coefficient ratio, β_{nw}/β_{an}, was taken equal to one. The exponent m in the expression for K_{rn} was taken equal to 3. The results, obtained numerically, by using the MATLAB program, are presented as points denoted "*", "+" and "o" in Figs. 3 and 4.

Based on these results, the functions $\varphi(u)$ and $\tau(u)$, with good accuracy, may be described by the following approximate expressions:

$$\varphi(u) = \beta_0 u^{\frac{1}{\nu+1}}\left(u + u_0\right)^{\frac{\nu}{\nu+1}}, \tag{16}$$

and

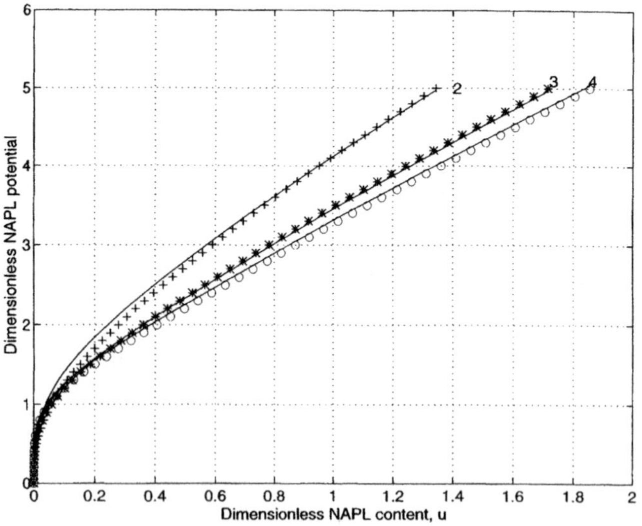

Fig. 3. Values of $\varphi(u)$ for $\nu = 2, 3, 4$

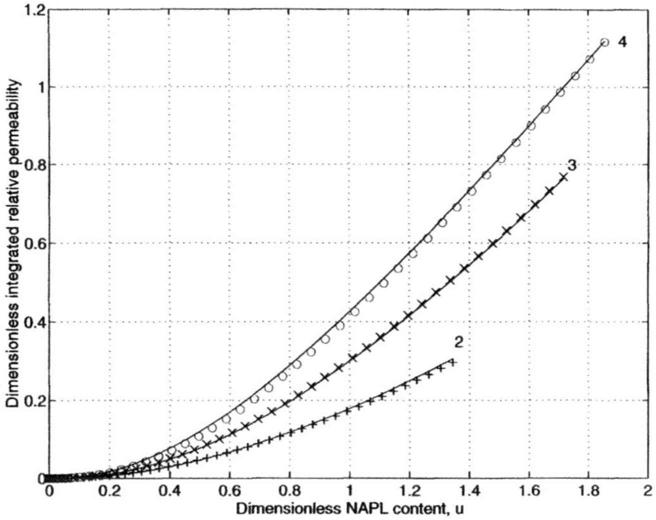

Fig. 4. Values of $\tau(u)$ for $\nu = 2, 3, 4$

$$\tau(u) = \beta_1 \frac{u^{\frac{m\nu+1}{\nu+1}}}{(u+u_1)^{\frac{\nu(m-1)}{\nu+1}}}. \tag{17}$$

Here, β_0, β_1, u_0 and u_1 are dimensionless constants, which depend on ν, m, and $\Delta\rho/\rho_n$.

The results of the calculations, made with Eqs. (16) and (17), are shown as solid lines in Figs. 3 and 4. They are compared with results obtained by direct integration (shown as points denoted by "*", "+" and "0").

4 Unsteady LNAPL Lens Movement over a Water Table

Let us now discuss some problems of LNAPL spreading over an initially planar watar table.

The initial conditions in all the problems are that $u(x,y,0)$ at $t = 0$ is a known function, $u_0(x,y)$, within some initial domain Ω_0 in the x,y-phane, and is zero outside of this domain.

The current lens domain, Ω_t, is bounded by a closed moving curve, Σ_t. This boundary is a "free boundary"; its shape and position vary continuously, in an a priori unknown manner. There is no flow of NAPL through the boundary Σ_t. This may be expressed by the boundary condition:

$$\lim_{u \to 0} \frac{\alpha Q_N}{\phi u} - \upsilon_N = 0 \quad \text{on } \Sigma_t, \tag{18}$$

where Q_N is a projection of the NAPL flow-rate vector on the normal to Σ_t, and v_N is the velocity of the boundary in the direction normal to its surface. Note that along the boundary $Q_N = u_N = 0$.

4.1 The Displacement of a NAPL Mound Along a Sloping Water Table

Consider first an LNAPL mound that was formed above a sloping water table. Immediately after its establishment, the mound starts moving along the water table. Figure 5 shows, schematically, the evolution of a mound of some arbitarily selected initial ($t = 0$) shape. The mound's displacement in the x-direction is described by Eq. (13), the x-axis being selected as the direction along the maximum slope.

If the dimensions of the lens are large in comparison with χ, and the value of i_0 is not too small, i.e. $\left|\chi \frac{\partial \varphi(u)}{\partial x}\right| << i_0$, we obtain from Eq. (13)

$$\frac{\partial u}{\partial t} - V_0 \tau'(u) \frac{\partial u}{\partial x} = 0, \tag{19}$$

where $V_0 = Ci_0$, $\tau'(u) = d\tau/du$.

The initial conditions corresponding to the lens movement problem may be written in the form:

$$t = 0, \quad u(x,0) = u_0(x), \quad 0 \le x \le L; \quad u = 0, \quad x < 0,\, x > L, \tag{20}$$

where $u_0(x)$ is a given initial distribution of u in the interval $0 \le x \le L$.

The first-order partial differential equation (PDE) Eq. (19) is similar to the well-known Buckley-Leverett equation for immiscible displacement in porous media (Buckley and Leverett 1942; Bear 1972; Barenblatt et al. 1990), except that the variable here is the dimensionless NAPL content in the mound, u, rather than saturation. According to the general theory, any value of u propagates at the velocity

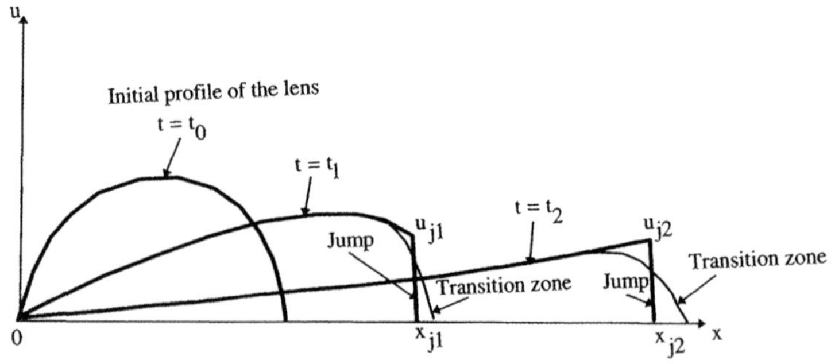

Fig. 5. The movement of an NAPL mound along an inclined water table and the formation of a jump

$$V_u(u) = \frac{dx}{dt} = V_0 \tau'(u), \tag{21}$$

increasing with u and tending to zero as $u \to 0$.

As the greater values of u propagate faster than the smaller ones, independent of the initial shape of the mound, eventually the leading edge of the mound becomes steeper, and the trailing edge flattens out. The evolution of the mound follows the general pattern shown in Fig. 5, which shows a sequence of positions of an initially convex mound.

As the function $\tau(u)$ is an increasing one, then with the initial conditions of Eq. (20), a jump in the function $u(x,t)$ forms at the leading front of the moving NAPL mound (Fig. 5). On the left (upper) side of the jump, $u = u_l = u_J$, while on the right side we have $u = u_r = 0$. The velocity of the jump, V_J, is defined by the following equation, obtained from the condition of the NAPL mass conservation:

$$V_J = \frac{dx_J}{dt} = V_0 \frac{\tau(u_J)}{u_J}. \tag{22}$$

Details are given by Bear (1972) and Barenblatt et al. (1990).

The velocity of the jump with "amplitude" u_J, is less than the velocity of propagation of the considered u_J-value, according to Eq. (21). Hence, the value of u_J decreases with time.

Using Eqs. (21) and (22), the evolution of a lens of any form may be calculated. The example of the evolution of a NAPL mound, having a rectangular initial form, is shown in dimensionless coordinates in Fig. 6.

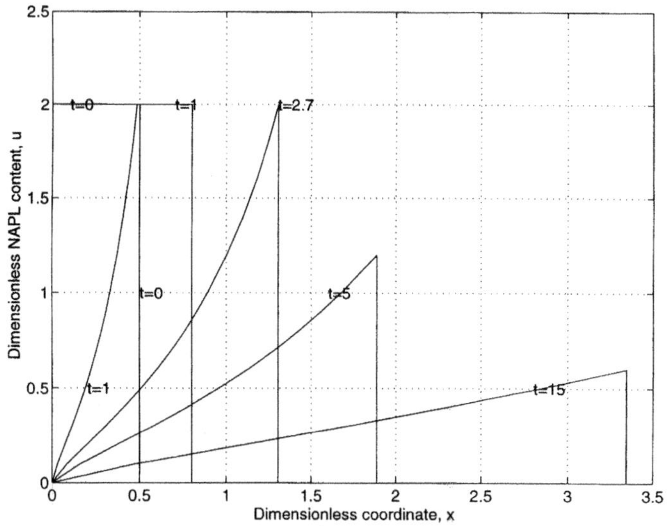

Fig. 6. Variations of NAPL content, u, along an inclined water table

If, for example, a mound of NAPL is initially concentrated in a narrow strip of 1 m width in the x-direction, the quantity of NAPL is 0.5 m^3 per 1 m length of strip, the soil parameters are the same as used in Section 3, $\nu = 2$ and $\alpha = 10\,m^{-1}$, one may calculate that the NAPL strip will widen in 1 day to the width 1.42 m and after 100 days to 10.2 m. The maximum (peak) values of u in this case are: 0.64 m in 1 day, and 0.089 m in 100 days.

The appearance of a jump in the solution of Eq. (19) is due to neglecting the capillary pressure, which is proportional to χ. Had we solved the problem of the LNAPL lens movement along an inclined water table for the "full" Eq. (13), we would have obtained a transition zone of small width, as shown in Fig. 5, instead of a jump.

4.2 Spreading of a NAPL Lens over a Horizontal Water Table with a Source

Let us consider now the spreading of a NAPL lens over a *horizontal* water table. Employing the approximation developed in Section 3, such spreading is described by Eq. (13) with $i_{0x} = i_{0y} = 0$.

At this stage, we shall add the feature of a NAPL source, augmenting the NAPL volume in the lens. We shall limit the discussion to the case of a point source, supplying NAPL at a constant rate, Q_n^0.

We assume that, initially, no NAPL exists floating on the horizontal water table. At $t = 0$, LNAPL is introduced through a source of a very small horizontal cross-sectional area, approximated as a point source. This source may be an approximate description of a leak from a tank or a pipeline. When the leaking NAPL reaches the water table (considered here as $t = 0$), a lens, floating over the water table, begins to form. Because the water table is horizontal, the lens is circular. Within the lens, the NAPL flow is radial, i.e., $u = u(r,t)$, where $r = \sqrt{x^2 + y^2}$.

For radial horizontal flow within the lens, Eq. (13) reduces to

$$\frac{\partial u}{\partial t} = \frac{a}{r}\frac{\partial}{\partial r}\left[r\tau(u)\frac{\partial}{\partial r}\varphi(u)\right], \quad a = C\chi. \tag{23}$$

The NAPL flow is radial in the domain $r_0 < r < r_t$, where $r_0 \to 0$, and $r_t = r_t(t)$ is the radius of the outer boundary of the lens, at $u = 0$. At $r = r_0 \to 0$, the boundary condition is

$$Q_n^0 = -\lim_{r\to 0}\left(2\pi r a\chi\tau(u)\varphi'(u)\frac{\partial u}{\partial r}\right) = \text{const}, \quad \varphi'(u) \equiv \frac{d\varphi}{du}. \tag{24}$$

The boundary condition at $r = r_t(t)$ expresses the absence of NAPL flow through the outer boundary of the lens, as formulated in Eq. (18). This condition transforms to:

$$\lim_{r\to r_t}\left(a\chi\frac{\tau(u)\varphi'(u)}{u}\frac{\partial u}{\partial r}\right) - \frac{dr_t(t)}{dt} = 0. \tag{25}$$

The problem as presented above has an exact (self-similar) solution, which, at least qualitatively, describes NAPL spreading over a horizontal water table. This solution belongs to the family of solutions developed by Barenblatt et al. (1990). The existence of self-similar solutions may be established by dimensional considerations. The value of u may be considered as a function of the dimensionless Boltzmann variable $\zeta = r/\sqrt{at}$.

Then Eq. (23) takes the form of an ordinary differential equation (ODE) in terms of $u = u(\zeta)$

$$\frac{1}{\xi}\frac{d}{d\xi}\left(\xi R(u)\frac{du}{d\xi}\right)+\frac{\xi}{2}\frac{du}{d\xi}=0, \quad R(u)\equiv \tau(u)\frac{d\varphi(u)}{du}. \tag{26}$$

The boundary conditions of Eqs. (24) and (25), take the form:

$$\lim_{\xi\to 0}\left(\xi R(u)\frac{du}{d\xi}\right)=-\tilde{Q}. \tag{27}$$

and

$$\lim_{\xi\to\xi_0}\frac{R(u)}{u}\frac{du}{d\xi}=-\frac{\xi_0}{2}, \quad u(\xi_0)=0, \tag{28}$$

where $\xi_0 = r_t/\sqrt{at}$. The value of ζ_0 is unknown, and must be defined in the course of numerical solution of Eq. (26).

Calculations for a typical example were carried out using MATLAB. For the function $\tau(u)$, $\varphi(u)$ and $R(u)$ Eqs. (16) and (17) with parameter values $\nu = 2$, and $m = 3$, were used. The calculated curves $u(\xi)$ for four values of $\tilde{Q}$ are shown in Fig. 7.

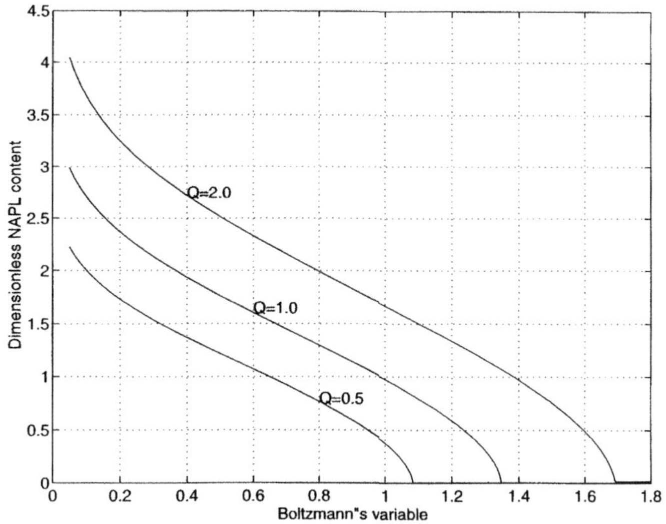

Fig. 7. NAPL content distribution around a source

The total flow rate of the NAPL source, Q_n^0, is equal to $2\pi va\chi\tilde{Q}$. If, for example, $a = 0.05\,\mathrm{cm}^2/\mathrm{s}$, $\chi = 10\,\mathrm{cm}$, then, if $\tilde{Q} = 1$, the value of Q_n^0 is about $0.25\,\mathrm{m}^3/\mathrm{day}$, and the radius of the NAPL "front" after a day of leakage is equal to 0.89 m, and is proportional to $\sqrt{t}$, where t is in days.

The self-similar solutions of the type described above may be obtained also for the case of an NAPL lens with a linear source.

4.3 Spreading of a LNAPL Mound over a Horizontal Water Table

Specific self-similar solutions of the problem of spreading of an NAPL mound over a horizontal water table were derived for an instantaneous source. By "instantaneous source" we mean that at the initial moment, $t = 0$, a finite quantity of the NAPL, concentrated in the form of the Dirac δ-function, is injected. A mound is created, which then continues to spread radially.

The "instantaneous source" is a good asymptotic approximation for the description of an NAPL lens that spreads out over time periods which are much larger than the time of initial NAPL application and when the initial (horizontal) area of source considered is much smaller than the diameter of the lens. The spreading from an instantaneous source is described by Eq. (23).

Analytical solutions of the instantaneous source type, may be obtained if we assume u to be small and that the power-law asymptotics apply for the functions $\tau(u)$ and $\varphi(u)$. Then Eq. (23) takes the form

$$\frac{\partial u}{\partial t} = \frac{a_*}{r}\frac{\partial}{\partial r}\left(r\frac{\partial u^{\lambda+1}}{\partial r}\right). \tag{29}$$

The constants λ and a_* may be expressed in terms of the constants defined by Eqs. (16) and (17), namely

$$\lambda = \frac{v(m-1)+1}{v+1}, \quad a_* = \frac{a\beta_0\beta_1}{v+1}. \tag{30}$$

The value of the exponent λ for soils with the values of the van Genuchten exponent v from 1.5 to 4 varies, approximately, in the range from 2 to 3.

Equation (29) is equivalent to that describing polytropic flow of a perfect gas in a porous medium studied by Barenblatt (see Chap. 3 in Barenblatt et al. 1990). For this equation, Barenblatt (1952) obtained a self-similar solution of the instantaneous source type, describing the process of the spreading of an initially concentrated (δ-function) gas mass.

Let U_0 be the total dimensionless volume of LNAPL that has leaked and reached the water table:

$$U_0 = \frac{2\pi\phi}{\chi}\int_0^\infty u r dr, \tag{31}$$

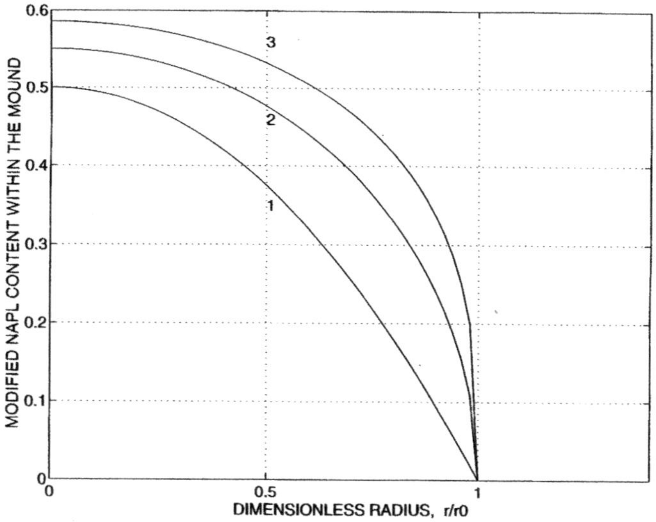

Fig. 8. NAPL content distribution within a mound

Recall that initially all the NAPL is assumed to be concentrated in the vicinity of $r = 0$.

The distribution of $u(r,t) = u^*(x,t)$ is then described by the following expressions:

$$u^* = \left(\frac{U_0}{2\pi a_* \phi t}\right)^{\frac{1}{\lambda+1}} \left[\frac{\lambda e_1}{4(\lambda+1)^2}\left(1-\frac{r^2}{r_0^2}\right)\right]^{\frac{1}{\lambda}}, \quad r < r_0(t), \tag{32}$$

and

$$r_0(t) = e_1^{\frac{1}{2}} \left[\frac{U_0^{\lambda} a_* t}{(2\pi\phi)^{\lambda}}\right]^{\frac{1}{\lambda+1}}, \quad e_1 = \frac{1}{\lambda}\left[2(\lambda+1)\right]^{\frac{\lambda+2}{\lambda+1}}. \tag{33}$$

It may be verified that for the solution described by Eq. (32), at $r = r_0$, a condition equivalent to that of Eq. (18) is satisfied.

The solution, analogous to Eq. (32), has also been obtained for one-dimensional linear flow.

The solution for the case $\lambda = 1$, corresponding to a water mound spreading over an impermeable horizontal surface, is well known and is presented in Bear (1972) and Barenblatt et al. (1990).

Dimensionless plots of $u^*(r)$, are shown in Fig. 8. for the values of λ = 1, 2 and 3.

From Eq. (32) it follows that the maximum height of a spreading mound decreases as $t^{-1/(\lambda+1)}$. The radius of the mound grows as $t^{1/(\lambda+1)}$. The total NAPL volume in the mound remains constant.

As a numerical example, let us consider the case of a radially spreading mound with $\nu = 2$, and $m = 3$, so that $\lambda = 8/3$. Using Eq. (32), we obtain an expression for $r_0(t)$, in terms of dimensional variables:

$$r_0(t) = 1.5415\chi\left(\frac{U_1}{2\pi\chi^3}\right)^{\frac{8}{11}}\left(\frac{Vt}{\chi}\right)^{\frac{3}{11}}. \tag{34}$$

Let us introduce $\chi = 20\,\text{cm}$, $\rho_n = 0.67\,\text{g/cm}^3$ and $V = 0.007\,\text{cm/s}$. If, say, $U_1 = 1000\,\text{l}$, then

$$r_{0d} = 18.8t^{3/11},$$

where r_0 is in meters and t in days.

4.4 LNAPL Spreading Above a Horizontal Water Table. The Case of Non-Zero Retained NAPL

The governing equations of Section 3 were obtained using the assumption that above the upper level of the lens, $z = b_2 + \eta$, and below the level of the water, $z = \eta$, NAPL saturation, S_n, is equal to zero. If in some (two-dimensional) domain, the NAPL content, $\mathcal{U}$, decreases, NAPL saturation at the levels $z = z_2$ and $z = \eta$ must be equal to some residual value, $S_n = S_{nr}$. Such a domain exists, for example, above and below the central part of a spreading NAPL mound, as shown in Fig. 9.

While spreading over a horizontal water table, an LNAPL lens (mound) becomes thinner and wider. Hence, in the central part of the lens, the upper boundary is lowered, while the lower one is rising, as shown in Fig. 9. Above and below the central part of the moving lens, the residual NAPL is left at a value that we shall refer to as "retained saturation", S_{nrt} ($\geq S_{nr}$) (shaded domain). At the same time, in the outer part of the lens, where it is spreading into a domain where initially no NAPL was present, the NAPL saturation above and below the lens is equal to zero. The "central" zone of the lens is characterized by the condition $\partial u/\partial t < 0$, and the "outer" zone by $\partial u/\partial t > 0$, where u is defined as (dimensionless) *mobile* NAPL content. The two zones are separated by a moving boundary.

Lenhard and Parker (1988) measured S_{nrt} in two- and three-phase displacement experiments.

For simplicity, we shall assume that the values of S_{nrt} at $z = b_2 + \eta$ and at $z = \eta$ are equal. If

$$S_n(z_2) = S_n(\eta) = S_{nrt} > 0,$$

the integration of the time-derivative term in the NAPL balance equation in the z-direction, using appropriate integration rules (Bear 1979), gives

$$\int_\eta^{z_2} \frac{\partial S_n}{\partial t}dz = \frac{\partial}{\partial t}\int_\eta^{z_2} S_n dz - S_{nrt}\frac{\partial(z_2 - \eta)}{\partial t} = \frac{1}{\phi}\frac{\partial U}{\partial t} - S_{nrt}\frac{\partial \zeta_2}{\partial t}. \tag{35}$$

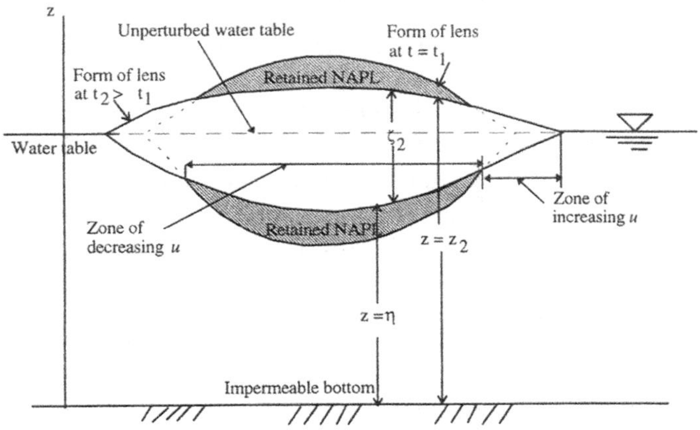

Fig. 9. Spreading of an NAPL mound showing the zones of residual phase

Theoretically, the value of S_{nrt} must be determined on the basis of the history of saturation changes, using hysteretic loops in the capillary pressure curve. Here, we shall use a simplified approach, assuming S_{nrt} to be proportional to the average, along the vertical, NAPL saturation, S^{n}_{av}, at the considered point of the lens.

If we use the van Genuchten expression for capillary pressure functions, at small values of $u = \mathcal{U}\alpha/\phi$, and lens thickness, the value ζ_2 may be described asymptotically as power-law function of u:

$$\zeta_2 = \frac{\kappa}{\alpha} u^{\frac{\nu}{\nu+1}}, \tag{36}$$

where κ is a dimensionless constant, depending on the parameters α and ν of the van Genuchten's curve. Then

$$S_n^{av} = \frac{u}{\alpha\zeta_2} = \frac{1}{\kappa} u^{\frac{1}{\nu+1}}, \quad S_{nrt} = \varepsilon S_n^{av} = \frac{1}{\kappa} \varepsilon u^{\frac{1}{\nu+1}}, \tag{37}$$

where ε is a constant.

The right-hand side of Eq. (13) remains the same, we merely use power-law approximations for the functions $\varphi(u)$ and $\tau(u)$. However, on the left-hand side, if $\partial u/\partial t < 0$, we must replace $\partial u/\partial t$ by the expression implied by Eq. (35).

In the case of radially symmetric flow, at small values of u, when the expressions of the power-law type are used for $\varphi(u)$ and $\tau(u)$, we arrive at the same Eq. (29), but with

$$a_* = a_1 \quad \text{at } \frac{\partial u}{\partial t} > 0, \quad a_* = a_2 = \frac{a_1}{1 - \dfrac{\varepsilon\nu}{\nu+1}} \quad \text{at } \frac{\partial u}{\partial t} < 0. \tag{38}$$

Our purpose is to construct self-similar solutions for Eq. (29) subject to the conditions of Eq. (38), describing an NAPL mound spreading over the horizontal water table.

For the case considered here, it follows from the general results of Barenblatt (1979) for nonlinear PDE of the parabolic type, with discontinuous coefficients, that the common continuous self-similar solution, corresponding to an instantaneous point source, described in Section 4.3, does not exist. Instead, for the initial conditions mentioned, the so-called self-similar solution of the second kind exists. The particular characteristic of the self-similar solutions of the second kind is that the dependent variable, as in the case of the common self-similar solution, is a function of the combination of independent variables r/t^{γ}. However, the constant γ is not determined by dimensional considerations alone. Instead, it depends on the boundary conditions. Kochina et al. (1983) obtained such a self-similar solution of the second kind, corresponding to the spreading of a water mound over an impermeable bottom, in the form of an instantaneous source, initially concentrated in an infinitesimal area around the origin of the coordinates. This solution is described by Barenblatt et al. (1990). For the problem of a mound spreading, γ is a function of the ratio a_1/a_2.

Searching for the self-similar solution of Eq. (29), subject to Eq. (38), yields an expression of the type:

$$u = t^{-\gamma} F(\xi), \quad \xi = \frac{r}{bt^{\delta}}, \tag{39}$$

where b, δ and γ are undefined constants.

A substitution of the Eq. (39) for u into Eq. (29), is compatible with dimensions of all variables only when $\gamma = (1 - 2\delta)/\gamma$. Then, after some transformations, one obtains from Eq. (29) the following ODE for the function $F(\xi)$:

$$\frac{d^2F}{d\xi^2} + \frac{\lambda}{F}\left(\frac{dF}{d\xi}\right)^2 + \frac{1}{\xi}\frac{dF}{d\xi} + \frac{K}{(\lambda+1)F^{\lambda}}\left[\frac{1-2\delta}{\lambda}F + \delta\xi\frac{dF}{d\xi}\right] = 0. \tag{40}$$

In Eq. (40) $K = 1$ at $dF/d\xi > 0$, and $K = a_2/a_1$ at $dF/d\xi < 0$. The constant b may be chosen in such a way that $F \equiv 0$ at $\xi \geq 1$ and $F \neq 0$ at $\xi < 1$. The problem of the mound spreading may now be formulated as the following boundary value problem for Eq. (40):

$$\text{at } \xi = 0, \quad F = 0, \quad \frac{dF}{d\xi} = 0, \tag{41}$$

(this condition corresponds to mound symmetry),

$$\text{at } \xi = 1, \quad \frac{dF^{\lambda}}{d\xi} = -\frac{K\delta\lambda}{\lambda+1}. \tag{42}$$

And, Eq. (42) is a consequence of continuity of total NAPL flow at the point $\zeta = 1$.

Hence, for an ODE of the second order, Eq. (40), we have a boundary problem with three conditions. This problem has a non-zero solution only for the special

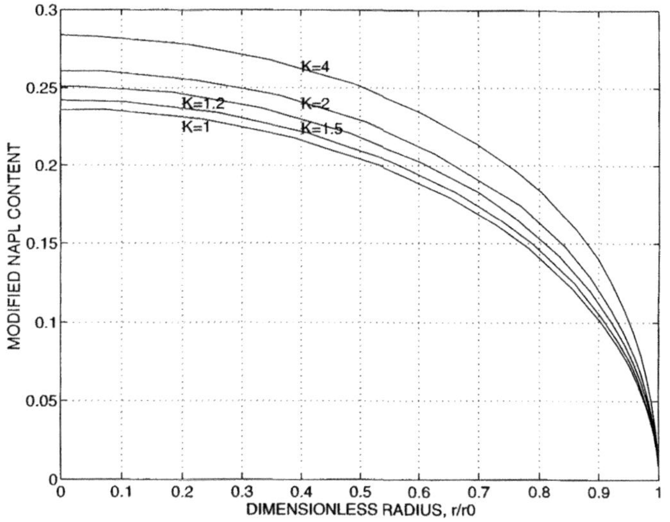

Fig. 10. Form of the NAPL mound in the radial case

values of the non-defined parameter δ, i.e. this is an eigen-value problem. The value of δ in this formulation depends on $K = a_1/a_2$. A solution of this problem was obtained numerically. The results of the calculations are presented in Fig. 10 as curves $F(\zeta)$ for $\lambda = 2$ and different values of K.

For example, if $\lambda = 2$, the values of δ are: at $K = 1$, $\delta = 1/6 = 0.167$, at $K = 1.5$, $\delta = 0.143$, at $K = 2$, $\delta = 0.130$, at $K \to 1$, $\delta \to 1/2(\lambda + 1)$.

Using the same approach, additional self-similar solutions of the second kind for the one-dimensional linear flow case were obtained.

In the case of spreading of an NAPL mound, with the formation of a retained NAPL zone, the total quantity of the mobile NAPL is

$$M_r = 2\pi \int_0^{R^*} urdr,$$

where the radius of the mound, R_*, does not remain constant, but decreases as a power-law function of t. If u is described by Eq. (39), we obtain the expression:

$$M_r = M_r^* t^{-\theta}, \tag{43}$$

where M_r^* is a constant with

$$\theta = \frac{1}{\lambda}\left(1 - 2\delta\left(\lambda + 1\right)\right). \tag{44}$$

The value of the exponent θ determines the rate at which NAPL transforms into the retained (trapped) state. If retained NAPL saturation, S_{nrt}, increases, the value of $K = a_2/a_1$, also increases. Consequently, the value of δ is reduced, and ϑ grows. Recall that at $K = 1$, i.e., with no retained phase, $\delta \to 1/2(\lambda + 1)$.

We have to emphasize that the self-similar solution, described above, being the self-similar solution of the second kind, is essentially incomplete, as the constants b or M_r^* cannot be determined from the solution of the ODE, Eq. (40). For their determination we must know the radius of the mound, R_*, or the mobile NAPL content in the mound, M_r, at some given moment of time t_0. These values may be found by solving (though only for some short initial interval of time), the full PDE Eq. (29) subject to Eq. (38).

5 Equilibrium and Movement of a DNAPL Lens over an Impermeable Bottom

When a denser than water non-aqueous phase liquid (DNAPL) infiltrates through the ground surface in a sufficiently large quantity, it will percolate through both the unsaturated and saturated zones, eventually reaching the impermeable bottom, forming a lens above it (Fig. 11). The lens contains both NAPL and water at different saturations.

To model the movement of such a lens and its spreading along the impermeable boundary, using the VE hypothesis, let the form of the impermeable boundary be described by:

$$z = \eta_*(x, y). \tag{45}$$

Under capillary equilibrium conditions, at a point of elevation z inside the lens, we have:

$$P_c(S_w) = p_n - p_w = \Delta\rho g(\eta_* + d - z), \tag{46}$$

where p_w and p_n are the water the NAPL pressures, respectively, $P_c(S_w)$ is the capillary pressure, assumed to be a function of water saturation, S_w, $\Delta\rho = \rho_n - \rho_w$, (here $\rho_n > \rho_w$,) d is the thickness of the lens, and H_0 the elevation of the water table (see Fig. 11).

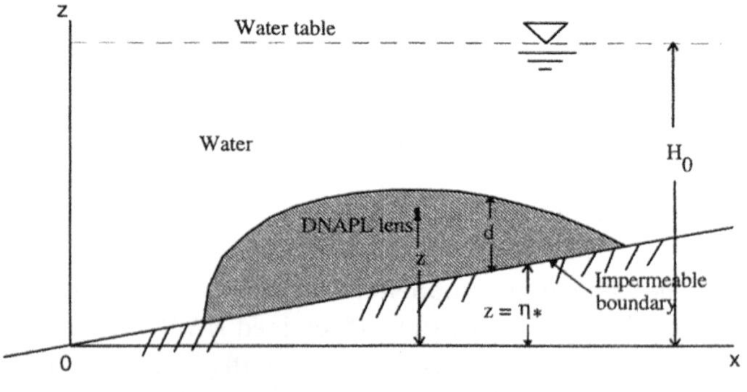

Fig. 11. Schematic view of a DNAPL lens in the subsurface

If $P_c(S_w)$ is a known function of water saturation, S_w, Eq. (46) describes static vertical saturation distribution, i.e., the function $S_w(z)$.

The governing 2-D equations for a moving lens are derived from the DNAPL and water balance equations, in the same way as in Section 3. The 2-D formulation of the flow model is obtained by integrating 3-D equations of flow over the thickness of the lens, assuming the pressures and saturation distributions to be hydrostatic (i.e., assuming flow within the lens to be horizontal). Also, the thickness of the lens is assumed to be small in comparison with that of the aquifer. Under this assumption, the water table is assumed to be unperturbed. Therefore, the mobility of water is assumed to be much greater than that of the DNAPL, and hence the water pressure is distributed hydrostatically throughout the (water) saturated zone (in both the horizontal and vertical directions). This assumption implies that the elevation of the water table, H_0, may be considered as a constant.

In deriving the balance equation for DNAPL, integrated along z from η_* to $\eta_* + d$ we assume also that the capillary pressure function is expressed by van Genuhten's expression, Eq. (3).

The integration of the 3-D equations of NAPL continuity and flow, under the VE hypothesis gives, after some transformations

$$\frac{\partial}{\partial t}\left[\vartheta-\Psi_v(\vartheta)\right]-a\frac{\partial}{\partial x}\left[\tau_v(\vartheta)\left(\frac{\partial\vartheta}{\partial x}-i_x(x,y)\right)\right]- \\ -a\frac{\partial}{\partial y}\left[\tau_v(\vartheta)\left(\frac{\partial\vartheta}{\partial y}-i_y(x,y)\right)\right]=0, \tag{47}$$

where $\vartheta = d/\chi_*$ and $\zeta = (z-\eta_*)/\chi_*$, with $\chi_* = P_{c0}/\Delta\rho g$, with Pc_0 denoting a characteristic capillary pressure, and $-i_x(x,y)$ and $-i_y(x,y)$ denoting the components of the vector $\nabla_{x,y}\eta_*$; in addition $a = (k/\mu_n)\chi_*\Delta\rho g$;

$$\tau_v(\eta)=\frac{1}{\chi_*}\int_{\eta^*}^{\eta^*+d}k_{rn}(S_n)dz,$$

and k_{rn} is the relative permeability for NAPL. The function $\Psi_v(\vartheta)$ is expressed by

$$\Psi_v(\vartheta)=\int_0^{\vartheta}\left(1+\xi^v\right)^{-\frac{v-1}{v}}d\xi. \tag{48}$$

The functions $\Psi_v(\vartheta)$ and $\tau_v(\vartheta)$ may be calculated numerically, and approximate expressions for them may be obtained and used in solving Eq. (47), as was done in Section 3. For the case $v = 2$, the analytical expressions for these functions exist.

In view of the assumptions made above, the water flow equations are replaced by the condition of hydrostatic water pressure distribution.

Equation (47) is similar to Eq. (13) of Section 3. The analytical solutions of this equation may be obtained in the same way as for the movement of an LNAPL lens above the water table. For example, if the impermeable bottom is a plane, the x-axis is directed along the maximum slope, and with the assumption $|\partial\vartheta/\partial z| \ll i_0$, we obtain as a "large scale approximation", instead of Eq. (47), the PDE of the first order,

$$\frac{\partial \vartheta}{\partial t} + a i_0 \frac{1}{1 - d\Psi_v / d\vartheta} \frac{d\tau_v}{d\vartheta} \frac{\partial \vartheta}{\partial x} = 0, \tag{49}$$

where $i_0 = i_x = \text{const.}$

The solutions of this equation for the case of NAPL lens movement along a sloping surface were described in Section 4.1.

We note that in the case of a DNAPL lens movement along an impermeable bottom, the boundary inclination, i_0, may be much greater than the possible inclination of the water table. Hence, the velocity of a DNAPL lens movement may be much greater than that of an LNAPL lens movement along an inclined water table.

For a radially symmetric (cylindrical) flow over a horizontal impermeable boundary, Eq. (47) takes the form:

$$\frac{\partial}{\partial t}\left[\vartheta - \Psi_v(\vartheta)\right] = \frac{a}{r}\frac{\partial}{\partial r}\left(r\tau_v(\vartheta)\frac{\partial \eta}{\partial r}\right). \tag{50}$$

If a source of DNAPL exists having the constant intensity Q_*, then, similar to the case of a LNAPL lens over a horizontal water table, the problem has a self-similar solution, i.e., $\vartheta(r,t)$ is a function of the variable $\xi = r/\sqrt{at}$, and of the parameter $\vartheta = Q_*/a$.

The self-similar solution of Eq. (50) for the case of a DNAPL source of constant intensity, Q_*, may be obtained in the same way as in Section 4.2. The self-similar solution of the same equation for the case of an instantaneous source of DNAPL is analogous to one described in Section 4.3 for the spreading of a LNAPL mound over a horizontal water table.

6 Conclusion

In this chapter, approximate mathematical models describing the movement of a relatively thin LNAPL lens over the water table and DNAPL lens over an impermeable bottom are proposed. The main assumptions are: (1) that "vertical equilibrium phase distribution" prevails in water, LNAPL and air in the unsaturated zone, and in DNAPL and water within a DNAPL lens; and (2) that the thickness of the NAPL lens is small in comparison with the depth of water-saturated zone. As a consequence, each model reduces to a unique 2-D PDE of the second order for the dimensionless NAPL content per area of the lens. Some exact analytical solutions were obtained for cases of a lens moving and spreading along a sloping or horizontal water table, and for a DNAPL lens moving and spreading along an impervious aquifer bottom.

Acknowledgment. Part of the work has been conducted in cooperation with Prof. Carol Braester, Technion, and Prof. V. Entov, Moscow. The first author wishes to acknowledge the support of the Fund for the Promotion of Research at the Technion. The second author wishes to acknowledge the support of the Water Institute, Technion.

References

Barenblatt GI (1952) On self-similar movement of compressible fluid in a porous medium. Prikl Mat Mekh 16:679–698 (in Russian)

Barenblatt GI (1979) Similarity, self-similarity and intermediate asymptotic. Plenum Press, New York

Barenblatt GI, Entov VM, Ryzhik VM (1990) Theory of fluid flows in natural rocks. Kluwer, Dordrecht

Bear J (1972) Dynamics of fluids in porous media. Elsevier, New York

Bear J (1979) Hydraulics of groundwater. McGraw-Hill, London

Blunt M, Dengen Zhou, Fenwick D (1995) Three phase flow and gravity drainage in porous media. Transport Porous Media 20:77–103

Buckley SE, Leverett MC (1942) Mechanism of fluid displacement in sands. Trans AIME 142:107–116

Corapcioglu MY, Tuncay K, Lingham R, Kambham KKR (1994) Analytical expressions to estimate free product recovery in oil-contaminated aquifers. Water Resour Res 30:3301–3311

Kochina IN, Mikhailov NN, Filinov MV (1983) Groundwater mound damping. Int J Eng Sci 21:413–421

Lenhard RJ, Parker JC (1987a) Measurement and prediction of saturation-pressure relationship in three-phase porous media system. J Contam Hydrol 1:407–424

Lenhard RJ, Parker JC (1987b) A model for hysteretic constitutive relations governing multiphase flow, 2. Permeability-saturation relations. Water Resour Res 23:2197–2206

Lenhard RJ, Parker JC (1988) Experimental validation of the theory of extending two-phase saturation-pressure relations to three fluid phase systems for monotonic drainage path. Water Resour Res 24:373–380

Parker JC, Lenhard RJ (1987) A model for hysteretic constitutive relations governing multiphase flow, 1. Saturation-pressure relations. Water Resour Res 23:2187–2196

Parker JC, Lenhard RJ (1990) Vertical integration of three-phase flow equations for analysis of light hydrocarbon plume movement. Transport Porous Media 5:187–206

Schiegg HO, Schwille F (1991) Hydrocarbons in porous media. NATO ASI Ser E202:69–201

Sleep BE, Sykes JF (1993) Compositional simulation of groundwater contamination by organic compounds. 1. Model development and verification. 2. Model applications. Water Resour Res 29:1697–1708, 1709–1718

Van Dam J (1967) The migration of hydrocarbons in a water-bearing stratum. In: Hepple P (ed) Joint problems of oil and water industries. Inst of Petroleum, London, pp 55–88

Van Genuchten MT (1980) A closed-form equation for predicting the hydraulic conductivity of unsaturated soils. Soil Sci Soc Am J 44:892–898

Yu-Shu Wu, Huyakorn PS, Park NS (1994) A vertical equilibrium model for assessing nonaqueous phase liquid contamination and remediation of groundwater systems. Water Resour Res 30:903–912

21 In situ Bioremediation of Residual Entrapped NAPL in Aquifers by a Biodegradable Nutrient-Surfactant Mix

U. Zoller

Contents

1 Introduction: The Problem

Anthropogenic organic compounds are ubiquitous in our technologically oriented society, due to the ever-increasing demands, use and consumption of these and related commercial products. Consequently, they constitute a substantial portion of the worldwide atmospheric, water and soil pollution profiles.

Subsurface soils, aquifers and groundwater are quite different from other environmental heterogeneous solid and aqueous systems. Typical characteristics are long residence times, low (relative to that of surface soil and water) temperature, low degrees of desorption, dispersion, mixing and dilution processes, decreasing aerobic conditions and, therefore, decreased microbial populations, and fewer – albeit occasionally more complex – operative biochemical organic transformations. The ultimate result is a slow environmental response. Consequently, aquifers and groundwater are susceptible to long-term contamination by hydrophobic organic substances, particularly those having a high degree of hydrophobe branching which is the main cause for their being biodegradation-resistant and, consequently, environmentally persistent organic pollutants (POPs). Fossil fuels, chlorinated hydrocarbons and low HLB branched-chain surfactants are the environmentally most abundant examples of such aquifer pollutants. Indeed, recent studies show, repeatedly, that the penetration of POPs into soil and groundwater is much more substantial than was previously believed (Muszkat et al. 1994; Goldenberg et al. 1996).

Department of Science Education – Chemistry, Haifa University – Oranim, Kiryat Tivon 36006, Israel

H. Rubin/N. Narkis/ J. Carberry
Soil and Aquifer Pollution

The increasing production and use of fossil fuels have substantially increased the probability of water resource contamination by them, particularly groundwater and aquifers. Indeed, aquifers and groundwater contamination by fuels and other non-aqueous phase-liquid (NAPL) compounds (e.g., chlorinated hydrocarbons) associated with industrial, commercial and military activities, already constitute an environmental issue of major concern worldwide (Sabatini et al. 1992; Rubin et al. 1994). In fact, nearly one quarter of the organic chemicals listed as priority pollutants by the US Environmental Protection Agency (US EPA) are organic NAPLs which, like the petroleum-originated products, are immiscible with water. Once they penetrate into the subsurface, they may be trapped as residual saturation in the unsaturated or saturated zones. Currently, entrapped NAPLs in aquifers seem to be one of the biggest existing remediation problems and the most intransigent of contaminated environmental systems (Oppelt and Kavanaugh 1996). The penetration of fuels into soil and groundwater is associated with accidents, continuous neglect, improper codes of fuel handling and disposal, as well as with the leaking of storage tank and fuel conveying systems. Thus, for example, significant number of these systems (about 2 million storage tanks in the US alone; Chen 1992) leak, and significant spillage of jet fuel associated with various military activities in Israel (due to a large number of aircraft per unit area) was and still is taking place. This underground leakage has contaminated soil and groundwater with low molecular weight aromatic hydrocarbons that can threaten human health, particularly if they get into drinking water supplies. Since contamination of the subsurface environment is still taking place, the cleanup of the contaminated soils and aquifers poses a major environmental challenge.

It turns out that the major problem in remediation of NAPL-contaminated aquifers is not the pumping out of either the high or low density (free) NAPLs which are trapped in the aquifers – immiscible with the water phase – below, above, or within the water table. Rather, the major difficulty is the removal of the residual entrapped NAPL which is strongly held by the capillary forces of the soil/porous matrix, causing the former to become virtually immobile. In Israel, for instance, the trapping of fuel in the upper layer of the coastal aquifer constitutes a special problem since the oscillations of the groundwater table there are particularly large. There are several cases of fuel contamination in the aquifers adjacent to the seashore in which the groundwater table is lower than the sea level. In order to avoid the penetration of sea water into the aquifer, it is necessary to raise the groundwater table. Such a raise results in the trapping of additional fuel in the saturated zone (Rubin et al. 1994).

The US EPA has recently concluded that the environmental "pump-and-treat" remediation technology that has been used for removing hydrocarbon NAPL contamination from aquifers is not working (Sabatini et al. 1992). Field demonstrations of surfactant flushing have illustrated potential problems (Nash 1988) and aquifers contaminated with dense nonaqueous phase liquids (DNAPLs) were shown to be extremely difficult to remediate with the standard "pump-and-treat" methods. Enhanced remediation methods, such as flushing with cosolvent or surfactant solutions, promise to reduce remediation times, but actually result in complex physicochemical systems of which we still lack both fundamental understanding and reliable process-based models (Okuda et al. 1996).

It appears that the difficulty in remediating contaminated aquifers with standard pump-and-treat methods is due, in part, to the slow dissolution rates of residual NAPLs (both DNAPLs and LNAPLS) into water (Bedient et al. 1994). Yet, since the solubility of hydrocarbons (as well as the magnitude of other organic solvents and substances) in water can be increased by several orders of magnitude by the addition of surfactants acting as emulsifiers/dispersants (Kile and Chiou 1989; Edwards et al. 1991), there is a growing interest (so far, particularly in the US) in the use of surfactants injected into the contaminated aquifers for enhancing the remediation of soils and aquifers contaminated by entrapped NAPLs (Sabatini et al. 1995).

Based on the state of the available data, accumulated experience and economical/feasibility considerations concerning the remediation of NAPL-contaminated aquifers, *the site-specific, in situ bioremediation of the residual entrapped NAPL by a biodegradable nutrient-surfactant mix* appears to be the approach of choice (Zoller 1996). Guided by the above approach, this chapter focuses on the according conceptualization/evaluation of the problem and the major issues involved, followed by its translation into a target-oriented active research program. As such, it constitutes a response to the following questions with respect to the emerging technologies for the remediation of NAPL-contaminated soils and aquifers:

Why in situ bioremediation? Why surfactant-enhanced? Why biodegradable (soft) surfactants? Why nutrient surfactants? And, why surfactant mix?

However, regardless of the particular technology that would be ultimately used in practice, the crucial and consequential question which should guide any future remediation endeavors of NAPL-contaminated aquifers is: what is the appropriate endpoint of the remediation treatment (or cleanup levels of the entrapped NAPL) given the intended future use of the remediated aquifer?

2 Factors and Mechanisms Affecting the State of NAPLs in Aquifers

The fate of the NAPL hydrocarbons in the subsurface environment is affected by (1) the character of the soil (grain size, porosity, permeability, surface area, organic matter content, cation exchange capacity, etc.); (2) the composition and physical and chemical properties of the NAPL fuel, e.g., its water solubility, soil adsorbtivity, vapor pressure, etc. (Chen 1992); and (3) the characterization (qualitatively and quantitatively) of the particular subsurface site in terms of available organic substrates and inorganic nutrients (e.g., N and P) required for biodegradation to take place and the specific microorganisms present (Steffensen and Alexander 1995). The migration of the NAPL materials may be caused by advection, dispersion or diffusion. Their fate in the subsurface can be: volatilization, adsorption to the soil or rock, dissolution in either the pore water or the water film surrounding the soil particles, trapping in the pores or fractures of rock; floating on, trapping under, or dissolution in the groundwater, and undergoing biochemical transformations or biodegradation. In view of the heterogeneous composition of the soil matrix and the complex nature of the pore network, the description of transport processes of NAPL in terms of mass transfer, adsorption and diffusion models is inherently difficult (Arocha et al. 1996). Typically, NAPL contaminants have low water solubilities and their mass transfer rates in weathered soils are small. These, together with the slow dissolution/elution rates of residual and pooled NAPLs into the aqueous soil solution and the accompanying diffusion and adsorption processes in the micropores are responsible for the persistence of the hydrophobic NAPL hydrocarbons in aquifer soils (Bedient et al. 1994; Arocha et al. 1996; Okuda et al. 1996). The ultimate in situ mineralization of residual entrapped NAPL via biodegradation is clearly the most desirable as far as the remediation of the NAPL-contaminated aquifers is concerned. This requires (1) accessibility of the NAPL substrates to microorganisms capable of degrading the former within the two-phase (solid soil/aqueous solution) subsurface system; and (2) mitigation of the toxicity to the microorganisms present by several components of the NAPL contaminant (Robertson and Alexander 1996). Surfactants have the ability of increasing the desorption, aqueous solubility (Zoller 1994) and mobility in soils (Goldenberg et al. 1996; Smith et al. 1991) of poorly soluble hydrophobic compounds and thus the potential of improving the accessibility to microorganisms of these substrates resulting, ultimately, in enhanced biodegradation (Rouse et al. 1994; Yeom et al. 1996). Therefore, the use of surfactant-based technologies in subsurface remediation (Sabatini et al. 1995) appears to be very promising.

Both enhancement and inhibition of biodegradation of organic compounds (present in NAPLs) in the presence of surfactants have been reported and correlation between inhibited biodegradation of hydrocarbons and surfactant concentrations above the critical micelle concentration (CMC) have been shown by studies involving commercial surfactants and mixed microbial cultures. Yet the mechanisms behind these phenomena are not well understood (Rouse et al. 1994). Significantly, however, surfactants below CMC concentrations have been shown to stimulate bioremediation of sorbed hydrocarbons in samples of aquifer sands (Aronstein and Alexander 1992). This is of particular importance when in situ surfactant-enhanced bioremediation of residual NAPL in aquifers is considered.

3 Bioremediation of NAPL-Contaminated Aquifers (NCAs)

In view of the high hydrophobicity and, consequently, the low water solubility of NAPLs (Table 1), their adsorption on soils and their resistance to chemical oxidative degradation processes under the natural conditions prevailing in the unsaturated and saturated zones of aquifers, *biodegradation* by living organisms is the main process leading, ultimately, to their mineralization. Without biodegradation,

Table 1. Biodegradation-related physical properties and characteristics of selected (A) bioremediable components[a] of NAPL; and (B) commercial surfactants used in enhanced bioremediation studies

Compound	Mol. wt.[b]	C-atoms	Water solubility (mg/l)	log K_{ow}	CMC[c] (g/l)
A					
Benzene	78	6	1790.0		
Toluene	92	7	579.0		
m-Xylene	106	8	215.0	3.20	
Naphthalene	128	10	30.6	3.36[d]	
Phenanthrene	178	14	1.2		
n-Hexane	86	6	9.0		
B					
Zoharpon LAS[e]	295	12	[f]		0.40[g]
SDBS[h]	348	18	[f]		0.40[i]
Triton X-100[j,k]	646	34	[f]		0.11[l]
Igepal CO 660[j,m]	660	35	[f]	2.37	0.06[n]
Zoharteric D[o]	502	44	[f]		0.16[g]

[a] Reported to have been successfully bioremediated.
[b] Round figures for single compounds; averaged mol. wt. for the commercial product mixtures.
[c] Critical micelle concentration.
[d] Nayyar et al. (1994).
[e] Straight chain, biodegradable ("soft") anionic surfactant (lauryl alcohol sulfate-sodium salt).
[f] Very high solubility: easily >100 g/l.
[g] Our work.
[h] Branched alkyl chain, nonbiodegradable ("hard") anionic surfactant (sodium dodecylbenzene sulfonate).
[i] Mukerjee and Myseles (1971).
[j] Branched alkyl chain, nonbiodegradable ("hard") nonionic surfactants.
[k] Octylphenol polyoxyethylene (10 ethylene oxide units).
[l] Yeom et al. (1996).
[m] Nonylphenol polyoxyethylene (10 ethylene oxide units).
[n] Approximate figure.
[o] Straight alkyl chain, biodegradable ("soft") amphoteric anionic surfactant.

NAPL chemicals can accumulate in the subsurface environment and persist there, reaching levels that may threaten human health directly via drinking the water pumped from NCAs or indirectly from bioaccumulation in organisms at higher levels of ecological food webs.

The enormous natural capacity of microorganisms to degrade NAPL organic compounds is the basis for bioremediation. However, since microorganisms cannot adequately decompose every NAPL component, particularly under natural subsurface conditions, bioremediation has limitations that must be recognized when considering its use (Atlas 1995). The optimization of NCA bioremediation is clearly needed, particularly because many of the NAPL constituents either inhibit microorganisms or completely prevent their growth and, in addition, their toxicity to microorganisms lengthens the acclimatization period prior to biodegradation (Robertson and Alexander 1996). Furthermore, since effective biodegradation is contingent on the availability of sufficient inorganic nutrients (e.g., nitrogen and phosphorus) and dissolved oxygen in the NCAs groundwater, the effectiveness of bioremediation is dependent on all the above limiting factors. Therefore although, in principle, bioremediation can treat both soil and groundwater in essentially one step without generating its own hazardous waste products, in situ bioremediation cannot deal with very high levels of organic compounds. With respect to the relatively small quantities of residual entrapped NAPL in aquifers, however, site specific in situ bioremediation has rosy future (Carberry et al. 1996).

Two major approaches have been taken by researchers with respect to the development of bioremediation based technologies. One is focusing on attempts to expand the range of microorganisms used for bioremediation in terms of better pollutant degradation kinetics, attacking a wider range of NAPL contaminants and doing so under extreme subsurface environmental conditions. The other is focusing on NAPLs' desorption (Hayes 1996) from the soil matrix and their dissolution (Diallo et al. 1994) into the aqueous phase of the aquifer. This should substantially increase the NAPLs accessibility and, therefore, availability – as nutritional substrates – to the microorganisms present and, thus, to facilitate effective biodegradation. In the first approach, the development of new microbial strains and the efforts to identify and genetically manipulate the enzymes active in biodegradation are followed by extensive work aiming at the optimization of the biochemical processes involved under a variety of NCA site conditions (Atlas 1995; Trombly 1995). In the second approach, the screening of surfactants having the highest potential for enhancing NCA remediation via desorption-dissolution of NAPL per se or (in addition) by enhancing its biodegradation is also followed by efforts to optimize these surfactant-enhanced processes under real field conditions (Sabatini et al. 1995; Sabatini, this Vol.). The technologies which evolved as a result of the above research efforts appear to be capable of cost effectively bioremediating NCAs.

NAPLs in the subsurface heterogeneous media seem to be one of the greatest bioremediation problems. However, in view of the recent trend to predetermine the endpoint (i.e., the targeted cleanup levels) for the specific site to be remediated (not necessarily restoration to the background levels), site-specific bioremediation is emerging as a very promising and economically feasible technology (Oppelt and Kavanaugh 1996). The cost of moving, transporting or removing large quantities of polluted material is at least several times that of biological treatment. Therefore, in situ bioremediation, the effectiveness of which depends on the degree and characteristics of the organic pollutants, is clearly an advantage, particularly with respect to the residual entrapped NAPL in aquifers. However, in situ bioremediation requires detailed site-specific study concerning the particular nature of the site (i.e., type, constituents/contents and relevant properties of the soil), its naturally occurring populations of microorganisms, and the actual effectiveness of the latter to biodegrade the present NAPL under the specific environmental site conditions; all the above in the presence of the particular agent(s) planned to be used for enhancing, directly or indirectly, bioremediation processes (e.g., surfactants). Since residual contamination may remain in NCAs when the NAPL concentration becomes too low to sustain a microbial population that would degrade it, the use of additional substance(s) as the source of carbon and energy to sustain microbial growth (co-metabolism) might be necessary. Therefore, the pre-determination of the site-specific acceptable endpoint of the bioremediation process is of crucial importance. Given the constraints of in situ bioremediation, the addition of surfactants to enhance the process makes sense only if the latter are *biodegradable.*

4 Surfactant-Enhanced Bioremediation of NCAs

Surfactant-enhanced bioremediation processes of NCAs, targeting the entrapped residual NAPL in particular, are based on the ability of surface active agents, due to their particular chemical makeup, to effectively increase the aqueous concentrations of poorly soluble compounds and the interfacial areas between immiscible fluids (e.g., oil/water). As a result of this surface activity, the NAPL components dissolved in the aqueous solution become more accessible to microorganisms present, leading, eventually, to the enhancement of the biodegradation-bioremediation process.

The hydrophilic/hydrophobic two part structure of the surfactants' molecules is the key to their effectiveness and performance under various conditions in different applications. The structures and formulae of commercially available products (always mixtures of isomers and/or homologues) are given in Fig. 1. These are also mostly the ones studied as candidates for use in surfactant-enhanced subsurface bioremediation.

The types of surfactants mostly used are the *anionics*, *nonionics*, *cationics* and *amphoterics*, classified by the charge on the larger part of the molecule (Chick 1967; Linfield 1976; Porter 1994). The current "workhorse" of the synthetic deter-

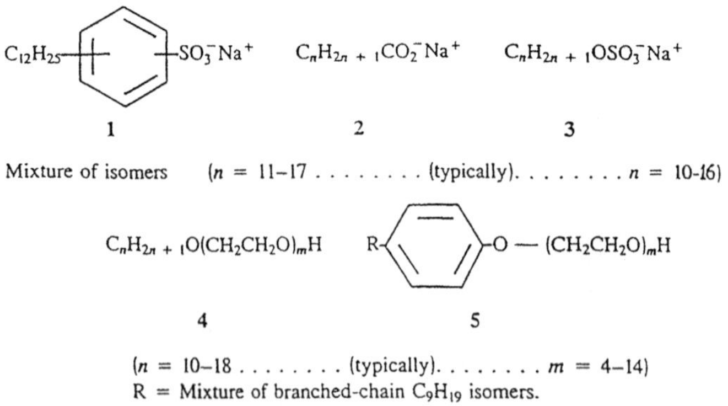

Fig. 1. Structures and formulae of representative, commercially available, surface active products (mixtures of homologues and/or structural isomers). (Zoller 1994)

gent industry (~35% in the United States) is the anionic linear alkylbenzene sulfonate (LABS or LAS), mainly in its sodium salt form (1). The other two anionic surfactants are the salts of carboxylic (fatty) acids (2) and (fatty) alcohol sulfate esters (3) (about 15% each). The two most important nonionic surfactants are the linear ethoxylated alcohols (EAs) (4) and the branched-chain alkylphenol ethoxylates (APEs) (5) (about 17% of the total output of nonionics in the United States). From the point of view of environmental contamination, anionic surfactants having the linear molecular structures of (1), (2), (3) and (4) are considered to be biodegradable (albeit at different rates and to different extents, dependent on the intimate environmental conditions) and classified as "soft", whereas those of the branched-chain type, like (5), are classified as "hard"; that is, nonbiodegradable. The branching, i.e., tertiary and quaternary carbon atoms, makes these surfactants resist biodegradation by microorganisms in natural environments (Swisher 1987). Clearly, the less biodegradable a surfactant is, the more of an environmental problem it constitutes. Thus, the degree of "hardness", or "softness" of any surfactant should be a major criterion for its possible use in surfactant enhanced bioremediation. Indeed, both enhancements and inhibitions of biodegradation of organic compounds in the presence of surfactants have been reported (Rouse et al. 1994).

The three general mechanisms by which surfactants can enhance the removal of residual (and free phase) hydrocarbons from the saturated zone are desorption-solubilization, mobilization and macroemulsification. The first is a straightforward process resulting in dissolved NAPL microemulsion, which operates whenever a surface active agent is added to an aqueous system, and is the easiest to apply from an engineering standpoint. The second, mobilization, refers to those special cases when the interfacial tension between the trapped residual phase and the flowing aqueous phase is reduced by several orders of magnitude, resulting in a separate phase of coalescable droplets of NAPL. As the interfacial tension decreases by the adsorption of the surfactant added at the hydrocarbon/water interface, the capillary number (dimensionless) decreases and mobilization of the hydrocarbon becomes easier. The design of an appropriate surfactant system for the ultra-lowering of the interfacial tension will require the matching of the hydrophilic/lipophilic balance (HLB) of the surfactant (more likely, the surfactant mixture) to the particular nature of the residual hydrocarbon phase (Rosen 1989). The third is the macroemulsification of NAPL, by the surfactant, to form a thermochemically meso-stable macroemulsion consisting of polydispersed NAPL droplets whose transport behavior can be significantly different from the bulk fluid. Since water/NAPL/surfactant(s) systems quite often exhibit complex phase behavior, the mechanisms of NAPL removal should be attributed to all three of these, albeit in different proportions in each case. All three should, therefore, be accounted for in mathematical modeling of surfactant-enhanced remediation of NCAs, if a significant deviation of the calculated prediction from the observed reality is to be avoided (Okuda et al. 1996).

Of the variety of processes affecting the efficiency of surfactant-enhanced remediation, some are associated with losses of the surfactant; e.g., precipitation, partitioning into trapped residual phases, or adsorption onto the aquifer's solid surface (Palmer et al. 1992; Sabatini et al. 1995; Hayes 1996). Each of these processes can reduce the availability of surfactant to remove the entrapped NAPL. Extensive studies have shown that (1) there is a selective sorption of the different homologues in a given surfactant product (Poisson distributed in mixtures of alkylphenol- and linear alcohol-based ethoxylates) on soil (Hayes 1996); and (2) the sorption behavior of nonionic surfactants is different from that of neutral organic contaminants such as xylenes (components of NAPL fuels) in that their sorption exceeds predictions based on K_{ow} values (Palmer et al. 1992). Thus, if a commercial surfactant product (containing broadly-distributed homologues and/or isomers) is to be employed within the remediation process, its specific (selective) sorption behavior should be taken into consideration. Furthermore, each of these processes can reduce the ability of the surfactant to remove the entrapped NAPL. Since in most cases soil particles are negatively charged, anionic surfactants are generally less adsorbed on soil than cationic or nonionic surfactants. Anionic surfactants, however, are susceptible to precipitation, mainly by abundant divalent cations like Ca^{2+} and Mg^{2+} (Shiau et al. 1995). In taking all of the above into consideration, twin-head anionic surfactants were sought as possible good candidates for subsurface remediation of NAPL while minimizing surfactant losses by sorption and precipitation processes. The research results demonstrated that indeed anionic disulfonates were significantly less susceptible to precipitation and sorption than monosulphonates and also less prone to sorption than nonionic surfactants (Rause et al. 1993). However, although the disulfonates studied exhibited greater solubilization of naphthalene (a common aromatic constituent of NAPL) than monosulfonates, their solubilization was slightly lower than that of nonionic surfactants. Thus, a *site specific* optimization of the surfactant-enhanced bioremediation process with respect to the particular surfactant, or surfactant mix, to be used is clearly essential in order to get the best possible results. Ethoxylated nonionic surfactants were the most studied with respect to surfactant sorption onto soils, the solubilization of NAPLs by them, their precipitation from the aqueous solution and phase partitioning in the water/NAPL system (Palmer et al. 1992; Diallo et al. 1994; Shiau et al. 1995; Hayes 1996; Okuda et al. 1996; Yeom et al. 1996). This is due, perhaps, to the commercial availability of these well-defined homologue and/or structural isomer product mixtures, enabling the selection and use, by the researcher, of the surfactant(s) with the desired properties (e.g., hydrophilic/lipophilic balance, HLB) in controlled laboratory experiments and simulations. Yet, a number of key questions related to the impact of sorption-desorption on surfactant-enhanced solubilization of NAPL have emerged and need to be addressed, particularly with respect to the surface chemical basis for the sorption/desorption behavior (Hayes 1996). The bottom line is, that in order to obtain in the lab results and data which are quantitatively applicable to the optimization and modeling of a real case of in situ surfactant-enhanced bioremediation of residual entrapped NAPL in NCA, the actual site aquifer material (the soil matrix as it is) should be used.

5 Problem Conceptualization and Rationale

Of particular importance, as far as in situ surfactant-enhanced remediation of NAPL-contaminated soils and groundwater is concerned, are: (1) the effect of the soil character on the partitioning of the NAPL hydrocarbons as well as the added surfactant (Shiau et al. 1995), or surfactant mix, between the stationary soil particles and the mobile water phase; (2) the water solubility of the NAPL contaminant in the absence and presence (of different types and in various concentrations in the mix) of (commercially) available (preferentially) biodegradable surfactants; (3) the surfactants losses due to precipitation and sorption processes in the subsurface; (4) the on site biodegradation rate and fate of both the various NAPL hydrocarbon components and the added surfactants in the subsurface (Zoller 1992, 1994); (5) the availability, in higher than the "limiting concentrations", of the necessary nutrients (e.g., N and P) to make the complete biodegradation of NAPL and the residual surfactants possible; and (6) the efficiency of the surfactant-enhanced desorption-extraction of the residual NAPL adsorbed onto the aquifer's porous soil/rock matrices.

Subsurface bioremediation cannot deal with very high levels of NAPL due to aquifer clogging (Nash 1988). Such high levels may result in poor economics in surfactant enhanced processes due to the high cost of surfactant(s) as well as of the added nutrients which are necessary for the co-metabolic transformation to occur if low levels of NAPLs are to be achieved at the endpoint. Since the lowest possible level of surfactant added is the most desirable from the bioremediation process, technology/engineering and economical points of view, the optimization of an effective cleanup process will, most probably, require the use of surfactant(s) in concentrations below their CMC (the aqueous concentration in which their surface activity is at maximum). This implies, most probably, the need of more than one surfactant in order to achieve effective bioremediation under these (low level surfactant concentration) constraints. Also, the most probable need to coadd, with the surfactant, adequate sources of essential nutrients – such as nitrogen and phosphorus – should be taken into consideration, particularly with respect to bioremediation of the residual entrapped NAPL in NCAs.

5.1 Why in Situ Nutrient-Surfactant Mix-Enhanced Bioremediation of NCAs?

Since the focus in this chapter is on the residual entrapped NAPL in NCAs, the site-specific approach to deal effectively with the problem, within the prior research, design and actual execution of the remediation process, is obvious. In situ enhanced bioremediation appears technologically and economically to be the simplest feasible and practical methodology to be applied in such cases. If a "nutrient surfactant" can be used to provide, simultaneously, the surface activity required for the desorption-solubilization of the entrapped residual NAPL and the essential nutrients required to sustain the biodegradation process under these circumstances, then, obviously, this is an advantage. The problem is that the commercially available (in bulk quantities) nutrient-surfactants, which contain in their structure the nitrogen and phosphorus needed for the bioremediation process, have a relatively low surface activity, particularly in the low level concentrations in the aqueous solution to be used. Therefore, the need for an appropriate surfactant mix for the designed bioremediation of NCAs is apparent.

In conclusion, our assertion that in situ bioremediation with the aid of biodegradable nutrient-surfactants is the method of choice for the removal of residual entrapped NAPL in aquifers, is thus based on fundamental biological-hydrological as well as technological-economical feasibility considerations, guided by well-established environmental imperatives.

In the preliminary steps of our study, which has evolved from the above conceptualization of the problem, we focused on the characterization of surfactants, and the nutrient-surfactant mix-enhanced release of soil-entrapped fuel, aiming at the development of surfactant-based, economically feasible technology for effective bioremediation of residual NAPL-contaminated aquifers. The approach is thus systemic, site-specific (Oppelt and Kavanugh 1996) and eclectic, dealing pragmatically with the relevant core problems in a search for the approximation of good practical solutions without looking for perfection. The laboratory results (Zoller and Rubin 1996), together with the accompanying modeling and the on-site field validation at pilot scale to follow, are expected to provide an applicable, efficient and feasible methodology/technology for surfactant-enhanced bioremediation of NCAs.

References

Arocha MA, Jackman AP, McCoy BJ (1996) Adsorption soil agglomerates: soil as a biporous sorbent. Environ Sci Technol 30:1500–1507

Aronstein BN, Alexander M (1992) Surfactants at low concentrations stimulate biodegradation of sorbed hydrocarbons in samples of aquifer sands and soil slurries. Environ Toxicol Chem 11:1227–1230

Atlas RM (1995) Bioremediation. Chem Eng News April:32–42

Bedient PB, Rifai HS, Newell CJ (1994) Groundwater contamination, transfer and remediation. PTR Prentice Hall, Englewood Cliffs

Carberry J, Kavanaugh M, Mulligan W, Weber W (1996) In: Trends and challenges; emerging frontiers. Environ Sci Technol 30(1):32–36

Cekirge HM, Gidmmona CP, Berlin J, Long C, Koch M (1994) Oil spill modeling using parallel computations. Spill Sci Technol Bull 1:61–68

Chen CT (1992) Understanding the fate of petroleum hydrocarbons in the subsurface environment. J Chem Educ 69(5):357–361

Chick MJ (ed) (1967) Nonionic surfactants. Marcel Dekker, New York

Diallo MS, Abriola LM, Weber WJ Jr (1994) Solubilization of nonaqueous phase liquid hydrocarbons in micellar solutions of dodecyl alcoholethoxylates. Environ Sci Technol 128:1829–1837

Edwards DA, Luthy RG, Liu Z (1991) Solubilization of polycyclic aromatic hydrocarbons in micellar nonionic surfactant solutions. Environ Sci Technol 25:127–133

Goldenberg LC, Melloul AJ, Zoller U (1996) The "shortcut" approach for the reality of enhanced groundwater contamination. J Environ Manage 46:311–326

Hayes KF (1996) Sorption-desorption studies. In: the use of surfactants to enhance the removal of chlorinated hydrocarbons from aquifer materials. Spring Progress Reports, The Great Lakes Mid-Atlantic Center for Hazardous Substance Research, The University of Michigan, Ann Arbor, pp 13–16, 39–40

Kile DE, Chiou CT (1989) Water solubility enhancement of DDT and trichlorobenzene by some surfactants below and above the critical micelle concentration. Environ Sci Technol 23:832–837

Linfield WM (ed) (1976) Anionic surfactants. Marcel Dekker, New York

Mukerjee P, Myssels KJ (1971) Critical micelle concentrations of aqueous surfactant systems. US Dept of Commerce, Washington, DC, NS RDS-NBS 36

Muszkat L, Raucher D, Magaritz M, Ronen D (1994) Groundwater contamination by organic pollutants. In: Zoller U (ed) Groundwater contamination and control. Marcel Dekker, New York, pp 257–271

Nash JH (1988) Field studies of in situ soil washing. US Environmental Agency, Cincinnati, Ohio, EPA/6000/2-87/110

Nayyar SP, Sabatini DA, Harwell JH (1994) Surfactant adsolubilization and modified admicellar sorption of nonpolar, polar and ionizable organic contaminants. Environ Sci Technol 28:1874–1881

Okuda I, McBride JF, Gleyzer SN, Miller CT (1996) Physicochemical transport processes affecting the removal of residual DNAPL by nonionic surfactant solutions. Environ Sci Technol 30:1852–1860

Oppelt ET, Kavanaugh MC (1996) In: trends and challenges; priority concerns. Environ Sci Technol 30(1):27A–41A

Oreskes N, Shrader-Frechette K, Belitz K (1994) Verification, validation and certification of numerical models in the earth sciences. Science 263:641–646

Palmer C, Sabatini DA, Harwell JH (1992) Sorption of hydrophobic organic compounds and nonionic surfactants with subsurface materials. In: Sabatini DA, Knox RC (eds) Transport and remediation of subsurface contaminants, chap 14. American Chemical Society, Washington, DC, pp 169–181

Porter MR (ed) (1994) Handbook of surfactants. Blackie Academic & Professional, Glasgow, pp 99–304

Robertson BK, Alexander M (1996) Mitigating toxicity to permit bioremediation of constituents of nonaqeous phase liquids. Environ Sci Technol 30:2066–2070

Rosen MJ (1989) Surfactants and interfacial phenomena, 2nd edn. Wiley-Interscience, New York

Rouse JD, Sabatini DA, Harwell JH (1993) Minimizing surfactant losses using twin-head anionic surfactants in subsurface remediation. Environ Sci Technol 27(10):2072–2078

Rouse JD, Sabatini DA, Suflita JM, Harwell JH (1994) Influence of surfactants on microbial degradation of organic compounds. Crit Rev Environ Sci Technol 24(4):325–370

Rubin H, Braxein A, Daniels H, Rouve G (1994) Migration of oil and petroleum pollutants in soils and groundwater. In: Zoller U (ed) Groundwater contamination and control. Marcel Dekker, New York, pp 355–375

Sabatini DA, Knox RC, Harrwell JH (1992) The potential use of edible surfactants for the remediation of contaminated groundwater supplies. Proc 3rd CESIO Int Surfactants Congr, London, June 1992, pp 349–358

Sabatini DA, Knox RC, Harrwell JH (eds) (1995) Surfactant-enhanced subsurface remediation; emerging technologies. American Chemical Society, Washington, DC

Shiau BJ, Sabatini DA, Harrwell JH (1995) Properties of food grade (edible) surfactants affecting subsurface remediation of chlorinated solvents. Environ Sci Technol 29(12):2929–2935

Smith JA, Tuck DM, Jaffe PR, Mueller RT (1991) Humics and soil. In: Baker RA (ed) Organic substances and sediments in water, vol 1. Lewis, Chelsea, Michligan

Steffensen WS, Alexander M (1995) The effect on competition for available nutrients in systems containing mixtures of substrates. Appl Environ Microbiol 61(8):2859–2862

Swisher RD (1987) Surfactant bioremediation. Marcel Dekker, New York

Trombly J (1995) Engineering enzymes for better remediation. Environ Sci Technol 29(12):560A–564A

Yeom IT, Ghosh MM, Cox CD (1996) Kinetic aspects of surfactant solubilization of soil-bound polycyclic aromatic hydrocarbons. Environ Sci Technol 30:1589–1595

Zoller U (1992) Distribution and survival of nonionic surfactants in the surface, sea and groundwater of Israel. J Environ Sci Health A27(6):1521–1533

Zoller U (1994) Groundwater contamination by surfactants. In: Zoller U (ed) Groundwater contamination and control. Marcel Dekker, New York, pp 273–292

Zoller U (1996) In situ remediation of residual entrapped NAPL in aquifers by a biodegradable nutrient-surfactant mix. Paper presented at the Int Worksh Soil and aquifer pollution NAPL contamination and reclamation, May 13–15. Water Research Institute, Technion, Haifa, Israel

Zoller U, Rubin H (1996) Development of methods for the enhanced remediation of fuel-contaminated aquifers by surface active nutrients. Annual Research Project Report, Technion Institute of Water, Technion, Haifa, Israel

22 Application of a Eulerian–Lagrangian Approach for the Simulation of Entrapped NAPL Dissolution and Transport

S. Lumelsky[1], S. Sorek[2], and H. Rubin[1]

Contents

1 Introduction

Multiphase flow and transport in porous media are very complicated and difficult for modeling. Two main problems should be considered: accurate mathematical description of the physics of phenomena and correct numerical implementation of the mathematical model.

In general, the governing equations that constitute the core of any model of transport and flow in a porous medium domain, are presented in their Eulerian

[1] Department of Civil Engineering, Environmental and Water Resources Engineering, Technion - Israel Institute of Technology, Haifa 32000, Israel

[2] Water Resources Research Center, J. Blaustein Institute for Desert Research Ben-Gurion University of the Negev, Sde Boker Campus 84990, Israel

H. Rubin/N. Narkis/ J. Carberry
Soil and Aquifer Pollution

formulation. Accordingly, the main group of the numerical models are based on this formulation and use finite elements or the finite difference approach. However, recently, the finite element approach has been more widely used. Abriola and Pinder (1985a,b) developed a 2-D finite element model for the flow of water, gas and NAPL in a heterogeneous porous medium domain that also takes into account volatilization and dissolution of NAPL. A similar finite element model was presented by Corapcioglu and Baehr (1987). A subsequent extension was to incorporate hysteretic constitutive relations, as was done by Parker et al. (1987), Parker and Lenhard (1987). Parker and Lenhard (1989) proposed the model for vertical integration of three-phase flow equations for light hydrocarbon (LNAPL) plume movement. Later Kaluarachchi et al. (1990) also developed a 2-D areal finite element model for the prediction of water and light hydrocarbon migration. A 2-D two-phase (NAPL and water) model was developed by Mayer (1994) using a finite element domain decomposition scheme. The behavior of variable density, aqueous-phase contaminant plumes in saturated, homogeneous 2-D and 3-D intermediate-scale aquifer models was investigated with the finite element code by Oostrom et al. (1994). The finite element model of air injection in groundwater in a homogeneous 3-D axially symmetric domain with emphasis on the developing steady state was presented by Van Dijke and van der Zee (1994). Mohtar et al. (1994) presented a 2-D finite element model for evaluation of the air flow distribution during steady air spreading.

At the same time, another class of numerical models based on the Lagrangian formulation of governing equations is widely used. In most cases, the reason for switching to the Lagrangian approach stems from a problem associated with numerical approximation of the advective flux term in the balance equations. The resolution of such equations by Eulerian methods may introduce spurious oscillations and artificial dispersion. Special Lagrangian schemes, based on formal decomposition of the transport equation into "advective" and "dispersive" parts (Neuman 1981, 1984), have been developed to treat this problem. This results in development of different Lagrangian adaptive methods. These methods exploit the hyperbolic characteristics and gain the advantage of being able to use less restricted temporal grids, while still obtaining accurate solutions. Documented models based on variants of the characteristic method approach were developed by Chiang et al. (1989), Cohen et al. (1989), Ryan and Cohen (1991), etc.

Another Lagrangian based method, namely the Eulerian-Lagrangian localized adjoint method (ELLAM), was devised in the late 1980s (Celia et al. 1987) and resolves the issues of the treatment of boundary conditions and of mass conservation, both difficult in standard Lagrangian techniques such as the method of characteristics. The method has been applied by various authors to single-phase contaminant transport in 1-D (see Binning and Celia 1994 for a review). A finite volume ELLAM has been developed by Binning and Celia (1994) to simplify integrals in the ELLAM formulation and has the desirable property of ensuring local contaminant mass balance.

Neuman (1981, 1984) and Neuman and Sorek (1982) proposed the particle tracking method for advection-dominated problems based on tracking a moving particle along a characteristic line, while the dispersive problem is solved in a fixed Eulerian grid system.

The major purpose of the model presented here is to study and simulate the fate of the residual NAPL trapped within the unsaturated and saturated zones for cases of non-uniform phreatic aquifers with respect to remediation procedure. The model presented here considers the flow of water and air, and solute mass exchange from the trapped immobile NAPL into these phases. Specifically, the model is aimed at the following objectives: (1) determination of the dominant physical processes which control the groundwater contamination by oil spills; (2) quantitative evaluation and simulation of the flow of the fluid phases in the unsaturated and saturated domain of porous media; (3) prediction of the oil concentration distribution in time and space with respect to different oil spill scenarios; (4) determination of the effectiveness of different remediation schemes in terms of the dissolution and transport of entrapped oil; and (5) application of the modified Eulerian-Lagrangian numerical method for the simulation of the dissolution and transport of the oil component through the porous media.

2 Conceptual Model

Conceptualization of this problem results in the following set of assumptions:

1. The domain incorporates a solid matrix, two flowing fluid phases and the immobile trapped NAPL.
2. The solid matrix is rigid and stationary.
3. The porous medium is non-uniform, so that the diffusion coefficient and the permeability are spatial functions.
4. Three phases occupy the porous medium pores:
   a) the non-aqueous interwetting phase (NAPL), N, which is immobile and can be dissolved in the aqueous phase and can also volatilize in the gaseous phase;
   b) the aqueous wetting fluid phase (water), w, which contains the dissolved NAPL component with the concentration, C_w^N. There is no mass transfer between air and water phases;
   c) the gaseous non-wetting fluid phase (air), g, which contains the volatilized NAPL with the concentration, C_g^N.
5. The NAPL, aqueous and gaseous phases are subject to incompressible flow conditions.
6. Concentrations of the NAPL within the aqueous and gaseous phases do not affect the densities of these phases:

$$\rho_w \neq \rho_w\left(C_w^N\right); \quad \rho_g \neq \rho_g\left(C_g^N\right),$$

where the concentration is expressed in terms of mass of the relevant component per unit volume of the fluid phase.

7. We also assume that there is no direct contact between the solid, gas (g), and NAPL (N) phases. The wetting phase, w, forms a slim water film around the solid granule. So, the NAPL can be adsorbed onto the solid surface only from the w-phase.
8. Isothermal conditions prevail.
9. Adsorption is described by a linear isotherm.

Assumption 6 allows the transport and the flow to be decoupled and solved separately. However, mass transfer from NAPL into the flowing phases affects the flow problem which, in return, influences the results by reduction of the NAPL saturation. Hence, actually, the developed numerical codes are built independently with respect to the flow and transport, and are affected by the mass balance of the NAPL.

3 Mathematical Formulation

3.1 Two-Phase Flow in Heterogeneous Porous Domain

The space occupied by the porous medium is referred to as the region Ω with boundary Γ. Darcy's velocity $\mathbf{V}_\alpha$ for the $\alpha(\equiv w,g)$ fluid phases, reads as follows:

$$\mathbf{V}_\alpha = -\lambda_\alpha \nabla \psi_\alpha, \tag{1}$$

where λ_α denotes the "mobility" of the α-phase, defined as follows:

$$\lambda_\alpha = \frac{kk_{r\alpha}}{\mu_\alpha}; \tag{2}$$

and ψ_α denotes the potential of the α phase as follows:

$$\psi_\alpha = P_\alpha + \rho_\alpha g z; \tag{3}$$

k and $k_{r\alpha}$ denote the intristic and relative permeabilities, respectively; ρ_α denotes the density of α-phase; P_α denotes the pressure in the α phase; μ_α denotes the viscocity of the α phase; g denotes the acceleration due to gravity; z denotes the altitude above some datum.

Following assumption 2, the mass balance equation for the α-fluid phases reads as follows:

$$\nabla \cdot \mathbf{V}_\alpha + \phi \frac{\partial S_\alpha}{\partial t} = q_\alpha, \tag{4}$$

where ϕ denotes porosity; S_α denotes saturation of α-phase; q_α denotes the total volumetric influx of α-phase. The gas and water pressures at a point within the medium differ by the capillary pressure P_c, as follows:

$$P_c = P_g - P_w = \psi_g - \psi_w + (\rho_w - \rho_g) g z. \tag{5}$$

When residual entrapped oil occupies the porous space, the oil saturation (S_N) also should be taken into account, i.e. $S_g + S_w + S_N = 1$. Assuming that changes in the entrapped NAPL saturation S_N are practically negligible during a small evolution step, we denote $S(P_c)$ ($\equiv S_w$) and refer to the following constitutive relation:

$$S + S_g \approx 1 - S_N = const. \tag{6}$$

Hence, in view of Eqs. (5) and (6), we obtain the following relation:

$$\frac{\partial S}{\partial t} = S' \frac{\partial}{\partial t}(\psi_g - \psi_w), \tag{7}$$

in which $S' \equiv dS/dP_c$.

Substituting Eq. (7) into Eq. (4) yields Eq. (8):

$$\begin{aligned} \nabla \cdot (\lambda_g \nabla \psi_g) &= -\phi S' \frac{\partial}{\partial t}(\psi_g - \psi_w) + q_g; \\ \nabla \cdot (\lambda_w \nabla \psi_w) &= -\phi S' \frac{\partial}{\partial t}(\psi_g - \psi_w) + q_g. \end{aligned} \tag{8}$$

Because λ_g and λ_w are functions of the relative permeabilities, Eqs. (7) and (8) become a set of non-linear differential equations.

In a practical situation, the potentials ψ_g and ψ_w may be nearly equal, causing difficulties of round-off error in calculation of the capillary pressure from (5) (Zienkiewicz 1977). Thus, we can define the following parameters:

$$\psi_s \equiv \psi_g + \psi_w \quad \lambda_s \equiv \lambda_g + \lambda_w \quad q_s \equiv q_g + q_w; \tag{9}$$

$$\psi_d \equiv \psi_g - \psi_w \quad \lambda_d \equiv \lambda_g - \lambda_w \quad q_w \equiv q_g - q_w. \tag{10}$$

By virtue of Eqs. (9) and (10), the set of equations (7) and (8) can be expressed as

$$\begin{aligned} &\nabla \cdot \left[\lambda_s(S) \nabla \left(\frac{\psi_s}{2} \right) \right] + \nabla \cdot \left[\lambda_d(S) \nabla \left(\frac{\psi_d}{2} \right) \right] = q_s \\ &\nabla \cdot \left[\lambda_d(S) \nabla \left(\frac{\psi_s}{2} \right) \right] + \nabla \cdot \left[\lambda_s(S) \nabla \left(\frac{\psi_d}{2} \right) \right] = -2\phi S' \frac{\partial \psi_s}{\partial t} + q_d \\ &\frac{\partial S}{\partial t} = S' \frac{\partial \psi_s}{\partial t}. \end{aligned} \tag{11}$$

The non-linear set of governing differential equations described in Eq. (11) can be solved for the three unknowns: ψ_s; ψ_d; S, subject to the following boundary conditions:

$$\psi_\alpha = \psi_{\alpha\Gamma_1} \quad on\, \Gamma_1, \tag{12}$$

and/or

$$V = \mathbf{n} \cdot V_{\alpha\Gamma_2} \quad on\, \Gamma_2, \tag{13}$$

where $\mathbf{n}$ denotes the unit outward normal vector; ψ_{Γ_1} and $\mathbf{V}_{\Gamma_2}$ are prescribed along the boundary segments Γ_1 and Γ_2, respectively.

3.2 Two-Phase Transport in a Heterogeneous Porous Domain

The immobile NAPL phase is entrapped in the porous media in the form of blobs. In this case the mass balance of NAPL in the α-phase with respect to assumption 2 is described by

$$\phi\frac{\partial\left(S_\alpha C_\alpha^N\right)}{\partial t}=\phi\nabla\cdot\left[S_\alpha \mathbf{D}_\alpha^N\nabla C_\alpha^N\right]-\nabla\cdot\left[\mathbf{V}_\alpha C_\alpha^N\right]+F_\alpha^N, \tag{14}$$

where F_α^N denotes the rate of mass transfer from NAPL into the α fluid phase; $\mathbf{D}_\alpha^N$ denotes the hydrodynamic dispersion tensor; C_α^N denotes the concentration of NAPL in the fluid α-phase.

In view of assumption 7 and Eq. (14), we can write the mass balance equations for NAPL in water and solid phases as follows:

$$\phi\frac{\partial\left(S_w C_w^N\right)}{\partial t}+(1-\phi)\frac{\partial\left(\rho_s K_d C_w^N\right)}{\partial t}=\phi\nabla\cdot\left[S_w \mathbf{D}_w^N\nabla C_w^N\right]-\nabla\cdot\left[\mathbf{V}_w C_w^N\right]+F_w^N, \tag{15}$$

and in the air phase,

$$\phi\frac{\left(\partial S_g C_g^N\right)}{\partial t}=\phi\nabla\cdot\left[S_g \mathbf{D}_g^N\nabla C_g^N\right]-\nabla\cdot\left[\mathbf{V}_g C_g^N\right]+F_g^N, \tag{16}$$

where, according to assumption 9, K_d denotes the constant partitioning coefficient and ρ_s denotes the constant density of the solid phase.

3.3 Mass Transfer Relationships

We follow the model for *non-equilibrium* dissolution of the NAPL in groundwater, proposed by Powers et al. (1991). Accordingly, the mass transfer term can be described as follows:

$$F_\alpha^N=-k_{fN}^\alpha\left(C_\alpha^N-C_{*\alpha}^N\right)a_\alpha^N, \tag{17}$$

where k_{fN}^α denotes the effective mass transfer coefficient; $C_{*\alpha}^N$ denotes the equilibrium concentration of the NAPL in the α phase, which is equal to the solubility of NAPL in α-phase; a_α^N denotes the specific interfacial area between NAPL blobs and the water phase. The specific interfacial area between NAPL and an adjoint phase, a^N, for the case of one mobile phase (water) is defined by

$$a^N=S_N\left(\frac{A}{V}\right)_b f; \tag{18}$$

in which $(A/V)_b$ denotes the surface area to volume ratio of a blob; f denotes the fraction of the blob surface area exposed to the mobile adjoined phase.

For the case of two flowing fluid phases (water and gas) we complete (17) by the additional relation for source term F_g^N with $\alpha(\equiv g)$:

$$F_g^N = -k_{fN}^g\left(C_g^N - C_{*g}^N\right)a_g^N. \tag{19}$$

In the case of two flowing fluid phases, the term f in expression (18) should be divided into two parts, accordingly. We assume that:

$$a^N = a_w^N + a_g^N \quad f = f_w + f_g. \tag{20}$$

The fraction of the blob surface that interacts with each of the α fluids is related to the S_α saturation of this fluid. We assume that this also reflects the fraction coefficient, namely

$$\frac{f_w}{f_g} = \frac{S_w}{S_g}. \tag{21}$$

3.4 Mass Balance for the NAPL Phase

The mass balance equation for the NAPL phase reads

$$\phi\rho_N \frac{\partial(S_N)}{\partial t} = -F_w^N - F_g^N, \tag{22}$$

in which ρ_N denotes NAPL density; S_N denotes saturation of the NAPL phase.

4 Numerical Solution

4.1 General Algorithm of the Numerical Solution

The algorithm was written for time level $t = t^{k+1}$. As the first approximation (for $iter = 0$-th iteration step) we impose NAPL saturation $S_N^{(iter=0),\,k+1} = S_N^k$.

1. The Flow Problem
   This includes:
   - Mass balance equations for the fluid phases;
   - Darcy's velocities;
   - Constitutive relations.

   From the flow problem solution we obtain Darcy velocities, V_w; V_g, and saturations, S_w; S_g, for water and gas phases, which serve as input for the transport problem.
2. The Transport Problem
   This includes:
   - The mass balance equation for component NAPL within each fluid phase;

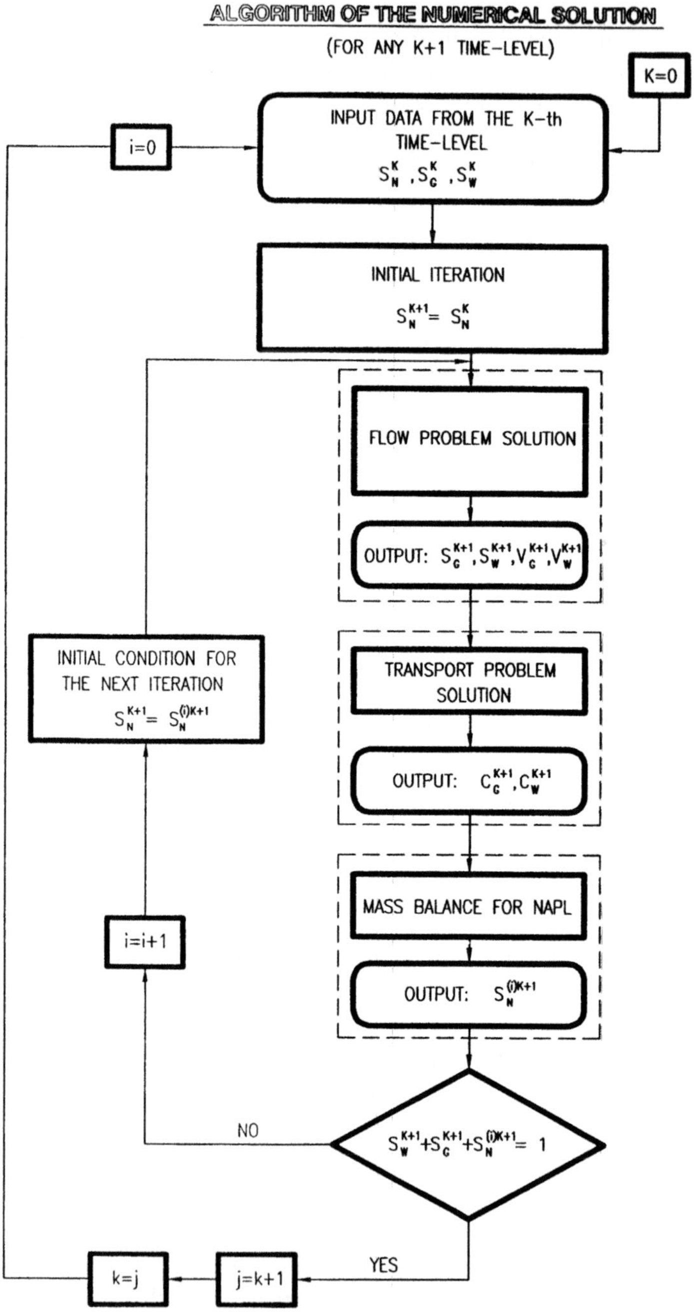

Fig. 1. General solution algorithm

- Relationships for the mass transfer description as given by Powers et al. (1991);
- Relationships for adsorption.

From the transport problem solution we obtain NAPL concentrations within the water and gas phases, C_w^N, C_g^N, and the mass transfer terms F_w^N and F_g^N which serve as input for the NAPL balance.

3. End of Time Step from t^k to t^{k+1}

The new NAPL saturation is calculated at each time level t^{k+1} by subtracting the volume of organic contaminant which was transferred to the aqueous and gaseous phases during the time step Δt from the immobile NAPL as follows:

$$S_N^{(iter),k+1} = S_N^k - f\left(C_w^N, C_g^N\right).$$

If $S_N^{(iter),k+1}$ complied with the condition $S_N + S + S_g = 1$ at any time level, then $S_N^{k+1} = S_N^{(iter),k+1}$, and we can proceed to the next evolution step; if not, we assign $S_N^{k+1} = S_N^{(iter),k+1}$, and repeat steps from (1) to (3). The scheme of the algorithm is presented in Fig. 1.

4.2 Flow Problem Solution

4.2.1 Finite Element Scheme

The vector of unknowns, $[\psi_s, \psi_d, S]^T$, is solved by the finite element method. We note that since λ_α is a function of relative permeabilities, which are non-linear functions of the saturation, we will also interpolate the λ_α values within each element by the same procedure as for saturation S as follows:

$$\lambda_\alpha(x,y,t) \simeq \hat{\lambda}_\alpha(x,y,t) \equiv \sum_{n=1}^{N} \lambda_{\alpha n}(t)\xi_n(x,y).$$

Applying to Eq. (11):

- Lagrangian interpolation (ξ_j denotes shape functions);
- Galerkin's weighted residual procedure;
- Integration by parts;
- Approximation of time derivative by forward difference as follows:

$$\frac{dS_j}{dt} = \frac{S_j^{k+1} - S_j^k}{\Delta t}; \quad \frac{d\psi_{dj}}{dt} = \frac{\psi_{dj}^{k+1} - \psi_{dj}^k}{\Delta t}, \tag{23}$$

the following set of non-linear algebraic equations was obtained:

$$\begin{aligned} &\mathbf{M}_s^{k+1}\psi_{sj}^{k+1} + \mathbf{M}_d^{k+1}\psi_{dj}^{k+1} = \mathbf{B}_s^{k+1} - \mathbf{Q}_s^{k+1}; \\ &\Delta t\mathbf{M}_d^{k+1}\psi_{sj}^{k+1} + \left[\Delta t\mathbf{M}_s^{k+1} - 4n\mathbf{I}^{k+1}\right]\psi_{dj}^{k+1} = \Delta t\mathbf{B}_d^{k+1} - \mathbf{Q}_d^{k+1} - 4n\mathbf{I}^k\psi_{dj}^k; \\ &\mathbf{W}S^{k+1} - \mathbf{I}^{k+1}\psi_{dj}^{k+1} = \mathbf{W}S^k - \mathbf{I}^k\psi_{dj}^k, \end{aligned} \tag{24}$$

where

$$\mathbf{M}_s = m_{s_{ij}} = \int\limits_{\Omega}\left[\sum_{j=n}^{N}\lambda_{sn}\xi_n\right]\nabla\xi_i\nabla\xi_j d\Omega; \quad \mathbf{I} = i_{ij} = \int\limits_{\Omega}\left[\sum_{j=n}^{N}S'_n\xi_n\right]\xi_i\xi_j d\Omega;$$

$$\mathbf{M}_d = m_{d_{ij}} = \int\limits_{\Omega}\left[\sum_{j=n}^{N}\lambda_{dn}\xi_n\right]\nabla\xi_i\nabla\xi_j d\Omega; \quad \mathbf{W} = w_{ij} = \int\limits_{\Omega}\xi_i\xi_j d\Omega;$$

$$\mathbf{B}_s = \int\limits_{\Gamma}\left[\lambda_s\frac{\partial\psi_s}{\partial n}\xi_i + \lambda_d\frac{\partial\psi_d}{\partial n}\xi_i\right]d\Gamma; \quad \mathbf{Q}_s = q_{s_{ij}} = 2\int\limits_{\Omega}q_{s_j}\delta_{ij}d\Omega;$$

$$\mathbf{B}_d = \int\limits_{\Gamma}\left[\lambda_d\frac{\partial\psi_s}{\partial n}\xi_i + \lambda_s\frac{\partial\psi_d}{\partial n}\xi_i\right]d\Gamma; \quad \mathbf{Q}_d = q_{dij} = 2\int\limits_{\Omega}q_{d_j}\delta_{ij}d\Omega.$$

To solve Eq. (24), we refer to the Levenberg-Marquardt algorithm which is a modification of Newton-Raphson method.

4.2.2 Verification Tests

To verify that the model of two-phase flow is properly solving the governing equations, it is necessary to compare computed results with analytical solutions. Because the results for the present multiphase formulation are not available for comparison, the verification problems that have been used are based on a subset of the present formulation. The first problem describes the behavior of the linear waterflood of a petroleum reservoir. The analytical solution was presented by Buckley and Leverett (1942). The second verification problem describes radial flow to a fully penetrating well in a confined aquifer. The analytical solution for this problem is presented by Theis (1935). A good agreement between numerical and analytical results was obtained consistently.

In addition to indicating the accuracy of the presented numerical model, these test problems also demonstrate the capability of the model to simulate conditions where two flowing phases are NAPL and water, or the flowing phase is water only. The test problems serve an additional purpose of providing benchmarks for alternative models using the proposed or similar governing equations and solution techniques.

4.3 Modified Eulerian-Lagrangian Method for Solving the Transport Problem

4.3.1 Theory

Application of the Eulerian-based methods for advection-dominated solute transport results in some range of numerical dispersion. On the other hand, the numerical dispersion is avoided when a Lagrangian formulation is used. Neuman and Sorek (1982), Neuman (1984), Sorek (1985a,b, 1988), Baptista (1987), Sorek and Braester (1988), among many others, have developed solutions of the Lagrangian formulation of the mass balance equation for the mass of a component in a fluid phase. In what follows, we extend this formulation to 2-D two-phase transport of NAPL by using the concept of an *apparent* or *modified velocity*. This concept introduces the modified Eulerian-Lagrangian (MEL) method.

Consider the general advection–dispersion problem, without source terms, given by Eq. (25)

$$\frac{\partial(\phi Se)}{\partial t} = -\nabla\cdot\left[\mathbf{V}e - \phi S\mathbf{D}\nabla e\right], \tag{25}$$

where $e(\mathbf{X}, t)$ denotes an intensive variable being a function of the spatial vector $\mathbf{X}(x, y)$ and time t; $\mathbf{V}(\mathbf{X}, t)$ denotes the Darcy's velocity vector; $\mathbf{D}(\mathbf{X})$ denotes a tensor associated with linear dependency between the dispersive flux and ∇e;

The Eulerian form (25) may be conformed to different Eulerian-Lagrangian presentations, depending on φ, as follows:

$$\frac{D_{V*}}{Dt}(\phi Se) = (1-\varphi)\nabla\cdot\left[\phi S\mathbf{D}\nabla e\right] + \varphi e S\mathbf{D}\nabla^2 e - e\nabla\cdot\mathbf{V}, \tag{26}$$

where the hydrodynamic derivative is defined by using the velocity, $\mathbf{V}^*$, as follows:

$$\frac{D_{V*}}{Dt} \equiv \frac{\partial}{\partial t} + \mathbf{V}*\cdot\nabla; \quad \mathbf{V}* = \mathbf{V} - \frac{\varphi}{\phi S}\nabla(\phi S\mathbf{D}), \tag{27}$$

in which $\mathbf{V}^*$ denotes a *modified velocity* due to spatial variations in ϕ, S, and $\mathbf{D}$, which serves as the *characteristic velocity* for the propagation of e. Note that in Eq. (26), e is interpreted as being carried, say by a particle, along a path line defined by the following expression:

$$\frac{D_{V*}\mathbf{X}}{Dt} = \mathbf{V}*. \tag{28}$$

In view of Eqs. (26)–(27), we may construct various solution schemes. The modified Eulerian-Lagrangian (MEL) scheme, proposed by Lumelsky and Sorek (1994), utilizes the modified hydrodynamic derivative incorporating $\nabla\cdot D$, whereas the conventional Eulerian-Lagrangian (EL) scheme is based on the standard hydrodynamic derivative. The various cases that may be obtained are described as follows:

$$\varphi = \begin{cases} 1, & \text{MEL} \\ 0, & \text{EL}. \end{cases} \tag{29}$$

Motivation. The EL formulation is reported to be effective in overcoming numerical dispersion phenomena when simulating problems are dominated by advection. In view of Eq. (27) for $\varphi = 1$, it is argued that the second term on the r.h.s. of Eq. (28) can also be influential in dictating an advection dominant situation. Note that this influence may occur for governing equations with no reference to $\mathbf{V}$, or even when $\mathbf{V} = 0$. We thus suggest that it may be preferable to simulate advection dominated problems by MEL formulation, described by Eq. (26) with $\varphi = 1$, rather than by the EL formulation with $\varphi = 0$.

Moreover, we often implement the EL formulation on the basis of presuming that advection fluxes are dominant. Such an assumption, however, may not always be justified. It is, therefore, preferable to implement the MEL scheme instead.

Decomposition of the Dependent Variable. Neuman and Sorek (1982) proposed splitting the dependent variable into an *advection part* $\bar{e}$ and a residual $\mathring{e}$, the latter considered as the *dispersion part*

$$e = \bar{e} + \mathring{e}. \tag{30}$$

The advection–dispersion equation, (25), can be formally decoupled into a purely hyperbolic "advection problem" defined in terms of $\bar{e}$ as follows:

$$\frac{D\bar{e}}{Dt} = 0, \tag{31}$$

and a predominantly parabolic "dispersion problem" defined in terms of $\mathring{e}$ and e.

So, in place of one governing Eq. (25) we obtain the system of two equations, (26) and (31), to be solved numerically.

MEL-Decomposed Equations. The governing equations, Eqs. (15) and (16) are decomposed in view of (27) and (30) with $e \equiv C_\alpha^N$ and $\varphi = 1$. The MEL form for the α-phase reads as follows:

$$\frac{D_{V_\alpha^\star}\left[\bar{C}_\alpha^N\right]}{Dt} = 0; \tag{32}$$

$$R_\alpha \frac{D_{V_\alpha^\star}\left[Co_\alpha^N\right]}{Dt} = \phi \mathbf{D}_\alpha^N S_\alpha \nabla^2 C_\alpha^N - C_\alpha^N \nabla \cdot \mathbf{V}_\alpha - \phi C_\alpha^N \frac{\partial S_\alpha}{\partial t} - F_\alpha^N, \tag{33}$$

in which the hydrodynamic derivative is given by Eq. (34)

$$\frac{D_{V_\alpha^\star}}{Dt} \equiv \frac{\partial}{\partial t} + \mathbf{V}_\alpha^\star \cdot \nabla, \tag{34}$$

with

$$\mathbf{V}_\alpha^\star = \frac{\mathbf{V}_\alpha - \phi \nabla\left(S_\alpha \mathbf{D}_\alpha^N\right)}{R_\alpha}; \tag{35}$$

and R_α defined by Eq. (36) for the water phase and Eq. (37) for the air phase:

$$R_\alpha \equiv R_w = \phi S_w + (1-\phi)\rho_s K_d \tag{36}$$

$$R_\alpha \equiv R_g \qquad = \phi S_g; \tag{37}$$

4.3.2 Numerical Scheme

The numerical solution includes two steps: advection by particle tracking techniques and dispersion by finite elements. The algorithm of the transport problem solution is depicted in Fig. 2.

1. **Advection by Particle Tracking Techniques.** In what follows, $C \equiv C_\alpha^N$. Let a particle, p, located at point (x_p, y_p) at time t^k, be associated with $\overline{C}_p^k \equiv C(x_p, y_p, t^k)$. At the end of the time step t^{k+1}, each particle, p, reaches a new forward

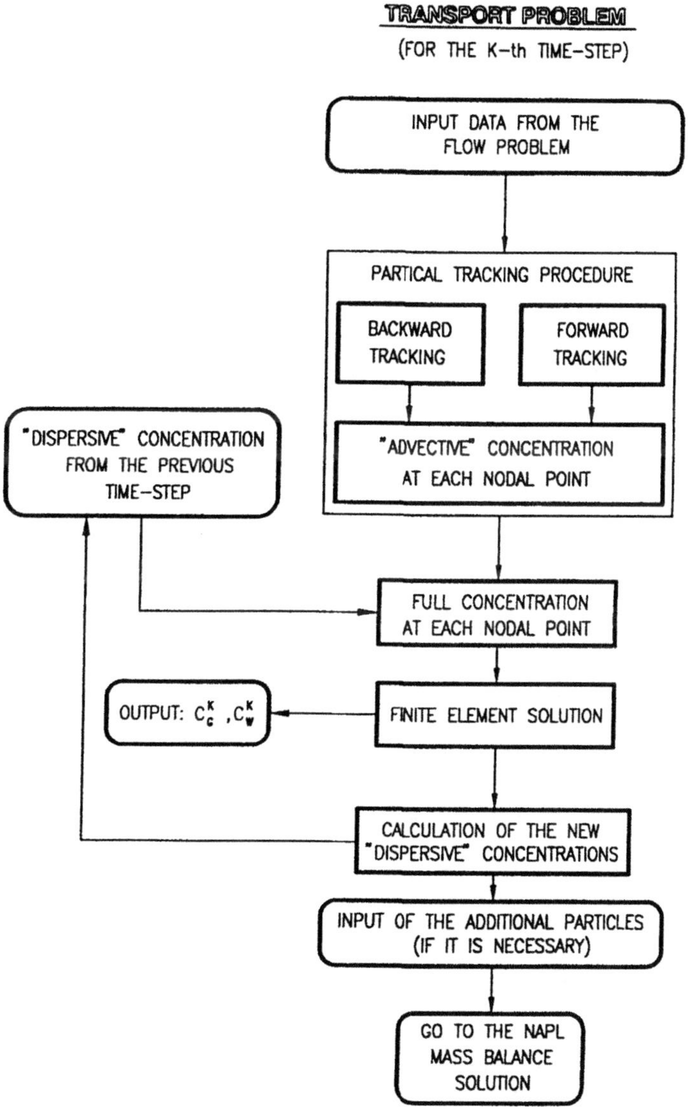

Fig. 2. Algorithm of the transport problem solution

position, (x_p^{k+1}, y_p^{k+1}). In order to obtain coordinates of the particle's new forward position it is necessary to solve the following Lagrangian equations for α-phase:

$$\frac{D_{V_\alpha^\star} \mathbf{X}_\alpha(x, y, t)}{Dt} = \mathbf{V}_\alpha^\star. \tag{38}$$

This describes the *forward particles tracking* procedure, which projects $\bar{C}_p^{k+1}$ [$= \bar{C}_p^k$, by virtue of Eq. (32)] onto nodal points. For nodes that are not covered by these clouds, i.e., for any node located at x_n, y_n such that

$$x_{p\,min}^{k+1} > x_n > x_{p\,max}^{k+1}, \tag{39}$$

$$y_{p\,min}^{k+1} > y_n > y_{p\,max}^{k+1}, \tag{40}$$

we usek a *backward particles tracking* procedure to project $C(\equiv {}^kC)$ from a backward location ${}^kx_p$ onto nodes obeying Eqs. (39) and (40). The ${}^kx_p$, ${}^ky_p$ location is obtained by solving Eq. (28).

2. **Dispersion by Finite Elements.** Equation (33) is solved with respect to $\alpha \equiv (w, g)$ for the residual problem by a finite elements method on a fixed mesh. We approximate $C(x, y, t)$, $V_\alpha(x, y, t)$, $S_\alpha(x, y, t)$. and $D(x, y)$ using spatial Lagrangian interpolation functions $\xi_j(x_i, y_i)$.

The hydrodynamic time derivative over $\mathring{\hat{C}}$ may be approximated by the backward difference for each $\Delta t (= t^{k+1} - t^k)$ as follows:

$$\frac{D\mathring{\hat{C}}}{Dt} \simeq \frac{\mathring{\hat{C}}^{k+1} - \mathring{\hat{C}}^{k}}{\Delta t} = \frac{\mathring{\hat{C}}^{k+1} - \left({}^k\hat{C} - \hat{\bar{C}}^{k+1}\right)}{\Delta t} = \frac{\hat{C}^{k+1} - {}^k\hat{C}}{\Delta t}, \tag{41}$$

where ${}^k\bar{C}$ is obtained as a result of the *particle tracking procedure.*

The resulting global algebraic set after integration by parts, in view of Eq. (41) and accounting for the forward and backward particle projection, for $\alpha (= w, g)$-phases becomes:

$$\begin{aligned}
&\sum_{j=1}^{N} C_j^{k+1} \int_\Omega \Bigg\{ \frac{R^{k+1}}{\Delta t} \xi_i \xi_j + \phi \sum_{l=1}^{N} \left(S_{\alpha l}^{k+1} \xi_l\right) \sum_{l=1}^{N} \left(D_l \xi_i\right) \frac{d\xi_i}{d\mathbf{x}} \frac{d\xi_j}{d\mathbf{x}} + \\
&\phi \left[\sum_{l=1}^{N} (D_l \xi_l) \sum_{l=1}^{N} \left(S_{\alpha l}^{k+1} \frac{d\xi_l}{d\mathbf{x}} \right) + \sum_{l=1}^{N} \left(S_{\alpha l}^{k+1} \xi_l\right) \sum_{l=1}^{N} \left(D_l \frac{d\xi_l}{d\mathbf{x}} \right) \right] \xi_i \frac{d\xi_j}{d\mathbf{x}} - \\
&\left[\sum_{l=1}^{N} \left(V_{\alpha l}^{k+1} \frac{d\xi_l}{d\mathbf{x}} \right) + \frac{\phi}{\Delta t} \sum_{l=1}^{N} \left(S_{\alpha l}^{k+1} \xi_l\right) + k_f^\alpha a_\alpha \right] \xi_i \xi_j \Bigg\} d\Omega = \\
&\int_\Omega \frac{R^k}{\Delta t} \left(\sum_{j=1}^{N_1} {}^kC_j \xi\left({}^kx_p, {}^ky_p\right) + \sum_{m=N_1+1}^{N} \bar{C}_m^{k+1} \delta_m \right) \xi_i d\Omega - \\
&\sum_{j=1}^{N} C_j^k \frac{\phi}{\Delta t} \int_\Omega \left[\sum_{l=1}^{N} \left(S_{\alpha l}^k \xi_l\right) \right] \xi_i \xi_j d\Omega - \int_\Omega \left(k_f^\alpha a_\alpha C_{\star\alpha} \delta_j\right) \xi_i d\Omega,
\end{aligned} \tag{42}$$

where N_1, denotes the nodes that are covered by clouds of particles.

4.3.3 Test Problems

We consider a 2-D porous medium domain spanning $x \in [0, L_1 = 34]$, and $y \in [0, L_2 = 24]$, during a time interval of $t \in [0, \tau]$. The time step was chosen to be $\Delta t = 1.0$. The domain was discretized by a finite elements rectangular and irregular mesh as shown in Fig. 3a. In order to verify the presented numerical scheme and illustrate its capability, the following cases have been incorporated:

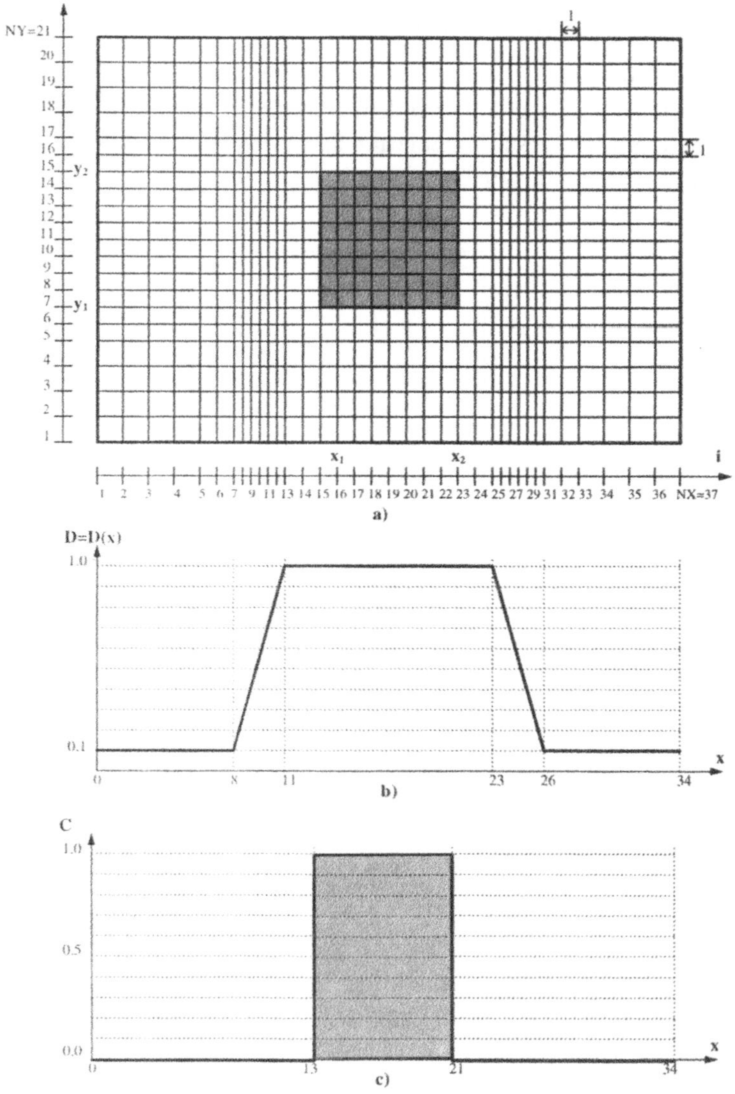

Fig. 3. Test problem for 2-D MEL scheme. **a** Finite elements mesh; **b** diffusion coefficient distribution; **c** initial conditions

Case A. We solved Eq. (14) for *a single phase* ($S = 1$) with zero velocity ($\mathbf{V} = 0$) and without a mass transfer term. This results in the heat transport equation (HTE). The numerical solution of the HTE is subjected to the following boundary and initial conditions (see Fig. 3c):

$$C(x, y=0, t)=0; \quad C(x=0, y, t)=0;$$
$$C(x, y=L_2, t)=0; \quad C(x=L_1, y, t)=0. \tag{43}$$

$$C(x,y,0)=\begin{cases} 1, & \text{if } \{x_1 \le x \le x_2\} \cap \{y_1 \le y \le y_2\}; \\ 0, & \text{otherwise.} \end{cases} \tag{44}$$

- $\mathbf{D} = const.$
  This concerns a flow problem (parabolic equation) with a known analytical solution. The resulting Eulerian finite element (EFE) numerical solution did not deviate from the analytical one.

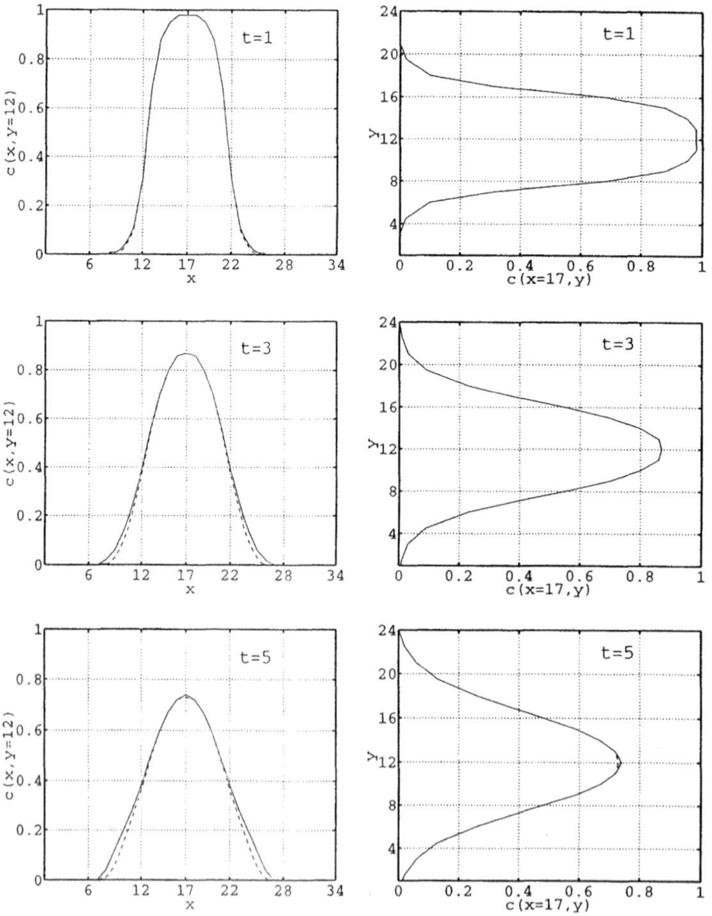

Fig. 4. Comparison between Eulerian (—) and MEL (- -) solutions of HTE for $\mathbf{D}(x, y)$

- $\mathbf{D} = \mathbf{D}(x, y)$.
  The dispersion coefficient is described in Fig. 3b.
  The flow problem is solved by the EFE and the MEL schemes. Comparisons are depicted in Fig. 4, proving deviations at the $\nabla\mathbf{D} \neq 0$ zone. With further grid refinement at this zone, we note that the EFE solution conforms to the MEL one. Simulations for different distributions of **D** are depicted in Fig. 5, and show that steeper gradients of **D** result in more deviations between EFE and MEL schemes.

Case B. This example applied non-zero fluid velocity **V** to the conditions of case A and results in the solution of an advection–diffusion equation (ADE). Here, the MEL scheme was compared with the Eulerian finite element (EFE) and conventional Eulerian-Lagrangian (EL) schemes for the following cases:

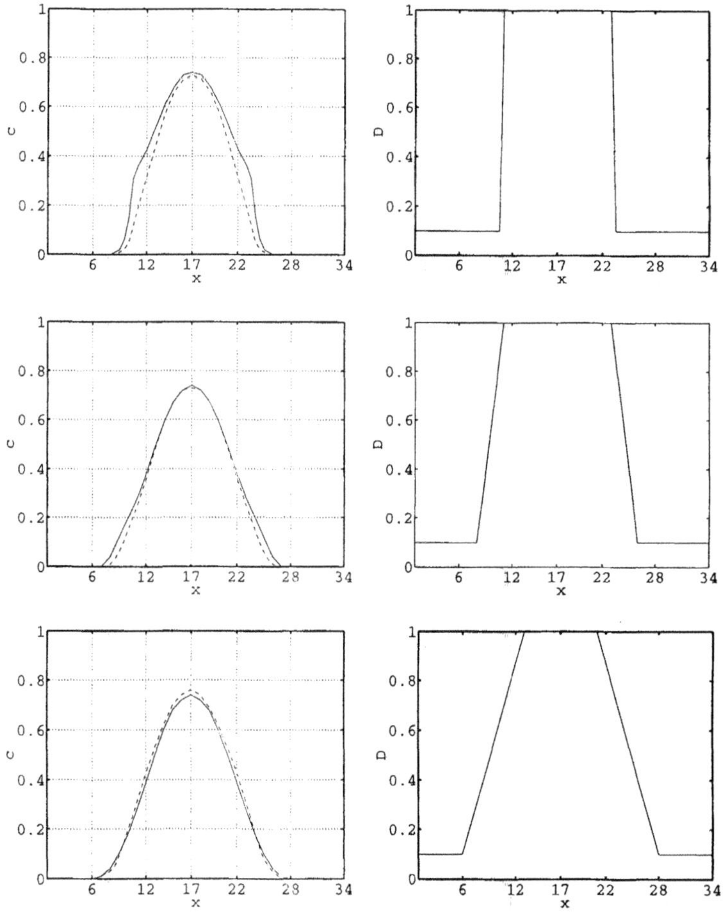

Fig. 5. Comparison between EFE and MEL solutions of HTE for different $\mathbf{D}(x, y)$

1. Constant velocity:
   a) $V_x = 0$; $V_y = 1$ (Fig. 6);
   b) $V_x = 1$; $V_y = 0$ (Fig. 7).
2. Non-uniform field of velocities: flow from source at $x = 0$; $y = 0$ (Fig. 8), with

$$V_x = 25\frac{x}{x^2 + y^2}; \quad V_y = 25\frac{y}{x^2 + y^2}.$$

From the comparison, it follows that the MEL scheme results in a steeper front of plume propagation because it takes into account the particle's slowing down in zone of **D**-coefficient variations.

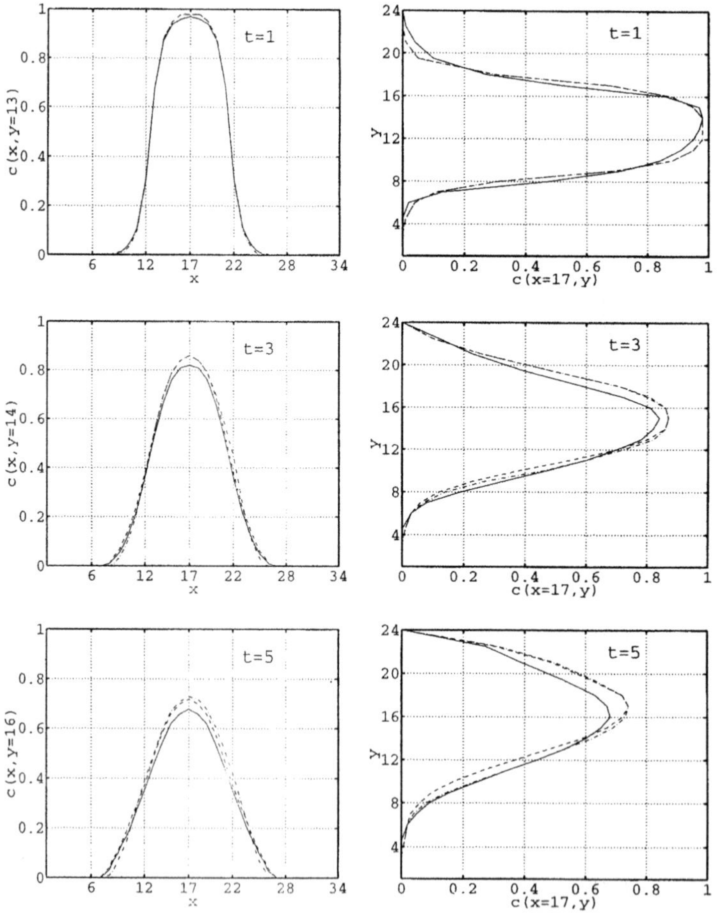

Fig. 6. Comparison between EFE (—), EL (- -), and MEL (-··-) solutions of ADE for **D**(x, y), $V_x = 0$; $V_y = 1$

4.4 NAPL Balance

NAPL balance is calculated from Eq. (22) on the same finite element mesh as for the flow and transport problems. The NAPL saturation (S_N^{k+1}) within the *i-th* element at the end of each time step, Δt, is calculated by subtracting the volume of contaminant which is transferred to the water and gas phases during the time step from the NAPL volume present at the beginning of the time step. The resulting NAPL saturation is expressed in Eq. (45)

$$S_N^{k+1} = \left(S_N^k - \frac{k_{fN}^w a_w^N \left[C_{\star w}^N - C_{w_{avg}i}^N \right] \Delta t}{\phi \rho_N} - \frac{k_{fN_i}^g a_g^N \left[C_{\star w}^N - C_{g_{avg}i}^N \right] \Delta t}{\phi \rho_N} \right), \tag{45}$$

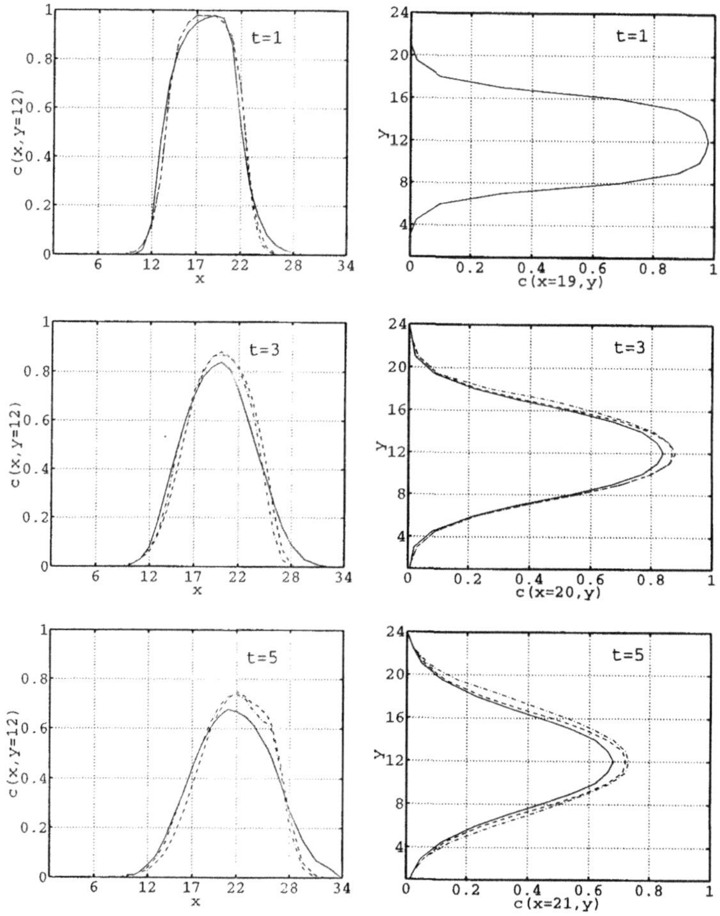

Fig. 7. Comparison between EFE (—), EL (- -), and MEL (-·-) solutions of ADE for D(*x*, *y*), $V_x = 1$; $V_y = 0$

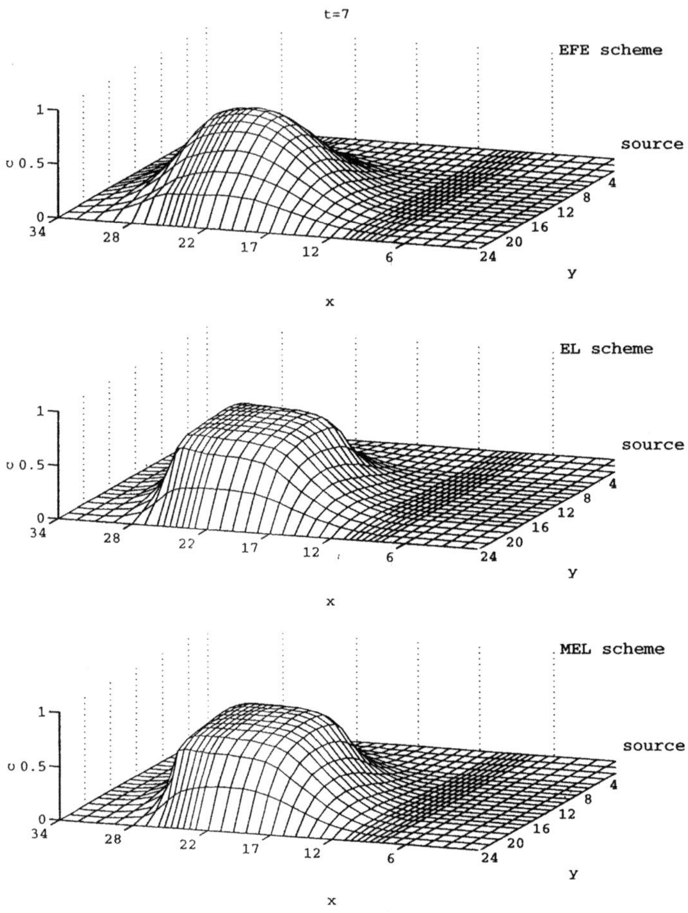

Fig. 8. Comparison between EFE, EL, and MEL solutions of ADE for **D**(x, y), non-uniform field of velocities

where Δt denotes the time step and $C^N_{\alpha_{avg}i}$ denotes the average concentration of NAPL in the α-phase within *i-th* element. $C^N_{w_{avg}i}$ is determined in the model by integrating $C^N_{\alpha i}$ over each element and averaging over the element square. The resultant average concentration is then variably weighted in time.

5 Conclusions

The presented mathematical model considers a real engineering problem of groundwater contamination by oil spills, and its numerical solution predicts displacement of the dissolved NAPL by the flowing water and air phases. Another task of this model is to determine the effectiveness of different remediation schemes in terms of the dissolution and transport of entrapped oil.

Application of the modified Eulerian-Lagrangian method improved the efficiency and accuracy of the numerical solution. The following conclusions can be summarized concerning the MEL scheme: (1) the MEL scheme may be superior in producing much less numerical dispersion when we are dealing with the first-order approximation of hyperbolic terms in the governing equations; (2) the MEL scheme provides solutions of the governing equations on the rough mesh without losing some information about the influence of the non-uniformity of the medium between nodal points; and (3) the MEL scheme, however, is a non-conservative scheme, so the mass balance must be checked at each time step of computation, and such a scheme may result in incorrect solutions when some coefficients described by discontinuous functions are used.

References

Abriola LM, Pinder GF (1985a) A multiphase approach to the modeling of porous media contamination by organic compounds, 1. Equation development. Water Resour Res 21:11–18

Abriola LM, Pinder GF (1985b) A multiphase approach to the modeling of porous media contamination by organic compounds, 2. Numerical simulation. Water Resour Res 21:19–26

Baptista AE (1987) Solution of advection dominated transport by Eulerian-Lagrangian method, using the backward method of characteristics. PhD Diss, MIT, Cambridge, Massachusetts

Binning P, Celia MA (1994) Two-Dimensional Eulerian-Lagrangian localised adjoint method for the solution of the contaminant transport equation in the saturated and unsaturated zones. Proc Computational methods in water resources, Heidelberg, Germany, pp 165–172

Buckley SE, Leverett MC (1942) Mechanism of fluid displacement in sands. Petrol Dev Technol pp 146:107

Celia MA, Ahuja LR, Pinder JF (1987) Orthogonal collocation and alternating-direction procedure for unsaturated flow problems. Adv Water Resour 4(10):178–187

Chiang CY, Wheeler MF, Bedient PB (1989) A modified method of characteristics, technique and mixed finite elements method for simulation of groundwater solute transport. Water Resour Res 25:1541–1549

Cohen Y, Ryan PA, Malina JJ (1989) NAPL and contaminant transport in the soil matrix. Environmental engineering. Proc 1989 Specialty Conf on Environmental Eng, Austin, Texas, 10–12 July 1989, pp 702–709

Corapcioglu MY, Baehr AL (1987) A compositional multiphase model for groundwater contamination by petroleum products. 1. Theoretical considerations. Water Resour Res 23:191–200

Kaluarachchi JJ, Parker JC, Lenhard RJ (1990) Modeling multicomponent organic chemical transport in three-fluid phase porous media. J Contam Hydrol 5:349–374

Lumelsky S, Sorek S (1994) Modified Eulerian-Lagrangian method for flow and transport in a heterogeneous aquifer. Proc Computational methods in water resources, Heidelberg, Germany, pp 956–965

Mayer AS (1994) Application of domain decomposition techniques for multiphase groundwater problems. Proc Computational methods in water resources, Heidelberg, Germany, pp 951–958

Mohtar RH, Wallace RB, Segerlind LJ (1994) Finite element simulation of oil spill cleanup using air sparging. Proc Computational methods in water resources, Heidelberg, Germany, pp 967–974

Neuman SP (1981) A Eulerian-Lagrangian numerical scheme for the dispersion-convection equation using conjugate space-time grids. J Comp Physiol 41(2):270–294

Neuman SP (1984) Adaptive Eulerian-Lagrangian finite element method for advection, dispersion. Int J Num Math Eng 20:321–337

Neuman SP, Sorek S (1982) Eulerian-Lagrangian methods for advection–dispersion. Proc 4th Int Conf on Finite elements in water resources, Germany, 14.41–14.68

Oostrom M, Leijnse A, Roberson KR (1994) Simulation of two- and three-dimensional dense solute plume behavior with the METROPOL-3 code. Proc Computational methods in water resources, Heidelberg, Germany, pp 967–974

Parker JC, Lenhard RJ (1987) Model for hysteretic constitutive relations governing multiphase flow. 1. Saturation–pressure relations. Water Resour Res 23:2187–2196

Parker JC, Lenhard RJ (1989) Vertical integration of three-phase flow equations for analysis of light hydrocarbon plume movement. J Transport Porous Media 5:187–206

Parker JC, Lenhard RJ, Kuppusamy T (1987) Physics of immiscible flow in the porous media. Gov Rep Announce Index (GRA&I), Issue 22

Parker JC, Katyal AK, Kaluarachchi JJ, Lenhard RJ, Johnson TJ (1991) Modeling multiphase organic chemical transport in soils and ground water. Gov Rep Announce Index (GRA&I), Issue 22

Powers SE, Loureiro CO, Abriola LM, Weber JJ (1991) Theoretical study of the significance of nonequilibrium dissolution of nonaqueous phase liquids in subsurface systems. Water Resour Res 27:463–477

Ryan PA, Cohen Y (1991) One-dimensional subsurface transport of a nonaqueous phase liquid containing sparingly water soluble organics: a front-tracking model. Water Resour Res 27:1487–1500

Sorek S (1985a) Eulerian-Lagrangian formulation for flow in soil. Adv Water Resour 8:118–120

Sorek S (1985b) Adaptive Eulerian-Lagrangian method for transport problems in soils. Scientific Basis for Water Resources Management. IASH Publ 153:393–403

Sorek S (1988) Eulerian-Lagrangian method for solving transport in aquifers. Adv Water Resour 11(2):67–73

Sorek S, Breaster C (1988) Eulerian-Lagrangian formulation of the equations for groundwater denitrification using bacterial activity. Adv Water Resour 11(4):162–169

Theis CV (1935) The relations between lowering of the piezometric surface and the rate and duration of discharge of a well using groundwater storage. Trans Am Geophys Union 2:519–524

van Dijke MIJ, van der Zee SEATM (1994) Multi-phase flow modeling of air injection in ground water. Proc Computational methods in water resources, Heidelberg, Germany, pp 933–940

Zienkiewicz OC (1977) The finite element method, 3rd edn. McGraw-Hill, New York

23 Groundwater Contamination Originating from Continuous Oil Leakage from an Underground Source

A. Pistiner[1] and M. Shapiro[2]

Contents

1 Introduction

Transport of pollutants is an important engineering and environmental problem. Groundwater contamination may result from contaminant leakages from various sources, including storage tanks, pipelines, etc. This study is devoted to investigation of the contamination pattern which may result from such leakages.

Various experimental studies reveal (Abriola 1983) that oil contained in a porous medium, fully saturated with water (e.g., a saturated aquifer), is partitioned between two zones: the first is the immiscible zone, containing mostly oil blobs, which are normally larger than the average pore size and held in pore spaces by capillary forces. The extension of the immiscible zone depends on the strength of the capillary forces, which is inversely proportional to the pore size. This zone is larger in porous media containing finer pores (Pistiner et al. 1990).

Adjacent to the immiscible zone is the dispersion zone, containing oil drops which are much smaller than the pore size. The oil blobs present in the immiscible zone constitute obstacles to the water flow. On the contrary, in the dispersion zone, the oil–water emulsion flows freely within the aquifer according to the mass transport advection–dispersion laws (Bear 1988). The oil blobs present in the immiscible zone serve as sources for contamination in the dispersion zone. Field studies of oil transport processes (Fried et al. 1979) show that the contaminant is distributed in the aquifer, within the two zones, as described above.

[1] Ministry of Environment, Water Quality Division, 2 Chury St., Haifa 33045, Israel
[2] Faculty of Mechanical Engineering, Technion-Israel Institute of Technology, Haifa 32000, Israel

H. Rubin/N. Narkis/ J. Carberry
Soil and Aquifer Pollution

Here, we treat the problem of contamination of an aquifer, resulting from a continuously leaking source by modelling oil exchange between the immiscible and dispersed zones. The goal of this work is to determine the extension of the immiscible zone for different flow conditions.

Several models were developed, aimed at description of the oil transfer process between the two zones. Spielman and Su (1977) assumed that the oil transfer rate between these zones depends on the droplet size, local flow rate, surface tension parameters, chemical environment and the local geometry of the porous medium. Spielman and Goren (1972) treated the problem of oil transfer between the two zones by assuming that blobs, held by the capillary forces in the porous medium, form a network of channels. These channels are sufficiently well connected to enable viscous oil flow. The oil in the immiscible zone prevails in a local equilibrium with the solid and water phase according to Laplace's law. When the blobs held in the pores are much larger than the average pore size, large local water pressure gradients force the trapped blobs through the pores, thereby causing transport of the trapped oil. This process is accompanied by the irreversible rupture of the transported blobs with subsequent formation of small droplets, which are further transported in the dispersed zone.

Determination of the extension of the immiscible zone is important for various practical purposes. Towards this goal Pinder and Abriola (1986) developed a 2-D non-steady numerical model by assuming the iso-saturation line $S = 0.1$ in the immiscible zone to be a surface source of the dissolved oil in the dispersion zone. As a result, they calculated evolution of the immiscible zone.

In various hydrological applications the inverse problem of groundwater contamination is of crucial importance. It means that the field data taken in several observation wells are used to assess the global extension of the contamination, locate its source and evaluate the evolution of the contamination pattern for given ground water flow conditions. For this purpose, a simple analytical model is needed, which enables evaluation of the influence of the oil discharge rate, groundwater flow rate, and porous structure of the aquifer on the extension of the immiscible zone. This work is devoted to the latter purpose. We will use the equations of two-phase oil-water transport in the immiscible zone to calculate the steady-state oil distribution, and, in particular, the location of the boundary between the immiscible and the dispersed zones.

2 Physico-Mathematical Model

Consider a two-dimensional homogeneous, generally anisotropic porous layer (e.g., aquifer), fully saturated with water, in the regional scale, i.e., over a planar horizontal extent much larger than its thickness. Consider also a line source of oil of the dimension L^*, located in the layer (see Fig. 1) which discharges oil with a constant volumetric rate $\dot{Q}^*$.

We will refer to L^* as the extension of the source perpendicular to the average water flow direction. The extension of the source along the flow is unimportant

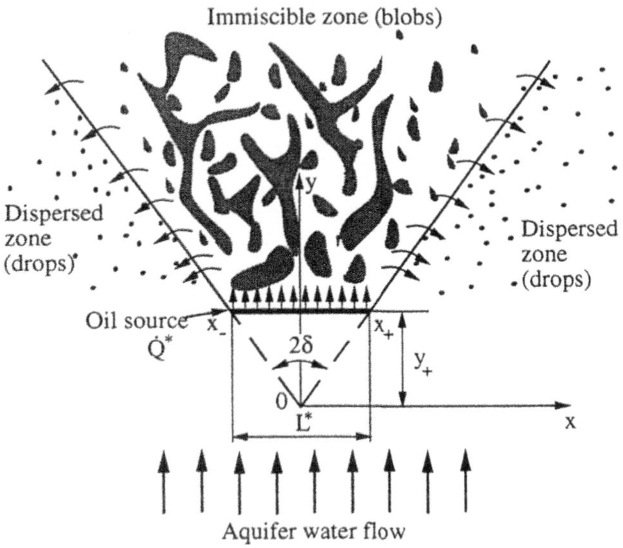

Fig. 1. Schematic description of two-dimensional flow of oil-water mixture in the aquifer with the presence an oil source

since the solution to be developed will be shown to be valid far downstream from the source. In this situation we will show that it is the extension of the line source in the x-direction and the total discharge $\dot{Q}^*$ which, in fact, affect the oil saturation distribution. The respective specific discharge vectors $\mathbf{q}_w$ and $\mathbf{q}_o$ of the water and oil in the immiscible zone in the absence of the gravity force, are described by Darcy's law. This can be written for each fluid in the Cartesian system, using the double index summation convention, ($x_1 = x$, $x_2 = y$) in the form

$$q_{wi} = -\kappa_{ij} \frac{K_{rw}(S)}{\mu_w} \frac{\partial P_w}{\partial x_j} \quad i = 1, 2 \tag{1}$$

and

$$q_{oi} = -\kappa_{ij} \frac{K_{ro}(S)}{\mu_o} \frac{\partial P_o}{\partial x_j} \quad i = 1, 2, \tag{2}$$

where S is the macroscopic (averaged over the layer height) oil saturation in the immiscible zone ($0 < S < 1$), $K_{rw}(S)$, $K_{ro}(S)$ are the relative permeabilities, dependent only on the oil saturation. Other quantities appearing in Eqs. (1) and (2) are: κ_{ij}, the Cartesian components of the permeability tensor; μ_o, μ_w, the oil and water viscosities, respectively; P_o, P_w, the oil and water pressures, respectively.

P_o and P_w are related via the capillary pressure P_c as follows:

$$P_c(S) = P_o - P_w. \tag{3}$$

The conservation of mass for each fluid in a rigid porous medium of porosity ϕ yields

$$\frac{\partial q_{wi}}{\partial x_i} - \phi \frac{\partial S}{\partial t} = 0 \quad i = 1, 2 \tag{4}$$

and

$$\frac{\partial q_{oi}}{\partial x_i} + \phi \frac{\partial S}{\partial t} = 0 \quad i = 1, 2, \tag{5}$$

where t is the time variable.

Combining Eqs. (1)–(5), one obtains (Pistiner et al. 1990)

$$\phi \frac{\partial S}{\partial t} = \frac{\partial}{\partial x_i}\left(\Psi(S)\kappa_{ij} \frac{\partial S}{\partial x_j} - q_i F(S) \right), \tag{6}$$

where

$$q_i = q_{wi} + q_{oi} \quad i = 1, 2 \tag{7a}$$

and

$$F(S) = \left[1 + \frac{K_{rw}(S)\mu_o}{K_{ro}(S)\mu_w} \right]^{-1}; \quad \Psi(S) = \frac{K_{rw}(S)}{\mu_w} \frac{\mathrm{d}P_c(S)}{\mathrm{d}S} F(S). \tag{7b,c}$$

We will further assume that the flow is unidirectional, hence, $q_y = q$, $q_x = 0$. In addition to this, we will set (Bear 1988) $\kappa_{xy} = \kappa_{yx} = 0$.

The fractional flow curve $F(S)$ and the function $\Psi(S)$ can be represented by power-law relationships, usually valid for low oil saturation (Ramakrishnan et al. 1988):

$$F(S) \simeq F_0 S^n; \quad \Psi(S) \simeq \frac{\gamma_{ow}}{\mu_w} \Psi_0 S^{n-m}, \tag{8a,b}$$

where γ_{ow} is the oil–water surface tension coefficient. Moreover, one can deduce from Eqs. (7) and (8) that F_0 is a dimensionless factor of order one, and Ψ_0 is of the order $\Delta P_c/\gamma_{ow}$, where ΔP_c is the characteristic capillary pressure. Parameters m and n were found to depend on the oil properties; in particular, viscosity and also on the pore size index of the porous medium (Pistiner et al. 1990). From the experimental data on fractional flow rates and permeabilities it was found that m and n vary in the following limits: $1 < n < 4$, $1 < m < 4$.

3 Mathematical Formulation

We define dimensionles variables

$$\hat{x} = x/L_x; \quad \hat{y} = y/L_y; \quad \hat{t} = t/T, \tag{9a,b,c}$$

where the characteristic convection lengths, L_x, L_y and the characteristic convection time, T, are:

$$L_x = \frac{\Psi_0 \gamma_{ow} \sqrt{\kappa_{xx}\kappa_{yy}}}{qF_0\mu_w}; \quad L_y = \frac{\Psi_0 \gamma_{ow}\kappa_{yy}}{qF_0\mu_w} \tag{9d,e}$$

$$T = \frac{L_y \phi}{qF_0}. \tag{9f}$$

Substituting Eqs. (9a–e) into Eqs. (6) and (8a,b) to obtain

$$\frac{\partial J_x}{\partial \hat{x}} + \frac{\partial J_y}{\partial \hat{y}} + \frac{\partial S}{\partial \hat{t}} = 0, \tag{10a}$$

where J_x, J_y are the dimensionless oil flux components:

$$J_x = -S^{n-m}\frac{\partial S}{\partial \hat{x}}; \quad J_y = -S^{n-m}\frac{\partial S}{\partial \hat{y}} + S^n. \tag{10b,c}$$

We will assume that a steady oil saturation distribution is established a sufficiently long time after the beginning of oil release ($t \gg T$). In such a case, $\partial S/\partial \hat{t}$ on the right-hand side of Eq. (10a) vanishes and Eqs. (10a,b,c) describe a steady-state distribution of the oil saturation $S(\hat{x},\hat{y})$ within the aquifer.

The volumetric discharge $\dot{Q}^*$ of the pollutant may be expresed in the dimensionless form by integrating across the flow direction using (9b) and (10c):

$$\dot{Q} = \frac{\mu_w \dot{Q}^*}{\gamma_{ow}\Psi_0\sqrt{\kappa_{xx}\kappa_{yy}}} = \int_{-L/2}^{+L/2} J_y\Big|_{\hat{y}=\hat{y}_+} d\hat{x}, \tag{11a}$$

where the dimensionles parameter L is defined as

$$L = L^* \frac{\mu_w q F_0}{\gamma_{ow}\Psi_0\sqrt{\kappa_{xx}\kappa_{yy}}}. \tag{11b}$$

Far from the source, the oil saturation and the sizes of oil blobs decrease down to the value $S = 0$, where the immiscible zone terminates. Beginning at this location, the oil prevails only in the dispersed phase. In fact, the boundary between the immiscible zone and the dispersion zone is not clearly defined. Rather, a transition zone exists which includes oil drops of different sizes (Soo and Radke 1986). However, we will assume that the latter transition zone is very narrow and immiscible and dispersion zones are separated by boundaries

$$\hat{y} = Y(\hat{x}) \quad \hat{x}_+ < \hat{x} < \infty \tag{12a}$$

$$\hat{y} = -Y(\hat{x}) \quad -\infty < \hat{x} < \hat{x}_-, \tag{12b}$$

where $\hat{x}_\pm = \pm L/2$ in accordance with the coordinate system chosen (see Fig. 1). This assumption is consistent with various field observations (Pinder and Abriola 1986) and implies that the minimal iso-saturation line $S = 0$ serves as a boundary between the two zones.

In accordance with the above, the saturation is subject to the following boundary conditions

$$S\left(\hat{x},Y\left(\hat{x}\right)\right)=0 \quad \hat{x}_+ < \hat{x} < \infty \tag{13a}$$

$$S\left(\hat{x},-Y\left(\hat{x}\right)\right)=0 \quad -\infty < \hat{x} < \hat{x}_+. \tag{13b}$$

We will place the origin of the coordinate system at the point located at a distance $\hat{y}_+$ upstream from the leakage source. This distance is to be determined in the course of the solution, together with the boundary separating the immiscible and the dispersed oil zones.

All iso-saturation lines are symmetrical about the flow direction (i.e., the $\hat{y}$ direction). Hence, the problem may be posed in the domain $0 < \hat{x} < \infty$, $\hat{y}_+ < \hat{y} < \infty$ with an additional boundary condition imposed at $\hat{x} = 0$

$$\frac{\partial S}{\partial \hat{x}}=0 \quad \text{on} \quad \hat{x}=0. \tag{14}$$

Early on, when the oil transport is affected by initial saturation distribution, Eqs. (10a,b,c) should be solved for the function $S(\hat{x},\hat{t})$ which involves, in particular, calculation of the boundary $Y(\hat{x},\hat{t})$, separating the continuous and discontinuous oil zones [see (13a,b)]. This requires formulation of an additional boundary condition, expressing continuity of the oil flux across the boundary (Spielman and Su 1977). As a result, this problem should be solved both in continuous and discontinuous zones.

We will look for a long-term solution of Eqs. (10a,b,c), where the saturation reaches a steady state form $S(\hat{x},\hat{y})$. This means that the oil discharged per unit time by the source is equal to the total oil flux across the transition zone $Y(\hat{x})$. Estimates performed in the discussion section show that such a steady saturation distribution in sand aquifers is reached during about a six month period. It turns out that at such long times the solution in the discontinuous zone is unnecesary, since both the boundary $Y(\hat{x})$ and the continuous oil saturation $S(\hat{x},\hat{y})$ in the immiscible zone are determined from the symmetry properties (similarity nature) of the problem. An additional boundary condition, required in order to obtain an unambiguous solution, is the integral relation expressed in Eq. (11a) between the oil discharge rate and the strength $\dot{Q}$ of the oil contamination source.

In the following section we will develop a solution which is valid at large distances ($\hat{x},\hat{y} >> L$) from the pollution source, i.e., where it may be viewed as a point source. The similarity solution developed here satisfies the integral Eq. (11a).

In the case where the real oil flux distribution along the line source is consistent with the solution $J_y(\hat{x},\hat{y}_+)$ [see Eq. (39a)], the similarity solution developed in the following section is valid everywhere within the domain $-L/2 < \hat{x} < +L/2$, $\hat{y} > \hat{y}_+$.

In fact, the oil discharge flux distribution is generally unknown. It is the total discharge $\dot{Q}^*$ which is of interest in hydrological applications. In this case, the solution obtained here is valid for large distances $\hat{y} > \hat{y}_+$ from the source.

4 The Self-Similar Solution

We assume that at the steady state, the distribution of the residual oil is derived from a certain universal behavior, which is investigated in this study by the similarity method. Towards this end, we will look for a transformation that reduces the nonlinear PDE (10a,b,c) into an ODE. Explicitly, we will construct a similarity solution of the form:

$$S(\hat{x},\hat{y}) = u(\xi)^{1/(n-m+1)} / \hat{y}^{1/(m-1)}, \xi = \hat{x}/\hat{y}, \tag{15a,b}$$

where $u(\xi)$ is a similarity function. This implies that the boundary $Y(\hat{x})$ in Eq. (12a) is $Y(\hat{x}) = \hat{x}/\xi_b$, where a constant ξ_b is to be determined. Substituting Eqs. (15a,b) into Eqs. (10a,b,c), one obtains

$$\frac{u'(1+\xi^2)}{n-m+1} = g(\xi) - \xi u^{\frac{n}{n-m+1}} - \frac{2}{m-1} u\xi \tag{16a}$$

$$g' = \frac{n-m+1}{m-1}\left(-u^{\frac{n}{n-m+1}} + \left(\frac{1}{n-m+1} - \frac{1}{m-1}\right)u\right). \tag{16b}$$

This system may be solved (e.g. numerically) in $0 < \xi < \xi_b$ for any m and n subject to the appropriate boundary conditions derived from Eqs. (13a,b) and (14):

$$u'(0) = 0; \quad u(\xi_b) = 0. \tag{17a,b}$$

We will consider a particular case $m = n/2 + 1$, where a closed form analytical solution exists, enabling the influence of various physical parameters to be examined. With the above relationship Eqs. (15a,b) reduces to

$$S(\hat{x},\hat{y}) = \left(u(\xi)/\hat{y}\right)^{2/n}, \xi = \hat{x}/\hat{y}, \tag{18a,b}$$

and the system of Eqs. (16a,b) adopts the form

$$\frac{2}{n}\left(u(1+\xi^2)\right)' = g(\xi) - \xi u^2, \tag{19a}$$

$$g' = -u^2. \tag{19b}$$

This is to be solved in the region $0 < \xi < \xi_b$ subject to the boundary conditions of Eqs. (17a,b). Combining Eqs. (18a,b) and integration yield the following expression for g:

$$g = \frac{u(1+\xi^2) - 2\lambda/n}{(n/2)\xi}, \tag{20}$$

where λ is an integration constant which will be determined below.

Introduction of Eq. (20) into Eq. (19a) yields

$$u' = -\frac{u(\xi^2-1) + 2\lambda/n}{\xi(\xi^2+1)} - \frac{n\xi}{2(\xi^2+1)}u^2. \tag{21}$$

In order to integrate this equation we apply the Ricatti transformation (Kamke 1959)

$$u = \frac{2(\xi^2+1)}{n\xi}\frac{1}{w}\frac{dw}{d\xi}, \tag{22}$$

in terms of the yet unknown function $w(\xi)$. Introducing Eq. (22) into Eq. (21) one obtains the following linear ODE:

$$w'' + \frac{2(\xi^2-1)}{\xi(\xi^2+1)}w' + \frac{\lambda}{(\xi^2+1)^2}w = 0. \tag{23}$$

Further, a new variable $z(\xi)$ is defined as follows:

$$z = \frac{1}{2}\left(1 - \sqrt{\frac{1}{\xi^2+1}}\right) \quad 0 < z < z_b < 1\ 2 \tag{24a}$$

where $z_b = z(\xi_b)$ with the inverse function $\xi(z)$ given by

$$\xi = \frac{2\sqrt{z(1-z)}}{1-2z}. \tag{24b}$$

Introduction of Eq. (24a) into Eq. (23) then yields the hypergeometric equation

$$z(z-1)\frac{d^2w}{dz^2} + (0.5-z)\frac{dw}{dz} - \lambda w = 0 \tag{25}$$

possessing the general solution

$$w(z) = C_1F_1(z) + C_2 z^{3/2}F_2(z). \tag{26}$$

In Eq. (26)

$$F_1(z) = F\left(-1-\sigma, -1+\sigma, -\frac{1}{2}, z\right); \quad F_2(z) = F\left(\frac{1}{2}-\sigma, \frac{1}{2}+\sigma, \frac{5}{2}, z\right), \tag{27a,b}$$

are expressed via the hypergeometric functions (Abramowitz and Stegun 1970) and $\sigma = \sqrt{1+\lambda}$, C_1 and C_2 are constants to be determined later. Using the properties of the hypergeometric series, we rewrite Eqs. (27a,b) in the following forms

$$F_1(z) = 2\left(\frac{1}{2} - z\right)\cos\beta + 2\sigma\sqrt{z(1-z)}\sin\beta \tag{28a}$$

and

$$F_2(z) = -\frac{3z^{-3/2}}{8\lambda\sigma}\left(2\sigma\sqrt{z(1-z)}\cos\beta + (2z-1)\sin\beta\right), \tag{28b}$$

where

$$\beta = 2\sigma\left(\arcsin\sqrt{z} + \pi\ell\right) \quad \ell = 0,1,2,3\ldots \tag{29}$$

Introduction of Eqs. (28a,b) into Eq. (26) yields

$$w(z) = \sin\beta\left(2C_1\sigma\sqrt{z(1-z)} - \frac{3C_2(2z-1)}{8\lambda\sigma}\right) + \cos\beta\left(C_1(1-2z) - \frac{3C_2}{4\lambda}\sqrt{z(1-z)}\right). \tag{30}$$

Substituting Eq. (24b) in Eq. (22), the function u becomes

$$u(z) = -\frac{2z-1}{n}\frac{1}{w}\frac{\mathrm{d}w}{\mathrm{d}z}. \tag{31}$$

Differentiation of Eq. (30) with respect to z yields the following expression

$$\frac{\mathrm{d}w}{\mathrm{d}z} = 2C_1\lambda\cos\beta + \frac{3C_2}{4\sigma}\sin\beta. \tag{32}$$

Introduction of Eqs. (30) and (32) into Eq. (31) yields the following solution for u:

$$u(\xi(z)) = \frac{2\lambda}{n}\frac{(2z-1)(3\tan\beta - 8\sigma\lambda C)}{(2z-1)(3\tan\beta + 8\sigma\lambda C) + 2\sigma\sqrt{z(1-z)}(3 - 8\sigma\lambda C\tan\beta)}, \tag{33}$$

where $C = C_1/C_2$. Using Eqs. (22) and (24a), $\mathrm{d}u/\mathrm{d}\xi$ can be expressed as follows:

$$\frac{\mathrm{d}u}{\mathrm{d}\xi}(z) = -(2z-1)^2\sqrt{z(1-z)}\left(2\frac{(u+2u^2/n)}{2z-1} - \frac{n(2z-1)}{4}\frac{1}{w(z)}\frac{\mathrm{d}^2w}{\mathrm{d}z^2}\right), \tag{34}$$

wherein $\mathrm{d}^2w/\mathrm{d}z^2$ is obtained by differentiation of Eq. (32):

$$\frac{\mathrm{d}^2w}{\mathrm{d}z^2} = \frac{-2C_1\lambda\sigma\sin\beta + \frac{3}{4}C_2\cos\beta}{\sqrt{z(z-1)}}. \tag{35}$$

Introducing Eqs. (34) and (35) into Eq. (17a) using Eq. (33), one obtains C in the form

$$C = \frac{3}{8\lambda\sqrt{1+\lambda}\tan\left(2\sqrt{1+\lambda}\pi\ell\right)} \qquad \ell = 0,1,2,3\ \ldots \tag{36}$$

Substituting Eq. (36) into Eq. (33) we finally obtain the following expression for $u(z)$:

$$u(z) = \frac{2\lambda(2z-1)}{n}\frac{1}{2z-1+2\sqrt{1+\lambda}\sqrt{z(1-z)}\tan\left(2\sqrt{1+\lambda}\,\mathrm{arc\,sin}\sqrt{z}\right)}. \tag{37}$$

This solution is independent of the integer l. The boundary between the capillary zone and the finely dispersed oil zone can be obtained by using Eq. (17b)

$$2\sqrt{1+\lambda}\ \arcsin\sqrt{z(\xi_b)} = \frac{\pi}{2} \tag{38a}$$

where

$$\xi_b = \tan\delta$$

and δ is the angle of the boundary between the immiscible and the dispersed zones. From Eq. (38a) it follows that

$$\delta = \frac{\pi}{2\sqrt{1+\lambda}}. \tag{38b}$$

Introducing Eqs. (18a,b), into Eq. (10c) and using Eqs. (19a,b), we obtain the following flux equations:

$$J_y = \hat{y}^{-2}\left(\frac{2}{n}(u\xi)' + u^2\right) \tag{39a}$$

$$J_x = \hat{y}^{-2}\frac{2}{n}u'. \tag{39b}$$

Calculating the oil discharge $\dot{Q}$ at a distance $\hat{y} = \hat{y}_+$ from the origin, using Eqs. (10c) and (11a), Eqs. (18a,b) and (39a) and using the symmetry property of u, we obtain

$$\dot{Q} = 2\hat{y}_+^{-1}\int_0^{L/2}\left(\frac{2}{n}u\xi - g\right)' \mathrm{d}\xi = -2\hat{y}_+^{-1}\left[\frac{u-2\lambda/n}{n\xi/2}\right]_0^{\xi_b}. \tag{40a}$$

It can be shown that

$$\frac{u-2\lambda/n}{n\xi/2} \to 0 \quad \text{as} \quad \xi \to 0. \tag{40b}$$

Further, by introducing $\hat{y} = \hat{y}_+$ in Eq. (40a), using Eq. (40b) and $L/2 = \xi_b\hat{y}_+$ we obtain the following expression for the total discharge:

$$\dot{Q} = \frac{16\lambda}{Ln^2}. \tag{41}$$

This yields the value of the yet unknown coefficient λ:

$$\lambda = \frac{\dot{Q}Ln^2}{16} = \frac{\tilde{Q}n^2}{16}, \tag{42}$$

where $\tilde{Q} = \dot{Q}L$. According to the latter, $\tilde{Q}$ can be identified as the modified strength of the oil source. By introduction of Eq. (42) into Eq. (39b), one obtains

$$\delta = \frac{\pi}{2\sqrt{1+\tilde{Q}n^2/16}}. \tag{43}$$

One can show that the total discharge in the water flow direction at any location $\hat{y} > \hat{y}_+$ is

$$\dot{Q}(\hat{y}) = \hat{y}^{-1} \frac{\tilde{Q}}{2\tan\left[\dfrac{\pi}{2\sqrt{1+\tilde{Q}n^2/16}}\right]},$$

Which shows that the total contaminant rate decays as $1/\hat{y}$.

5 Results and Discussion

Figure 2 depicts the angle δ of the boundary between the immiscible and dispersed zone as a function of the dimensionless parameter $\tilde{Q} = \dot{Q}L = \dot{Q}^*L^*qF_0\mu^2{}_w/\gamma^2_{ow}\Psi^2_0\kappa_{xx}\kappa_{yy}$. This parameter measures the influence of both the total pollutant discharge rate $\dot{Q}$ and the length L as a single product. Therefore, at large distances $\hat{y}$ where our solution is valid, the total contamination may be viewed as a result of a point source $L \to 0$ with infinite discharge rate $\dot{Q} \to \infty$. This is in contrast with the comparable transfer process governed by *linear* transport equations, since in the latter case the saturation distribution is governed by $\dot{Q}$ only. It can be observed that as $\dot{Q}$ increases, the angle δ of the immiscible zone decreases. This phenomenon reveals the importance of oil advection in the direction of the flow. With increasing $\tilde{Q}$, this oil transport mechanism becomes more significant, compared with the capillary dispersion along the sidestream ($\hat{x}$) direction. This nonlinear effect of $\tilde{Q}$ may be rationalized by Eq. (10c), wherein the convective part is given by S^n. Clearly, the oil advection transport "velocity" (or specific discharge) is proportional to S^{n-1}. For the case $n \geq 1$ considered here, with increasing $\tilde{Q}$ and the concomitant increase in S, the advection is augmented.

Figure 3 depicts saturation profiles for three values of $\tilde{Q}$. This figure may be useful for calibration of the model solution, which can be achieved by matching

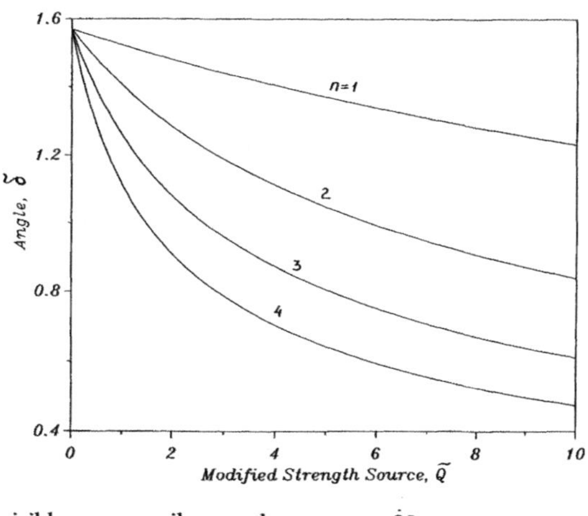

Fig. 2. The value of the angle δ of the immisible zone vs. oil strength parameter $\dot{Q}L$

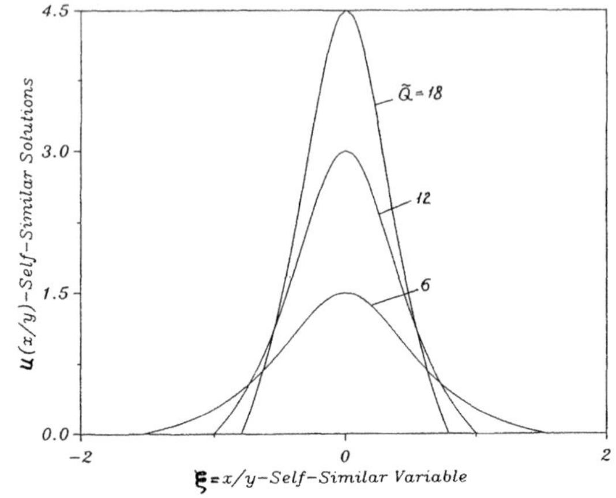

Fig. 3. Self-similar solutions for saturation distribution ($n = 2$)

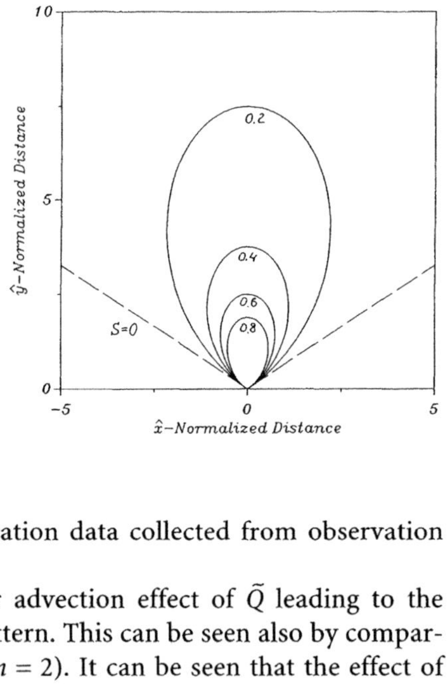

Fig. 4. Distribution of the oil saturation ($\tilde{Q} = 6$, $n = 2$)

various self-similar profiles with oil saturation data collected from observation wells in the aquifer.

One can clearly observe the nonlinear advection effect of $\tilde{Q}$ leading to the downstream plume-like contamination pattern. This can be seen also by comparing Figs. 4–6 obtained for various $\tilde{Q}$ (for $n = 2$). It can be seen that the effect of increasing $\tilde{Q}$ is to reduce the value of the angle δ and to increase the downstream values of saturations, as was also shown in Fig. 3.

Below, we perform estimations of the characteristic dimensions of the pollution zone, as obtained from the similarity solution for the water-oil mixture ($\mu_w = 10^{-3}$ kg/ms, $\gamma_{ow} = 0.03$ N/m) of water flowing with characteristic velocity 0.01 m/day in a coarse sand aquifer with capillary pressure of 2 cm H_2O, and permeability of 10^{-8} cm^2. One obtains from Eq. (9e) the characteristic length $L_y \sim 1.7$ m and the characteristic time scale given by Eq. (9c) of $T \sim 50$ days. Taking $L^* = 1$ m as a

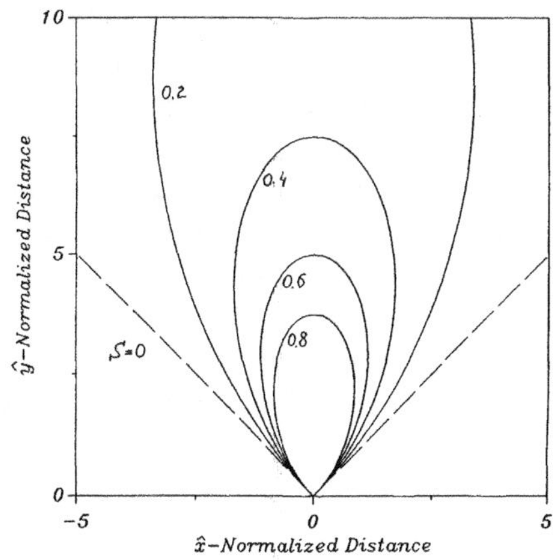

Fig. 5. Distribution of the oil saturation ($\tilde{Q} = 12$, $n = 2$)

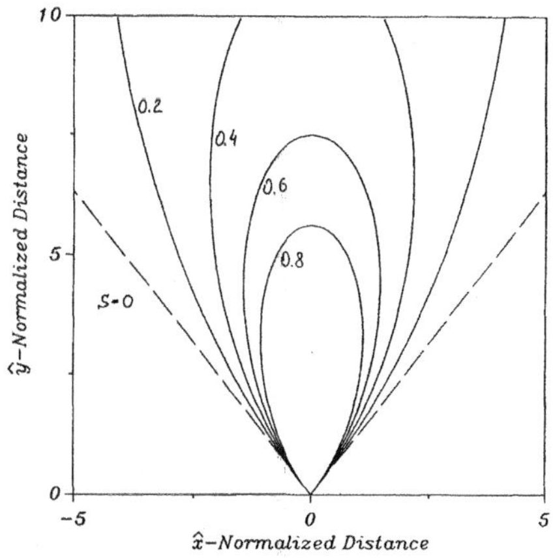

Fig. 6. Distribution of the oil saturation ($\tilde{Q} = 18$, $n = 2$)

characteristic extension of the oil leakage source, $F_0 \sim 1$, and the leakage rate $\dot{Q}^* = 0.36\,m^2/day$ which corresponds to the data of Fig. 5, one obtains the extension of the plume calculated in Fig. 5 to be about $(10–20)L_y \sim 17–34\,m$. Moreover, the time in which the steady oil distribution is achieved if $t \gg 50$ days, is about 0.5–1 year. These estimates are generally consistent with several field studies (Testa and Winegardner 1991).

6 Summary

The distribution of immiscible oil contaminant in a saturated porous medium is analyzed and described by the basic two-dimensional equations of the two-phase flow. The self-similar solution obtained here pertains to steady-state oil transport. The influences of the leakage rate from an oil source and its length are lumped in a single parameter $\dot{Q}L$ describing the source strength. It is found that increasing $\dot{Q}L$ reduces the value of the immiscible zone angle δ and increases the downstream oil saturations. The model developed here may be used jointly with the data collected from several observation wells to evaluate the extension of the immiscible oil zone and to locate the source and strength of the leak.

Acknowledgments. This research has been supported by the Technion V.P.R. Fund - B. and I. Green Research Fund.

References

Abramowitz M, Stegun IA (1970) Handbook of mathematical functions. Dover, New York

Abriola LM (1983) Mathematical modeling of the muliphase migration of organic compounds in a porous medium. PhD Thesis, Princeton University, Princeton, New Jersey

Bear J (1988) Dynamics of fluids in porous media. Dover, New York

Fried JJ, Muntzer P, Zilliox L (1979) Ground-water pollution by transfer of oil hydrocarbons. Ground Water 17(6):586–595

Kamke E (1959) Differentialgleichungen Lösungsmethoden und Lösungen. Chelsea Publication, New York

Pinder GF, Abriola LM (1986) On the simulation of nonaqueous phase organic compounds in subsurface aquifer. Water Resour Res 22(9):109s–119s

Pistiner A, Shapiro M, Rubin H (1990) Similarity solutions for immiscible phase migration in porous media: an analysis of free boundaries. Transp Porous Media 5:491–515

Ramakrishnan TS, Wilkinson D, Dias M (1988) Effect of capillary pressure on the approach to residual saturation. Transp Porous Media 3:51–79

Soo H, Radke CJ (1986) Filtration model for the flow of dilute stable emulsions in porous media. Chem Eng Sci 41:263–272

Spielman LA, Goren SL (1972) Theory of coalescence by flow through porous media. Ind Eng Chem Fund 1:66–73

Spielman LA, Su YP (1977) Coalescence of oil in water suspensions by flow through porous media. Ind Eng Chem Fund 16:272–282

Testa SM, Winegardner DL (1991) Restoration of petroleum contaminated aquifers. Lewis, Chelsea, Michigan

Part V

Field Studies

24 Surfactant-Enhanced NAPL Remediation: From the Laboratory to the Field

D. A. Sabatini[1,3], R. C. Knox[1,3], J. H. Harwell[2,3], and B. J. Shiau[1]

Contents

1 Introduction/Surfactant Fundamentals

Pump-and-treat remediation has routinely been prescribed for subsurface nonaqueous phase liquid (NAPL) contamination. However, pump-and-treat systems have proven incapable of remediating contaminated aquifers in a timely and economical manner. Cost overruns of 80% and cleanup times as much as three time longer than original estimates are typically reported (Olsen and Kavanaugh 1993). Several factors have been identified that contribute to the inefficiency of pump-and-treat remediation, including: (1) diffusion limitations for contaminants from low conductivity zones to high conductivity zones; (2) hydrodynamic isolation, such as dead end zones; (3) rate-limited desorption of contaminants from solid surfaces; and (4) liquid–liquid dissolution of residual NAPLs (Bouwer et al. 1988; Keely 1989; Mackay and Cherry 1989; Haley et al. 1991; Schmelling et al. 1992; NRC 1994). When residual NAPL is present, subsurface contamination can be categorized into three zones: the source/residual zone, the concentrated plume (center of mass of the groundwater plume), and the dilute groundwater plume.

[1] School of Civil Engineering and Environmental Science, The University of Oklahoma, Norman, Oklahoma 73019, USA
[2] School of Chemical Engineering and Materials Science, The University of Oklahoma, Norman, Oklahoma 73019, USA
[3] The Insitute for Applied Surfactant Research, The University of Oklahoma, Norman, Oklahoma 73019, USA

H. Rubin/N. Narkis/ J. Carberry
Soil and Aquifer Pollution

Innovative technologies are necessary for residual phase materials as pump-and-treat alone will be extremely inefficient in this zone.

Chemical amendments that have been suggested for expediting pump-and-treat remediation include the following: complexing agents, cosolvents, surfactants via solubilization (micellar partitioning) and mobilization (ultra-low interfacial tension via middle phase microemulsion), oxidation-reduction agents, precipitation-dissolution regaents, and ionization reagents (Palmer and Fish 1992). Surfactant enhanced subsurface remediation, the subject of this chapter, has received increasing attention as a viable technology for remediating NAPL source zones (EPA 1995; Sabatini et al. 1995).

Surfactant (*surf*ace-*act*ive-*agents*) molecules have two distinct regions (hydrophobic - water disliking, and hydrophilic - water liking); this dual nature causes surfactants to accumulate at interfaces (e.g., air-water, oil-water, water-solid). Surfactants can be classified in numerous ways, including by charge (cationic, anionic, nonionic, zwitterionic), source (e.g., biosurfactants from plant or microbial production, synthetic surfactants), and regulatory status (e.g., indirect or direct food additive status, acceptable for use in pesticide formulations, biodegradable in waste treatment processes).

Above a certain concentration, surfactant molecules self-assemble into aqueous phase aggregates (known as micelles). The concentration above which micelles form is referred to as the critical micelle concentration (CMC). Micelles have a hydrophilic exterior and a hydrophobic interior; the former makes them highly soluble in water, while the latter makes them an excellent sink for organic contaminants. The greater the surfactant concentration above the CMC, the larger the solubility enhancement anticipated; this is referred to as the solubilization mechanism.

Surfactant mobilization (formation of a middle phase microemulsion with concomitant ultra-low interfacial tensions - IFTs) results in bulk displacement of the residual oil phase. A middle phase system occurs on the continuum between aqueous phase micelles (as described above) and reverse micelles (having a hydrophobic exterior and a hydrophilic interior, and located in the oil phase). The middle phase has a density between that of the water and oil, resulting in its designation as a middle phase microemulsion. The transition between these three phases can be realized by various methods. For ionic surfactants, variations in salinity can produce the middle phase system (as in enhanced oil recovery), while temperature variations can produce the same result for nonionic surfactants. However, these approaches may not be feasible for aquifer remediation. Surfactant mixtures (combinations of mixed surfactants/hydrotropes/alcohols) can produce similar results without requiring salinity or temperature variations (Shiau et al. 1994, 1996a).

Surfactant-based remediation technologies have advanced significantly, based on environmental research conducted over the past 5 years, while benefiting also from prior research on surfactant-enhanced oil recovery (SEOR); (e.g., Valsaraj and Thibodeaux 1989; Vignon and Rubin 1989; Edwards et al. 1991; Jafvert and Heath 1991; Smith and Jaffe 1991; Abdul et al. 1992; Fountain 1992; West 1992;

West and Harwell 1992; Pennel et al. 1993; Baran et al. 1994; Sabatini et al. 1995). The success of surfactant-based technologies will depend upon their technical, economical, and regulatory viability. This chapter summarizes research contributions of our group that have aided in the development of these technologies, including results from economic analyses, laboratory studies, and field demonstrations. Space will not allow a thorough treatment of each of these areas; the interested reader is directed to the cited references for additional details.

2 Process Development: Economic, Laboratory and Modeling Results

In this section, the results of economic and laboratory studies will be summarized. The laboratory studies include initiatives aimed at minimizing surfactant losses (physicochemical and hydrodynamic losses), evaluating solubilization and mobilization using high performance and direct food additive surfactants, and developing unit processes and design criteria for contaminant–surfactant separation and surfactant reuse.

2.1 Economic Viability

It is one thing to conclude that surfactants can be used to reduce remediation times; it is another thing to conclude that one can afford to use surfactants for this purpose. Two efforts have considered the question of economic viability for surfactant enhanced subsurface remediation. In the first study (Krebs-Yuill et al. 1995) the optimized net present cost of micellar solubilization was compared to the optimized net present cost of conventional pump-and-treat. In the second study (Krebs-Yuill et al. 1996), the optimized net costs of solubilization and of mobilization were compared to a fixed cost for conventional pump-and-treat, with the latter being based on the experience of DuPont.

Krebs-Yuill et al. (1995) showed that surfactant solubilization was more economical than conventional pump-and-treat under two scenarios. In the first scenario, sites from 0.25 acre to 10 acres were considered, with a mixture of tetrachloroethylene (PCE), trichloroethylene (TCE), and 1,2-dichloroethylene (DCE) as the contaminants and using an arbitrary cleanup deadline of 7 years. The two most important conclusions were: (1) that up-front surfactant costs constitute the single largest cost in a surfactant-enhanced remediation process, and (2) that surfactant recovery from project extraction wells and reinjection must be implemented. The cost of surfactant alone exceeds the cost of 25 years of pump-and-treat on a discounted cash flow basis for sites of 10 acres. Surfactant recovery costs were estimated to be less than 10% of the annual surfactant replacement costs. A third important conclusion was that the cost of surfactant-enchanced pump-and-treat is highly sensitive to the size of the zone of residual contamination. As the surfactant injection is localized to a smaller and smaller zone (the zone of residual contamination), the cost of the surfactant required to fill the zone with micelles drops geometrically. For sites of 1 acre or less, significant reductions in the total project costs were realized. It was concludee that surfactant-enhanced remediation will outperform conventional pump-and-treat if the source zone is 1 acre or less, surfactant adsorptive losses are minimized, an efficient surfactant recovery and reuse scheme is implemented, the surfactant-enhanced remediation can be completed in less that 7 years, and if the conventional pump-and-treat remediation would require 15 or more years.

A subsequent economic study was initiated by DuPont for the purpose of evaluating surfactant-enhanced pump-and-treat for residual zones of 2 acres or less (Krebs-Yuill et al. 1996); standard DuPont economic assumptions were used in this study. DuPont estimated that the break-even point for surfactant-enhanced pump-and-treat was a total project net present cost of $5 million to offset 5 years of $1 million annual pump-and-treat costs. It was demonstrated that for PCE residual saturations of 1 to 10%, an economic optimum is reached by trading off the cost of surfactant versus the costs of additional wells and higher flow rates as less surfactant is used. Of greatest significance, however, was that the target cost of $5 million was met for all sites of less than 1 acre in size, and for 1-acre sites with residual saturation of 5% or less. Only for sites of 2 acres was solubilization unable to beat the cost of conventional pump-and-treat in addition to completing the cleanup within the 5-year framework. In fact, for 0.25 acre sites with residual saturations of 5% or less, the cost of solubilization remediation was only 40% of the cost of pump-and-treat, with a total project cost of less than $2 million. The minimum cost for the project was encountered when the total remediation time varied between 1 year and 2 years.

Krebs-Yuill et al. (1996) also compared mobilization to solubilization. In the comparison it was assumed that, while sub-pore volume flushes may be able to recover significant contaminant mass (e.g. 90%), a total of ten pore volumes of surfactant injection would be necessary to achieve treatment goals. While little cost difference was found for solubilization versus moilization for smaller sites and low residual saturations, as the site size and the residual saturation increased, mobilization became more attactive. For the 0.5 acre site at 10% residual, mobilization remediation net present cost increased to only $2.1 million, while solubilization remediation net present costs soared to $3.9 million, both below the break-even point of $5 million for conventional pump-and-treat.

2.2 Minimizing Surfactant Losses

Due to the fact that surfactant costs constitute a major capital expenditure in surfactant enhanced subsurface remediation, every effort should be made to minimize surfactant losses. Two forms of surfactant losses will be considered in this section: physicochemical losses (e.g., precipitation, adsorption, phase separation) and hydrodynamic losses (e.g., into uncontaminated portions of the aquifer).

Physicochemical Surfactant Losses. A major factor in surfactant selection is minimizing physicochemical losses (e.g., precipitation, sorption, phase separation). Impacts of these losses range from simply decreasing the efficiency of the system, and thus negatively impacting the economics of the system, to significantly reducing the conductivity of the formation and rendering the process ineffective.

While nonionic surfactants have lower CMC values and less potential for precipitation, they are susceptible to higher levels of sorption and may be more subject to phase separation than anionic surfactants (Rouse et al. 1993, 1996; Shiau et al. 1996a). Precipitation-resistant anionic surfactants may thus be ideal candidates for environmental remediation. Alkylbenzene sulfonates and alkyl sulfates are common classes of anionic surfactants; however, they are both susceptible to elevated levels of precipitation. Increasing the hydrophilicity of these surfactants would likely reduce their losses. The Dowfax series accomplishes this by adding a second sulfonate group (alkyl diphenyloxide disulfonate – DPDS), and the Steol series accomplishes this by adding ethylene oxide groups (EOs). It has been hypothesized that these alterations would reduced the potential for precipitation and sorption while maintaining a high solubilization potential (Rouse et al. 1993, 1996); data substantiating these hypotheses are presented below.

Precipitation studies compared the Dowfax series to their monosulfonate analogs and the Steol series to their nonethoxylated analogs. For the Dowfax series, the C10-, C12- and C16-DPDS surfactants evidenced no precipitation for calcium concentrations up to 0.1 M [$\log Ca(\mu m) = 5$], whereas sodium dodecyl benzene sulfonate (SDBS) realized precipitation at 0.001 M concentrations (Rouse et al. 1993). For the Steol series (ethoxylated dodecyl sulfate), the precipitation region decreased with increasing ethoxylation. The precipitation region for CS-130 (one EO) was not substantially different from sodium dodecyl sulfate (SDS – nonethoxylated), while the precipitation region became much smaller for CS-230 (two EOs). No precipitation was observed for CS-330 (three EOs) up to 0.1 M Ca. Thus, it is observed that the Dowfax and Steol surfactants are in fact significantly less susceptible to precipitation than their monosulfonated and nonethoxylated equivalents.

Sorption studies have also been conducted to compare the Dowfax series to monosulfonate analogs (e.g., SDBS) and the Steol series to nonethoxylated analogs (e.g., SDS). For comparison with SDBS, C12-DPDS was evaluated for its sorption on the same medium. Sorption parameters are summarized in Table 1; linear distribution (K_d) values apply at low concentrations and the Langmuirian coefficient (q_{max} – maximum or plateau sorption value) is applicable at high concentrations. For comparison purposes the sorption parameters for other surfactants are included in Table 1. From the sorption parameters in Table 1 it is seen that SDBS is significantly more susceptible to sorption than the Dowfax surfactants (K_d values

Table 1. Surfactant sorption losses with subsurface material. (After Rouse et al. 1993; Shiau et al. 1995)

Surfactant	K_d (cm^3/g)	q_{max} (mg/g)
SDBS (C12)	8.3	11.4
C12-DPDS	3.1	1.6
SDS	–	7.21
CS-130, CS-230, CS-330	8.3, 6.3, 4.6	2.90, 1.26, 0.92
T-MAZ 20	–	7.0
CO 620[a]	41.1	–
Triton X-100[b]	–	11.9

[a] Alkylphenolethoxylate, same medium.
[b] Alkylphenolethoxylate, different medium.

three times greater for SDBS than C12-DPDS; q_{max} about seven times higher for SDBS than C12-DPDS). The sorption differences for nonionic surfactants evaluated in other research are equally if not more significant, approaching an order of magitude in some cases (see Table 1).

Likewise, the Steol surfactants have been compared to SDS for sorption potential. Looking at the q_{max} values reported in Table 1, it is apparent that the level of sorption decreases with increasing ethoxylation (i.e., from SDS to CS-330). Again, the maximum Steol sorption amount is as much as an order of magnitude less than for certain other surfactants, while being similar in value to the Dowfax surfactant (C12-DPDS).

In addition to the economic advantages, there is also a technical advantage of reduced sorption. The sorbed surfactants, which can exist as a bilayer (admicelle) of surfactant molecules, having a hydrophobic interior, can act as an organic sink for contaminants (much as organic matter does). While this phenomenon can be exploited in sorbing barriers and unit separation processes (Smith and Jaffe 1991; Sun and Boyd 1993; Nayyar et al. 1994), its occurrence in surfactant-enhanced subsurface remediation can serce to delay aquifer restoration. This is especially true as elution of the adsorbed surfactant decreases as the surfactant concentration declines (a natural outgrowth of the nonlinear sorption isotherm). Thus, again the advantage of low sorbing surfactants is apparent.

Having demonstrated that the Dowfax and Steol surfactants are less susceptible to losses (precipitation and sorption), their solubilization potential is obviously of concern. Micellar-water partitioning coefficients (K_m) for naphthalene and the alkyl DPDSs were slightly higher than their monosulfonated equivalents ($\log K_m$ of 4.04 for SDBS and 4.32 for C12-DPDS, with C16-DPDS having a $\log K_m$ of 4.41 – Rouse et al. 1993). This indicates that the Dowfax surfactants are equally if not more efficient in enhancing the solubility of contaminants relative to their monosulfonated equivalents. Likewise, naphthalene solubilization by the Steol surfactants was comparable to slightly higher than for SDS ($\log K_m$ of 4.25 for SDS, and 4.33, 4.40 and 4.44 for CS-130, CS-230, and CS-330, respectively; Rouse et al. 1996). Thus, the Dowfax and Steol surfactants appear very promising for use in subsurface remediation, based on their ability to minimize surfactant losses while maintaining high removal efficiencies.

Hydrodynamic Losses. In addition to physicochemical losses, hydrodynamic processes may also result in increased surfactant demand. Simply replacing water with a surfactant solution in a conventional pump-and-treat system will not be ideal. The same hydrodynamic processes (e.g., heterogeneity of hydraulic conductivity) that negatively impact pump-and-treat contaminant removal efficiencies will only be amplified in surfactant-based remediation. Moreover, well geometries and pumping rates for simple pump-and-treat remediation schemes are often designed to hydraulically impact (i.e., drawdown or upcone) a miximum volume of an aquifer. This "brute force" approach to achieving hydraulic control would be very inefficient for systems using chemical enhancements.

The technical and economic efficiency, as well as regulatory approval for use of surfactants, will all be highly dependent on the ability to achieve hydraulic control of the fluids in the subsurface. Concurrently, it will be necessary to achieve hydraulic control while maximizing hydraulic efficiency. Hydraulic efficiency can be increased by: (1) minimizing the volume of injected surfactant solution; (2) minimizing the volume of fluid to be pumped to the surface (reducing treatment costs); (3) targeting injected chemicals to the contaminated zones of the aquifer; (4) preventing the movement of injected fluids toward clean portions of the aquifer; and (5) maximizing capture of resulting water-surfactant-contaminant mixtures (Chen et al. 1995).

The vertical circulation well (VCW) uses simultaneous injection and extraction within a single borehole to create a circulating flow pattern of fluids emanating from the borehole. Pumping the extraction interval at a higher rate can result in complete capture of fluids from the injection interval. The "closed" nature of the VCW hydraulic system allows for complete capture and subsequent reuse of surfactants injected into the formation. Physical (sand tank) modeling studies have shown that the VCW system is more efficient than the traditional multiple well injection/extraction system in terms of both contaminant removal and capture of injected surfactant solutions (Chen et al. 1995).

Surfactant-enhanced solubilization using the VCW system was recently demonstrated at a field site in Traverse City, Michigan. The highly conductive nature of the shallow groundwater formation (hydraulic conductivity = 10^{-3} cm/s) required an injection/extraction (I/E) ratio of 1 : 10 to approach complete capture of the injected surfactant solution. The low I/E ratio resulted in significant dilution of the recovered surfactant solution. Figure 1 shows the concentration breakthrough curve for surfactant in the VCW effluent. Although the relative concentrations (C/Co) stay below 0.3, mass recovery approaches 100% near the end of the test. It is important to note that dilution of the surfactant solution occurs within the borehole (not in the formation) due to the excessive groundwater removed.

The hydraulic efficiency of surfactant-enhanced remediation processes can be increased by strategically locating and operating impermeable (e.g., grout

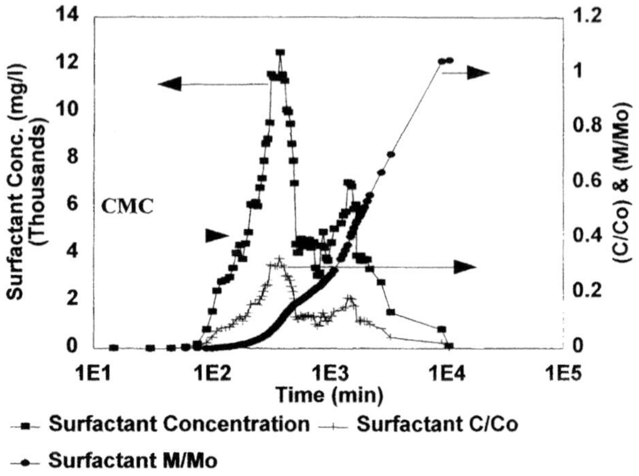

Fig. 1. Traverse City vertical circulation well results: surfactant concentration and relative surfactant concentration/mass recovered during test. (Adapted from Knox et al. 1996)

curtains, slurry walls, sheet piling) and/or hydraulic (e.g., injection wells, infiltration galleries) barriers. Gupta and Knox (1996) used numerical simulations to assess the relative effectiveness of hydraulic and impermeable barriers for improving the efficiency of a surfactant-enhanced process for DNAPL remediation. Sensitivity analyses were performed relative to barrier location, barrier length, and injection-deflector well injection rates. Results of the study showed the use of a dual (hydraulic and impermeable) barrier system to be most efficient in terms of reducing the volume of fresh ground water migrating through the contaminated zone and controlling migration of injected surfactant solutions.

2.3 Solubilization Versus Mobilization

In solubilization, the more hydrophobic compounds will realize the greatest relative enhancement (higher K_m indicates greater contaminant partitioning into micelles); thus, PCE will be enhanced to a greater relative extent than DCE. K_m is observed to vary more between contaminants for any surfactant than between surfactants for a given contaminant (Shiau et al. 1994). Thus, for the solubilization mechanism, surfactant selection is relatively independent of the contaminant(s) and will most likely be made based on factors such as cost, susceptibility to losses, toxicity, etc. At the same time, when estimating (extrapolating) K_m values or when modeling the solubilization process, it should be recognized that micellar solubilization (and thus K_m) varies as a function of contaminant type (nonpolar versus polar/ionic) and aqueous contaminant concentrations below their water solubility (Nayyar et al. 1994; Rouse et al. 1995). Aqueous contaminant concentrations below water solubility may be experienced due to nonideal solubilization, mixed NAPL phases, etc. (Pennell et al. 1993; Rouse et al. 1995; Soerens 1995).

Initial efforts in our laboratories to achieve chlorocarbon middle phase microemulsions, without consideration of surfactant structure, were unsuccessful. The hydrophilic-lipophilic balance (HLB) of the surfactant systems was varied from 2.1 to 40. Although phase inversion was observed in this HLB range (type II to type I), a clear middle phase was not achieved in the transition (instead a mesophase was realized). Using the branched surfactant Aerosol OT (AOT) and sodium mono- and dimethyl naphthalene sulfonate (SMDNS), a middle phase microemulsion was realized in the transition region (Shiau et al. 1994). Thus, it is observed that surfactant selection (structure) is critical to achieving middle-phase systems.

Mobilization with AOT and SMDNS was achieved by varying the SMDNS concentration while holding the AOT concentration constant. Figure 2 shows a phase diagram for 1,2-DCE using AOT and SMDNS as the surfactant system. At low SMDNS concentrations, a Type II system is realized (surfactant has partitioned into the oil phase). Increasing the SMDNS concentration diminishes the potential for AOT to partition into the oil, thereby allowing it to accumulate at the interface and produce a middle phase system. At yet higher SMDNS concentrations, the system is overoptimized and the surfactants reside in the water phase (Type I system). Middle-phase systems were achieved via this approach for PCE, TCE, and 1,2-DCE individually (Shiau et al. 1994) and in binary and ternary mixtures of these chlorocarbons (Shiau et al. 1996a). Regular solution theory proved better able to predict phase behavior in the ternary system than ideal solution theory (Shiau et al. 1996a). It is especially encouraging that, given analyses from binary systems, we can better predict results for ternary systems of variable DNAPL composition.

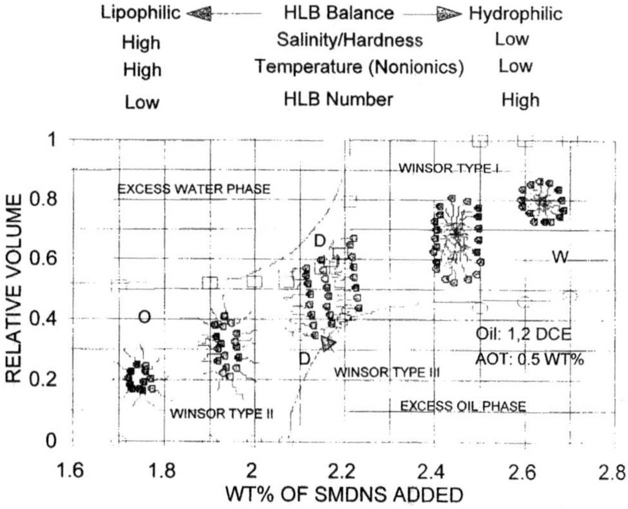

Fig. 2. Phase diagram showing surfactant systems (Winsor types I, II, and III) for 1,2-DCE (0.5 wt% AOT with varying SMDNS concentrations at 15 °C). (Adapted from Shiau et al. 1994)

Middle-phase systems are sensitive to aquifer conditions (temperature, hardness, etc.), contaminant composition, and surfactant structure; this should alert potential users to problems if proper surfactant selection and design efforts are not utilized. However, while deviations from ideal middle phase conditions will decrease the system efficiency, it will not negate the improved efficiency over solubilization. Shiau et al. (1996b) demonstrated that deviating just outside the middle phase regime had minor impacts on the removal efficiency (IFTs are still very low, just not minimum). This reinforces the concept that solubilization and mobilization are not discrete systems, but are rather points (regions) on the continuum of surfactant systems. Another concern regarding mobilization is that vertical migration of released residual may be realized and unacceptable. While recognizing these potential limitations, one must also remain cognizant of the fact that mobilization has the potential to be significantly more efficient than solubilization (as further demonstrated below) and thus should not be prematurely dismissed as a viable technology. In fact, vertical circulation wells can potentially mitigate and even take advantage of any tendencies for vertical migration.

In Fig. 3, solubilization and mobilization results from one-dimensional column studies are presented (Shiau et al. 1996b). Mobilization achieved higher concentrations and eluted the PCE more quickly than solubilization (>97% extracted in ca. three pore volumes). The tail on the solubilization curve indicates the reduced extraction rate and thus slow approach to complete PCE elution for solubilization (likely due to interfacial area constraints); while ca. 85% of the PCE is eluted within 10 pore volumes, less than 90% has been eluted by 30 pore volumes. Again, the potential advantages of mobilization over solubilization are apparent. These results corroborate batch data, which show mobilization to be two orders of magnitude more efficient than solubilization. Based on PCE solubility, ideal dissolution would require in excess of 400 pore volumes to dissolve the residual PCE. Considering nonideal dissolution (i.e., reduced concentrations with time) the actual number of pore volumes for water dissolution is expected to be much greater than 400

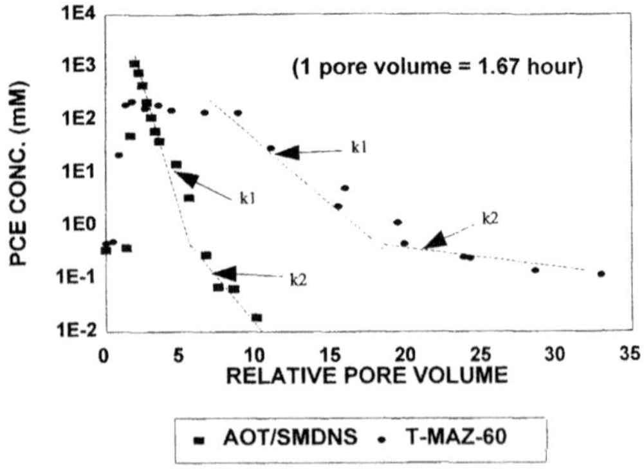

Fig. 3. PCE concentration versus pore volume of surfactant injected via solubilization and mobilization mechanisms in alluvial aquifer material. (Adapted from Shiau et al. 1996b)

pore volumes. Thus, the relative advantages of surfactant-enhanced remediation over water dissolution are apparent. It should be noted that the concentrations in Fig. 3 are still several orders of magnitude above the drinking water standard. Thus, while surfactant techniques significantly expedite mass removal, protracted amounts of time may be required to achieve drinking water standards (although much less than with water alone).

Nonideal surfactant solubilization has been attributed to interfacial mass transfer limitations (e.g., Pennel et al. 1993), dissolution fingering and a combination of the two (Soerens 1995); this phenomenon is evidenced in Fig. 3 by the two legs of the declining PCE elution. Dissolution fingering suggests that heterogeneous NAPL distribution occurs even within seemingly homogeneous systems, and that this heterogeneity results in preferential dissolution pathways. Whereas interfacial mass transfer cannot explain mass transfer zones that change with distance, dissolution fingering can account for this phenomenon while assuming equilibrium dissolution, or a combination of equilibrium and mass transfer-limited dissolution processes (Soerens 1995). Further understanding of these phenomena is necessary to accurately scale up these processes to full scale implementation.

2.4 Surfactant-Contaminant Separation and Surfactant Reuse

Economic analyses indicate that surfactant-contaminant separation and surfactant reuse are critical to successful implementation of this technology, especially for multiple pore volume flushes. There are actually two different separations that may be necessary; the first is the removal of contaminant from the concentrated surfactant stream, and the second is the removal of water from the surfactant stream to reconcentrate it for reinjection. The latter separation will be most important when groundwater extraction exceeds fluid injection (to achieve a desired capture zone).

The most critical question to be asked in selecting a unit process for surfactant decontamination is whether the contaminant is volatile or nonvolatile. While many options exist for removal of volatile contaminants, fewer technologies appear worthy of consideration for nonvolatiles. Candidate processes for treating volatile contaminants include air stripping with antifoams, membrane air stripping, vacuum stripping and vacuum distillation, pervaporation, membrane liquid extraction, and precipitation. Carbon adsorption and ion exchange are not included amongst the viable technologies because surfactants themselves will strongly adsorb on both hydrophobic and hydrophilic surfaces. Surfactant adsorption will result in concomitant sorption of the contaminant (surface adsolubilization, similar to aqueous solubilization – Nayyar et al. 1994), thereby negating the desired separation.

Conventional air stripping in packed columns can be used for surfactant solutions when the volatility of the contaminants is high. Lipe et al. (1996) evaluated packed tower air stripping for surfactant-contaminant separation of naphthalene and TCE from C12- and C16-DPDS. Air stripper efficiency declined with increasing surfactant concentration, for more hydrophobic contaminants (naphthalene versus TCE), and for surfactants with increased solubilization potential (C16-DPDS versus C12-DPDS). These results are as expected based on the reduced contaminant activity due to micellar solubilization. The air stripper design equations were re-derived to incorporate the micellar solubilization process. Model predictions agreed well with experimental results, supporting the hypothesis that existing mass transfer correlations are still valid in the presence of surfactants (i.e., the presence of the surfactant neither enhanced nor inhibited the overall mass transfer). Appropriate physical parameters did have to be updated to properly reflect surfactant-modified parameters (e.g., surface tension). The modified design equations allow proper design of air strippers to achieve the desired effluent concentrations, assuming that foaming is not a problem.

Low air flow rates may be used to avoid heavy foams in the column (e.g., Lipe et al. 1996). However, this may render the process even less efficient, and result in even larger tower heights. The use of antifoams should allow higher air flow rates, but implications of using an antifoam must be considered. For example, antifoam contamination of the reinjected surfactant solution may not be acceptable. In addition, there may still be an upper limit to the air flow rates – some kind of a foam point – that is below the conventional flooding limit. Vacuum stripping and vacuum distillation may extend the range of contaminants that can be removed, but again at the cost of using antifoams. Membrane air stripping (using hollow fiber columns) can mitigate the foaming problem (Lipe et al. 1996) without sacrificing efficiency and without antifoams. Control of the pressure difference across the porous hollow fibers in a hollow fiber membrane module can prevent weeping of the surfactant solution into the sweep air on the shell side of the module.

The separation of nonvolatile contaminants from surfactant solutions will most likely be done using membrane liquid-liquid extraction. Hollow fiber membranes will eliminate emulsion formation in the extraction step. Hasegawa et al. (1996) evaluated hollow fiber liquid-liquid extraction of naphthalene and phenanthrene from C12- and C16-DPDS; squalene was used as the extracting liquid. Liquid-liquid extraction efficiency was observed to decline with increasing surfactant concentration, for more hydrophobic contaminants (phenanthrene versus naphthalene), and for surfactants with increased solubilization potential (C16-DPDS versus C12-DPDS). These results are expected based on the reduced contaminant activity resulting from micellar solubilization. When the liquid-liquid extraction design equations accounted for micellar solubilization, model predictions agreed well with experimental results. The modified design equations allow proper design of the liquid-liquid extraction process to achieve the desired effluent concentrations.

Squalane was utilized in the above research due to its low water solubility, the ability to acquire it in a purified form, and its environmental friendliness (it is used in hand lotions). A systematic study of extracting solvents would be helpful as most solvents now associated with liquid-liquid extraction in the chemical process industries will not be usable because they would contaminate the surfactant solution. The solvent stream will have to be regenerated as well in this approach, and, again, care will have to be taken to prevent contamination of the surfactant solution.

If reconcentration of the surfactant solution is necessary for reinjection, ultrafiltration (UF) is an obvious approach for achieving this goal. The UF process will have an increased ability to retain surfactants due to the colloidal size of surfactant micelles. Ultrafilter membranes with 10000 molecular weight cutoff (MWCO) can efficiently retain surfactant micelles where monomers would not be retained. For example, Lipe et al. (1996) achieved 93 to 99% retention of surfactants using a 10000 MWCO membrane. While the UF process effectively retains surfactant micelles, monomeric concentrations will pass through with the permeate. If economics dictate, smaller filters may be used to capture this additional surfactant for reuse (Krebs-Yuill et al. 1995).

3 Process Scaleup: Field Demonstrations

3.1 Traverse City

A field demonstration study was conducted at a US Coast Guard Station located in Traverse City, Michigan. The facility had experienced previous leaks of JP-4 aviation fuel and a small volume spill (~55 gallons) of an industrial solvent (PCE). This site has undergone over a decade of remediation activities, including extensive bioremediation, ground water extraction and air sparging. The contaminant concentrations at the site have thus been significantly reduced. The background groundwater concentrations of PCE and TPH were low: 5 μg/l and 4 mg/l, respectively; soil TPH values of less than 1000 mg/kg were detected. For this reason, and because of the limited resources available for this project, the primary objectives of the fieldwork were maximizing surfactant recovery and evaluating the VCW system, and a secondary objective was enhancement in contaminant extraction. Complete mass balances, such as would be obtained from extensive pre- and post-test characterization activities (e.g., soil coring or partitioning tracer tests), were beyond the resources available for this site; such detailed characterization will be discussed relative to the second site.

Surfactant laboratory studies evaluated surfactant phase behavior relative to the field site. To ascertain that surfactant losses would not limit or negate the efficiency of the system, surfactant precipitation was evaluated in batch studies and surfactant sorption was evaluated in column studies. To evaluate the potential enhancement in contaminant extraction, surfactant solubilization studies were conducted in batch and column studies. Finally, surfactant foaming characteristics were evaluated; this was of special interest due to the use of air sparging in the vicinity of the field site (see Sabatini et al. 1996).

Modeling studies evaluated the surfactant concentrations resulting at downgradient wells if significant surfactant loss was realized; surfactant concentrations would be below acceptable levels (1–10 mg/l as assessed from biotoxicity testing) within 100 to 500 ft. of the VCW system. The results of the simulations, coupled with the available operational safeguards, were deemed sufficient and in January 1995 the Michigan Department of Natural Resources (MDNR) granted permission to implement the surfactant demonstration study; the field demonstration study was completed in June 1995. The results of the test are summarized below: for additional details see Knox et al. (1996).

Figure 1 shows the surfactant concentration and mass recovery over the course of the demonstration study; the erratic pattern of the surfactant concentration curve is attributable, in part, to the variations in pumping rates associated with changes in the system hydraulics. Mass recovery for the surfactant solution exceeds 95% over the duration of the test. The surfactant concentration in the treatment zone around the VCW exceeded the CMC (where the CMC ≈ 3600 mg/l) for almost 20 h, with the maximum C/Co value for the surfactant in the treatment zone exceeding 0.3 (12 000 mg/l versus 36 000 mg/l injected).

Figure 4 shows the elution curves for the conservative tracer (fluorescein), the surfactant, and the solubilized PCE. As expected, the surfactant solution appears later, shows a lower peak relative concentration and experiences greater spreading as compared to the conservative fluorescein. The surfactant retardation factor was in keeping with laboratory column results (~1.4). The increased PCE concentrations illustrate the solubilization potential of the surfactant solution; the effluent PCE concentrations increase by more than two orders of magnitude from an initial value of 5 μg/l, representing background concentration, to a high of 800 μg/l. The double peaks in the PCE elution curve are attributed to the heterogeneous distribution of PCE within the treatment zone. The delay in PCE elution relative to the surfactant is attributed to partitioning of the PCE between the aqueous micellar solution and the admicellar phase, as discussed previously (Nayyar et al. 1994). Since sorption of Dowfax 8390 is less than many other surfactants (Rouse et al. 1993), this admicellar effect would be likewise minimized.

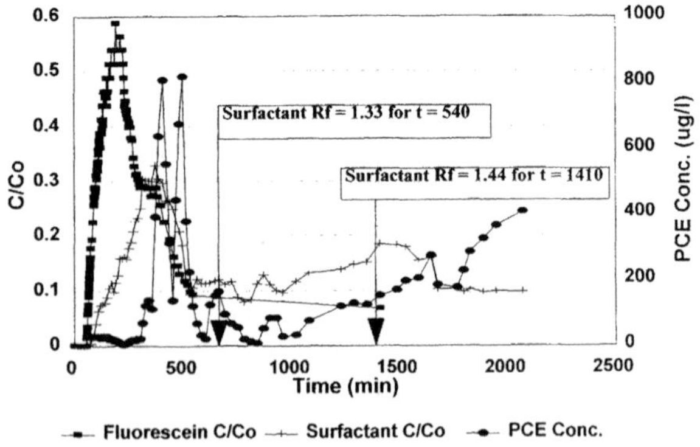

Fig. 4. Relative concentration of fluorescein and surfactant and absolute concentration of PCE during Traverse City field test. (Adapted from Knox et al. 1996)

Figure 5 depicts the mass of PCE removed during the demonstration study. The upper curves depict the total PCE mass removed (3354 mg) and the PCE mass removed by surfactants from the treatment zone (2757 mg); the bottom curves depict the total PCE mass that would have been removed by pumping water alone (654 mg) and the PCE mass that would have been removed from the treatment zone by pumping water alone (12 mg). The PCE mass removed within the treatment zone (region swept by micelles) was 40 times higher than what would have been removed using water only pump-and-treat remediation. This comparison assumes that the PCE concentration would have remained constant for pump-and-treat remediation; actually, it is expected that the PCE concentrations would continue to decrease during pump-and-treat remediation (as it was initially), further highlighting the micellar enhancement realized. For total hydrocarbons (TH), the surfactant-enhanced mass removal was almost 90 times what would have been removed by simple pump-and-treat remediation. The hydrocarbons extracted were methylated alkanes, which were relatively unimpacted by the bioremediation activities which had occured at the site; thus, the surfactants solubilized what appears to be the biorecalcitrant fraction of the jet fuel, illustrating the potential value of using multiple technologies.

The Traverse City demonstration study illustrated the high levels of surfactant recovery that can be realized (>95% for the high performance Dowfax 8390), and the significant enhancements in contaminant extraction that can be achieved, even when the initial contamination levels have been significantly reduced by past bioremediation activities. This study also illustrates the value of careful laboratory

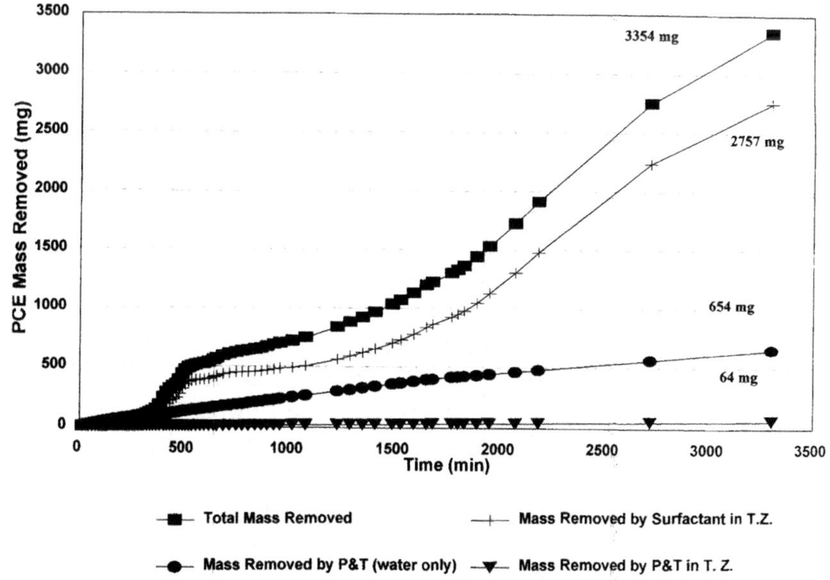

Fig. 5. PCE mass recovery curves for Traverse City field test. (Adapted from Knox et al. 1996)

testing, existing infrastructure, collaborative research, and cooperative regulatory personnel and site coordinators.

3.2 Hill Air Force Base

Surfactant enhanced demonstration studies are underway at Hill Air Force Base, which is located north of Salt Lake City, Utah. This base has been the site of military activities since 1920. The particular area of interest is Operable Unit 1 (OU1), which includes several landfills, chemical disposal pits and fire training areas. Principal contaminants of concern are LNAPLs (jet fuel, light lubricating oils), chlorinated VOCs, PCBs, etc. The soil of interest is predominantly sand and gravel, with less than 5% silt, clay and organic content.

A series of cells have been installed at OU1 for demonstration purposes. The 3×5 m cells consist of sheet piling driven down to a lower confining layer (8 to 10 m). The water table in these cells has been altered to incorporate heavily contaminated regions; as such, the saturated thickness ranges from 2 to 5 m. Our group has two cells at OU1; one is designated for solubilization while the second is designated for mobilization. To date, laboratory studies have been conducted to select the preferred solubilization and mobilization systems, and ongoing field studies are characterizing the background conditions in the cells. The surfactant demonstration studies are scheduled to occur over the next 4 to 6 months (summer 1996). Results from the efforts to date will be summarized below.

These cells have been/will be well characterized. Three-dimensional contaminant distributions will be assessed via soil corings within each cell and partitioning tracer tests (PTT – using partitioning alcohols, chromatographic separation can be used to back determine residual levels). Comparison of pre- and post-test coring and PTTs will provide an assessment of contaminant removed during each test; these results will be compared with mass extracted during the test. A grid of multilevel samplers within each cell allows three-dimensional characterization during the PTT; 60 sampling points result from the 3×4 grid with five depths at each grid point. The pre-coring has been completed in both cells, and the pre-PTT has been conducted in one cell. This level of characterization and data will allow for careful data interpretation and modeling analysis (currently UTCHEM is being used for this purpose).

Laboratory solubilization studies have considered a range of surfactants, including anionic and nonionic as well as direct food additive and high performance surfactants. The surfactants considered include Dowfax 8390, Steol CS-330, CO 660, Tween 80 and Witconol SN-120. Laboratory batch studies have evaluated the solubilization potential, sorption level and susceptibility to precipitation or phase separation using contaminated materials from the site. Based on these studies, Dowfax 8390, Steol CS-330 and Witconol SN-120 were selected for evaluation in column studies. Dowfax 8390 and Steol CS-330 outperformed Witconol SN-120 in the column studies; prior field experience with Dowfax 8390 at the Traverse City site suggests this surfactant as the candidate for use in the field demonstration study.

Identification of a mobilization system has been more challenging. Initial studies utilized JP-4, since this comprises the major fraction of the NAPL. However, middle phase compositions identified for JP-4 (e.g., 25 mM AOT and 1.2 wt% SMDNS) were not successful in achieving middle phase systems with the Hill AFB NAPL; apparently the constituents solubilized in the JP-4, although minor by fraction, significantly impact the NAPL behavior. Subsequent studies identified the following systems as candidates for implementing middle phase systems with the Hill AFB NAPL: 25 mM AOT, 1.12 wt% T-MAZ 60 and either 6500 ppm Na or 960 ppm Ca. Ongoing studies are further evaluating these systems in batch and column studies.

The pre-PTT test was conducted in the first cell in March of 1996. A 0.15 pore volume pulse of 1000 mg/l hexanol and 2,2-dimethyl-3-pentanol was injected, followed by roughly 8 pore volumes of water. Knowing the distribution coefficient of these alcohols with the NAPL, and quantifying the chromatographic separation of the tracers (via relative moments) at each point in the sampling grid allows back-determination of the residual level at each point. Preliminary analysis of the several thousand samples analyzed during the first pre-PTT suggests an overall residual saturation value of roughly 7%; the spatial trends in residual saturation from the pre-PTT are consistent with results of the coring studies.

4 Summary

Surfactant-enhanced subsurface remediation is one of several technologies being developed for remediation of subsurface NAPL contamination. Economic analyses indicate that this technology can be competitive with conventional pump-and-treat remediation if surfactant losses can be minimized, contaminant elution maximized, and surfactant-contaminant separation and surfactant reuse implemented. This chapter summarizes process development contributions of our group in each of these areas. This chapter also reports on the results of two field studies, one completed (surfactant solubilization of PCE and methylated alkanes at a Coast Guard facility in Traverse City, Michigan) and one in progress (solubilization and mobilization of JP-4 and other solubilized constituents at Hill AFB, north of Salt Lake City, Utah). The encouraging economic results, the process development work, and the successful field demonstration all point to the viability of surfactant-based remediation technologies, and point to the need for further development and implementation studies.

References

Abdul AS, Gibson TL, Ang CC, Smith JC, Sobczynski RE (1992) In situ surfactant washing of polychlorinated biphenyls and oils from a contaminated site. Ground water 30(2):219–231

Baran JR, Pope GA, Wade WH, Weerasooriyaa V, Yapa A (1994) Microemulsion formation with chlorinated hydrocarbons of differing polarity. Environ Sci Technol 28(7):1361–1366

Bouwer EJ, Mercer J, Kavanaugh M, DiGiano F (1988) Coping with groundwater contamination. J Water Pollut Control Fed August, 1415–1427

Chen Y, Chen LY, Knox RC (1995) Modeling the effectiveness of innovative measures for improving the hydraulic efficiency of surfactant injection and recovery systems. In: Sabatini DA, Knox RC, Harwell JH (eds) Surfactant-enhanced subsurface remediation: emerging technologies. ACS Symp Ser 594: American Chemical Society, pp 249–264

Edwards DA, Luthy RG, Liu Z (1991) Solubilization of polycyclic aromatic hydrocarbons in micellar nonionic surfactant solutions. Environ Sci Technol 25(1):127–133

Environmental Protection Agency (EPA) (1995) Surfactant injection for groundwater remediation: state regulators' perspectives and experiences. EPA 542-R-95-011, Cincinnati, Ohio

Fountain JC (1992) A comparison of field tests of surfactant flooding: examples of mobility control of DNAPL. In: Sabatini DA, Knox RC (eds) Transport and remediation of subsurface contaminants: colloidal, interfacial and surfactant phenomena. ACS Symp Ser 491:182–191

Gupta H, Knox RC (1996) Modeling the effectiveness of hydraulic and impermeable barriers for improving contaminant extraction or surfactant injection. PETRO-SAFE 96. ASME Petroleum Division and American Petroleum Institute, Houston, Texas

Haley JL, Hanson B, Enfield C, Glass J (1991) Evaluating the effectiveness of groundwater extraction systems. Ground Water Monit Rev Winter, 119–124

Hasegawa M, Sabatini DA, Harwell JH (1996) Liquid–liquid extraction for surfactant–contaminant separation. J Environ Eng Div – ASCE (in press)

Jafvert CT, Heath JK (1991) Sediment and saturated-soil associated reactions involving an anionic surfactant (dodecylsulfate). 1. Precipitation and micelle formation. Environ Sci Technol 25(6): 1031–1038

Keely JF (1989) Performance evaluation of pump-and-treat remediations. US Environmental Protection Agency, EPA/540/4-89/005, Ada, Oklahoma

Knox RC, Sabatini DA, Harwell JH, Brown RE, West CC, Blaha F, Griffin S (1996) A field demonstration of surfactant injection and recovery for subsurface remediation: II. Field demonstration results. Ground Water (in press)

Krebs-Yuill B, Harwell JH, Sabatini DA, Knox RC (1995) Economic considerations in surfactant-enhanced pump-and-treat. In: Sabatini DA, Knox RC, Harwell JH (eds) Surfactant-enhanced subsurface remediation: emerging technologies, ACS Symp Ser 594, American Chemical Society, Washington, DC, pp 265–278

Krebs-Yuill B, Harwell JH, Sabatini DA, Quinton GE, Shoemaker SH (1996) Economic study of surfactant-enhanced pump-and-treat remediation. 69th Annu Water Environment Federation Conf, Dallas, Texas, October 5–9

Lipe M, Sabatini DA, Hasegawa M, Harwell JH (1996) Micellar enhanced ultrafiltration and air stripping for surfactant-contaminant separation and surfactant reuse. Ground Water Monit Remediation, 16(1):85–92

Mackay DM, Cherry JA (1989) Groundwater contamination: pump and treat remediation. Environ Sci Technol 23, 6:630–636

National Research Council (NRC) (1994) Alternative for ground water cleanup. National Academy Press, Washington, DC

Nayyar SP, Sabatini DA, Harwell JH (1994) Surfactant adsolubilization and modified admicellar sorption of nonpolar, polar and ionogenic organic contaminants. Environ Sci Technol 28(11):1874–1881

Olsen RL, Kavanaugh MC (1993) Water Environ Technol, March, 42–47

Palmer CD, Fish W (1992) Chemical enhancements to pump and treat remediation. USEPA, EPA/540/S-92/001, Cincinnati, Ohio, 20 pp

Pennel KD, Abriola LM, Weber WJ Jr (1993) Surfactant-enhanced solubilization of residual dodecane in soil columns. I. Experimental investigations. Environ Sci Technol 27(12):2341–2351

Rouse JD, Sabatini DA, Harwell JH (1993) Minimizing surfactant losses using twin-head anionic surfactants in subsurface remediation. Environ Sci Technol 27(10):2072–2078

Rouse JD, Sabatini DA, Deeds NE, Brown E, Harwell JH (1995) Micellar solubilization of unsaturated hydrocarbon concentrations as evaluated by semi-equilibrium dialysis. Environ Sci Technol 29(10): 2484–2489

Rouse JD, Sabatini DA, Brown RE, Harwell JH (1996) Evaluation of ethoxylated alkylsulfate surfactants for use in subsurface remediation. Water Environ Res 68(2):162–168

Sabatini DA, Knox RC, Harwell JH (eds) (1995) Surfactant enhanced subsurface remediation: emerging technologies. ACS Symp Ser 594:312

Sabatini DA, Knox RC, Harwell JH, Soerens TS, Chen L, Brown RE, West CC (1996) A field demonstration of surfactant injection and recovery for subsurface remediation. I. Laboratory and modeling studies. Ground Water (in press)

Schmelling SG, Keely JW, Enfield CG (1992) Critical evaluation of treatment technologies with particular reference to pump-and-treat systems. In: Rees JF (ed) Contaminated land treatment technologies, Elsevier Applied Science, New York, pp 220–233

Shiau BJ, Sabatini DA, Harwell JH (1994) Solubilization and mobilization of DNAPLs using direct food additive (edible) surfactants. Ground Water 32(4):561–569

Shiau BJ, Sabatini DA, Harwell JH (1995) Properties of food grade surfactants affecting subsurface remediation of chlorinated solvents. Environ Sci Technol 29(12):2929–2935

Shiau BJ, Sabatini DA, Harwell JH, Vu D (1996a) Middle phase microemulsions of mixed chlorinated solvents using food grade (edible) surfactants. Environ Sci Technol 30(1):97–103

Shiau BJ, Sabatini DA, Harwell JH (1996b) Removal of chlorinated solvents in subsurface media using edible surfactants: column studies. J Environ Eng Div – ASCE (in press)

Smith JA, Jaffe PR (1991) Comparison of tetrachloroethane sorption to an alkylammonium-clay and alkyldiammonium and clay. Environ Sci Technol 25:2054–2058

Soerens TS (1995) Dissolution of nonaqueous phase liquids in porous media: mass transfer in aqeuous and surfactant systems. PhD Diss, University of Oklahoma, Norman, Oklahoma

Sun S, Boyd SA (1993) Sorption of nonionic organic compounds in soil water systems containing petroleum sulfonate-oil surfactants. Environ Sci Technol 27:1340–1346

Valsaraj KT, Thibodeaux LJ (1989) Relationships between micelle-water and octanol-water partition constants for hydrophobic organics of environmental interest. Water Res 23(2):183–189

Vignon BW, Rubin AJ (1989) Practical considerations in the surfactant-aided mobilization of contaminants in aquifers. J Water Pollut Control Fed 61(7):1233–1240

West CC (1992) In: Sabatini DA, Knox RC (eds) Transport and remediation of subsurface contaminants: colloidal, interfacial and surfactant phenomena. ACS Symp Ser 491:149–158

West CC, Harwell JH (1992) Surfactants and subsurface remediation. Environ Sci Technol 26:2324–2330

25 Case Study of Bioremediation of Soil Contaminated by Diesel Oil

M. Perle[1], H. Rubin[2], N. Narkis[2], C. Braester[2], P. Shoshani[1] and N. Heruti[3]

Contents

1 Introduction

This study originates from an accidental event of diesel oil contamination due to an overflow of a fuel storage tank. As a result of this accident, approximately 1300 m^3 of diesel oil was released to the environment and freely flowed as a small stream, contaminating the loamy soil in that site. The diesel oil was concentrated especially in three ponds along the stream path, and over 10 h it completely infiltrated into the soil. The total area contaminated by the diesel oil was about 7000 m^2. The maximum depth of the contamination was 5 m. The total petroleum hydrocarbon (TPH) concentration in the contaminated site was between 5000 to 50 000 mg/kg soil, and in two locations, the TPH concentration reached about 130 000 mg/kg soil.

Immediately after notification of the event, the IEC (Israel Electric Corp.), owner of the installation, pumped out some fuel accumulated in the three ponds to minimize the negative effect on the environment. Furthermore, a new ditch for surface runoff was excavated around the contaminated site to avoid the flow of surface water through that site. It should be noted that loamy soil quickly becomes

[1] Israel Electric Corp. P.O. Box 10, Haifa 31000, Israel
[2] Department of Civil Engineering, Environmental & Water Resources Engineering Technion – Israel Institute of Technology, Haifa 32000, Israel
[3] BIOTECH Environmental Company, P.O. Box 73, Ramat-Hasharon 47100, Israel

H. Rubin/N. Narkis/ J. Carberry
Soil and Aquifer Pollution

impermeable after its contact with water, whereas the diesel oil could penetrate much more quickly through that soil.

The Ministry of the Environment and The Water Commissioner's Office required complete reclamation of the polluted site. The IEC accepted responsibility to cover all expenses connected with the site reclamation. The government authorities had to determine the ultimate goal of the reclamation.

2 Selection of the Treatment Method

Several treatment methods were proposed and evaluated, based on the following major options: (1) excavation of contaminated soil and disposal with no treatment in an authorized disposal site in Ramat Hovav, (2) excavation of the contaminated soil to be incinerated by ovens of a power plant, located about 150 km north of the contaminated site, and (3) on site bioreclamation of the contaminated soil.

Owing to the large quantities of soil to be excavated and disposed, the first two options were rejected and, therefore, it was decided that on-site bioremediation would be the most appropriate method to be used.

The basic idea of the selected method was to excavate the contaminated soil, arrange it in several heaps in bioreactors, constructed near the contaminated site, and monitor the reduction of total petroleum hydrocarbon (TPH) concentration in the treated soil. The Ministry of the Environment and the Israel Hydrological Service determined the goal of the treatment. They demanded that the TPH content of 90% of the contaminated soil samples, taken from specified locations in the bioreactors, would not exceed the maximum level of 500 mg/kg soil.

Several companies responded to the tender issued by IEC, providing offers for the required reclamation project. BIOTECH-Environmental Company won the tender. Following the agreement between IEC and BIOTECH, a document specifying complete details of the reclamation method and the structure of the bioreactors was provided by the contractor.

3 Description of the Bioreactor

The objective of the reclamation project was to clean up the soil contaminated by diesel fuel. The selected bioremediation approach was based on use of an enriched culture of microorganisms, after their adaptation to degrade diesel fuel. The enhanced bioremediation would be carried out in controlled bioreactors, where a suitable environment of dissolved oxygen, nutrients and moisture would allow rapid breakdown of the diesel oil. Since these conditions could not be maintained in the original contaminated site, the soil had to be excavated and transferred to bioreactors specially built for the treatment purpose.

Major concern was associated with the nature of the local loamy soil. Loam has a very low hydraulic conductivity. It also becomes very quickly impermeable after its contact with water. The process of bioremediation requires the presence and activity of the biomass throughout the entire volume of the treated soil. Therefore, aqueous solutions of nutrients, acclimated microorganisms and dissolved oxygen should be evenly provided throughout the entire volume of the treated soil. Such conditions should be guaranteed during the entire remediation period. In order to enhance the supply of the aqueous solution of nutrients and oxygen, the soil permeability had to be increased. It was decided to condition the soil and mix it with hay as a bulking agent by using agricultural equipment, such as plows and discs.

3.1 Construction of the Bioreactors

The contaminated soil of a volume of 12000 m^3 was excavated and thoroughly mixed with hay on site. The hay was chopped and mixed with the contaminated soil by a disc plow. A tractor piled the soil and the hay mixture and loaded it on trucks to be transferred to the location of the bioreactor construction. Additional hay quantities were mixed with the next layer of soil and the same procedure was repeated several times, until most of the contaminated soil was transferred to the biological reactors. The mixture of soil and hay was larger by 30% than the original soil volume, and the total volume of the mixture was 17000 m^3. This material was arranged in several heaps in the bioreactors.

The bioreactor operation is associated with the addition of water, nutrients and oxygen. In order to prevent seepage of water into the clean soil beneath the reactor, the reactor was constructed on a flexible liner membrane made of low density polyethylene. On top of the polyethylene liner a drainage layer (5 to 10 cm thick) of small size gravel was placed. On top of the drainage layer, a layer of permeable

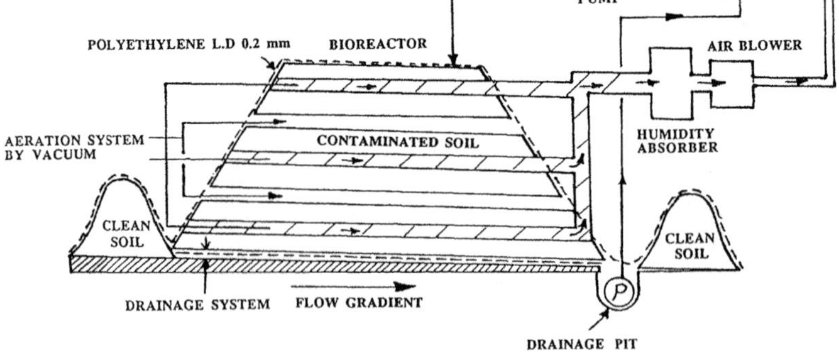

Fig. 1. Schematic description of the aeration and water systems

geotechnical textile was placed. A collection pipe was constructed along the downstream side of the reactor.

The mixture of soil and hay was arranged in five heaps in biological reactors. Two large biological reactors were built. The size of one biological reactor was 80 × 40 × 3 m. The size of the second reactor was 120 × 40 × 3 m. Figure 1 presents a schematic description of the bioreactor structure and equipment. The reactors were built on top of the low density polyethylene liners. Each reactor incorporated six layers of soil. During the placement of the soil layers, some water with microorganisms acclimated to diesel fuel biodegradation and nutrients were added to the mixture of soil and hay. Between two adjacent soil layers, perforated tubes, supplying air by vacuum pumps, were placed. The tubes were protected against soil clogging. A system supplying water, acclimated microorganisms and nutrients as an aqueous dispersion was installed. This system incorporated two separate lines of pipes with drippers. One pipeline was placed at a depth of 2 m below the top of the reactor. The other pipeline was placed on the top of the reactor. In order to reduce evaporation and wind erosion, each reactor was covered by low density polyethylene liners. Special pumps for recycling the liquids drained at the bottom of the reactors were installed and connected to the dripping system. The drained leachates were mixed with the nutrient aqueous solution and the seeding of a suspension of acclimated microorganisms.

3.2 Operation of the Bioreactors

The original plans were aimed at the operation of the bioreactors for about six months. During that time period all supplying units were continuously operated. Vacuum pumps drew air out of the soil and led to the introduction of fresh air with oxygen to the biomass. The water supply system was continuously fed from the mixing tank which contained any leachate pumped from the base of the reactor. Additional suspension of acclimated microorganisms and nutrients were supplied as required into the mixing tank, and guaranteed optimal performance of the bioremediation process. The air flow and leachate generation were measured regularly throughout the remediation period. Adjustment was made in the air flow, water delivery, acclimated bacteria and nutrient supply as needed.

3.3 Monitoring the Bioreactor Performance

The bioreactor performance was regularly checked by analysis of samples taken from specific locations of the bioreactor. Locations for soil sampling, and number of samples from each reactor were determined according to a sampling program required by The Ministry of the Environment and The Israel Hydrological Service. According to the results of the sample analysis, the final bioremediation objectives had to be determined. Soil samples were transferred to the chemical laboratory of The Israel Institute of Petroleum and Energy. The chemical test objective was to determine the residual concentration of the TPH. The method used by the chemical laboratory was based on EPA 418.1 (EPA 1986) procedure. According to this method, the soil is extracted by an organic solvent (freon, or methylene chloride) and is then analyzed by infra-red (IR) spectroscopy for the determination of TRPH (Total Recoverable Petroleum Hydrocarbons). Originally, this method was aimed to determine fuel quality. The method was modified for soil samples by using the solid extraction procedure from Method EPA 9071. The IR method specifies calibration by a reference fuel. In the present case, the reference was the original diesel fuel which contaminated the site. Some tests of TPH were performed by a private laboratory Spectrolab, according to method EPA 8015 (modified) (EPA 1986) which utilizes gas chromatography. The procedure for sampling by both laboratories is described in the following section.

After 6 months of operation, soil sample tests showed that the targeted TPH concentration of 500 mg/kg in 90% of soil samples was not achieved. Therefore, the bioreactor operation was extended for another time period of 6 months.

4 Results

The initial TPH concentration in soil samples taken from the reactors was between 5000 and 50 000 mg/kg soil. There were two small sites in which the TPH concentration was about 130 000 mg/kg soil. Soil with a TPH concentration higher than 500 mg/kg soil was considered as contaminated and required treatment. Such soil was excavated and transferred to the bioreactors. After the removal of the contaminated soil, samples of the exposed soil surface were taken and the TPH concentration was measured, to verify complete removal of contaminated soil. Figures 2 to 4 show the distribution of TPH concentration of the exposed soil surface, in all three contaminated sites, after the removal of the contaminated soil. These figures show the existence of several regions in which the TPH concentration exceeded the accepted criterion. In that region additional soil excavation was made.

Table 1 shows the data of residual TPH concentration distribution in the bioreactors after 3 months of operation. It indicates that TPH concentration was reduced to a level between 500 and 23 000 mg/kg soil, but further treatment was required.

After 6 months of bioreactors operation, 16 locations randomly selected in the large bioreactor and 4 locations in the smaller reactor were chosen for soil sampling. Table 2 shows the test results of residual TPH concentration of these samples. In addition, samples were taken from three different depths at the same location. All samples of the three depths from the same location were mixed together to obtain new samples, named "average samples". Table 3 shows the TPH concentration of these samples. Tables 2 and 3 indicate that the TPH concentration did not reach the accepted criterion of 500 mg/kg soil in 90% of the soil samples. Therefore, full operation of the bioreactors was extended for additional 6 months.

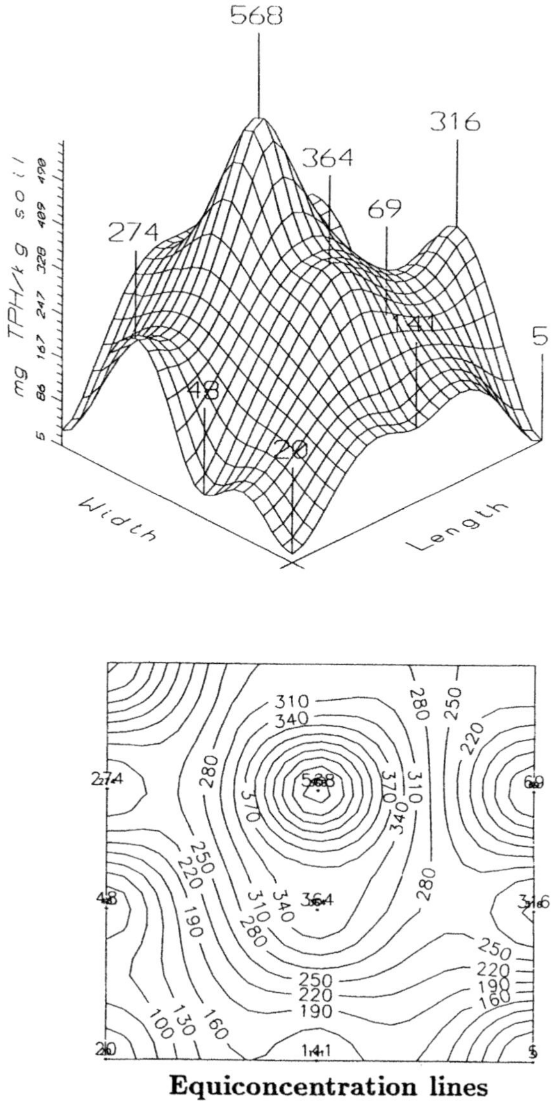

Fig. 2. Site A: TPH concentration of the remaining soil at the contaminated site, after excavating the upper part and transporting to the bioreactor

After 12 months of bioreactor operation, samples were taken from 20 different locations; 4 locations in the small bioreactor, and 16 in the large one. At each location, three samples from different depths were taken and mixed well together to form a single new integrated sample. Each integrated sample was divided into two samples. One of these two samples was sent to the chemical laboratory of the Israel Institute of Petroleum and Energy, which uses the IR method EPA 418.1. The

Table 1. Residual TPH concentration after 3 months of the bioreactor operation

Sample No.	Residual TPH concentration (mg TPH/kg soil)
1	3988
2	546.5
3	2562
4	12097
5	11078
6	6398
7	4638
8	1786
9	8765
10	23079
11	1885
12	8084
13	2788
14	3238
15	1851
16	2770
20	3117
24	1587
27	1082
29	480.3
31	2077
34	1468
38	1089

Table 2. Residual TPH concentration of the treated soil after 6 months of the bioreactor operation. Three random samples were taken from each of the six heaps in the bioreactor

Site	Random samples (mg TPH/kg soil)		
	Sample 1	Sample 2	Sample 3
1	2182	871	676
2	2009	2481	107
3	2919	2540	82
4	2134	2166	1523
5	998	111	80
6	49	5546	3820

second sample was sent to a private laboratory Spectrolab which uses the GC Method EPA 8015. In each one of the testing laboratories, each sample was divided into several doses and checked by three replications of the specific TPH concentration analytical procedure. Therefore, three replications of TPH concentration were

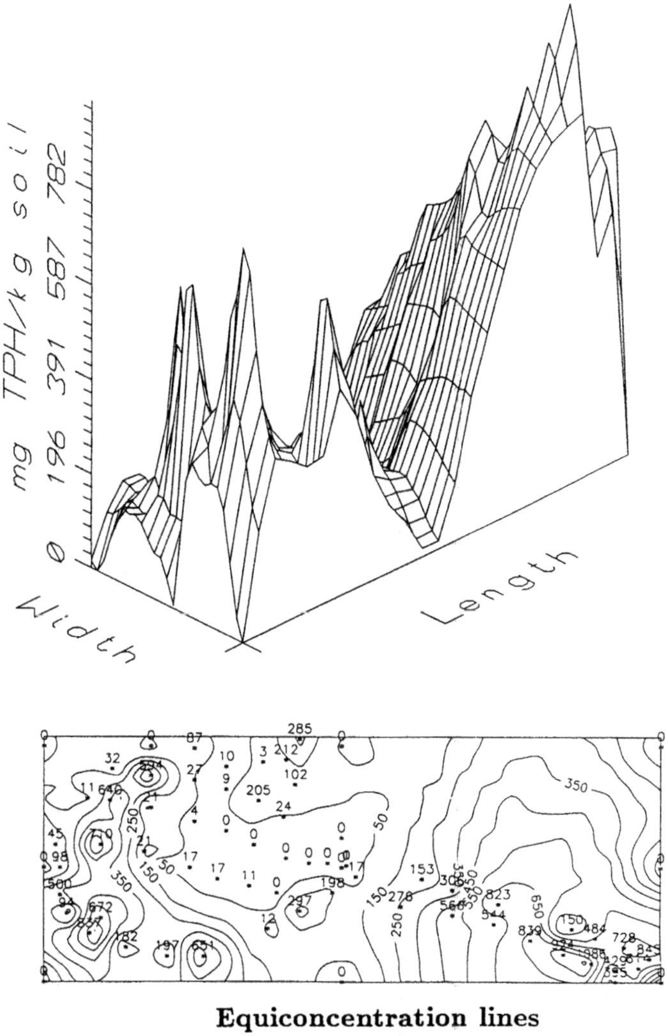

Fig. 3. Site B: TPH concentration of the remaining soil at the contaminated site, after excavating the upper part and transporting to the bioreactor

obtained by each chemical laboratory from each one of the 20 different sampling locations. Tables 4 and 5 show the average and standard deviation of TPH concentration values of both laboratories for each sampling location. TRPH results measured by the IR method in most case, showed higher values than those of the GC method, since the GC results reported only on BTEX (benzene, toluene, ethylbenzene, xylene) concentrations (Puri et al., 1994).

It should be noted that the project described in this paper was the first one in Israel of such a large scale. Therefore, some changes in initial decisions were made

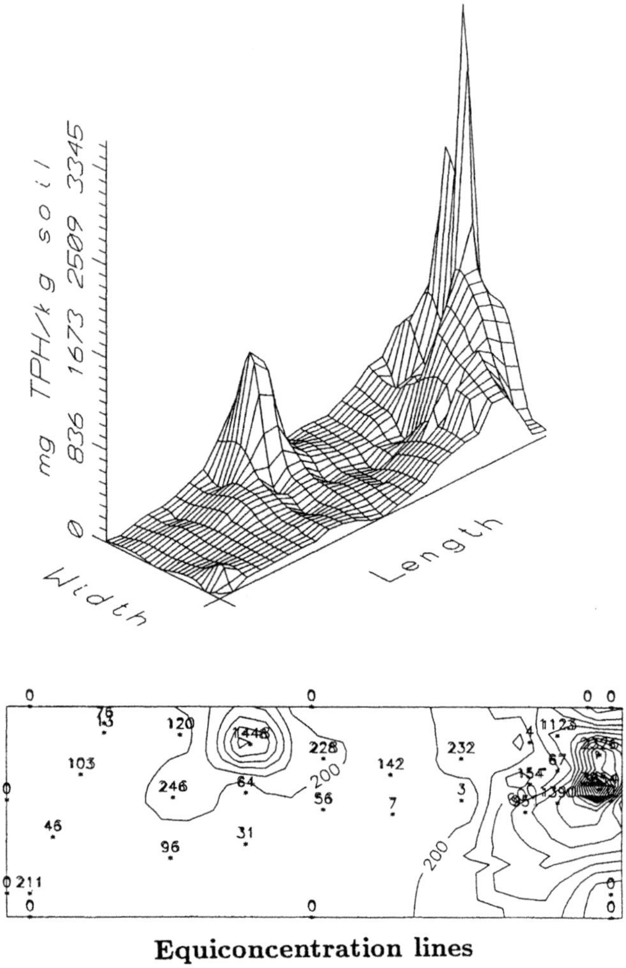

Fig. 4. Site C: TPH concentration of the contaminated upper part of the remaining soil, after excavating the upper part and transporting to the bioreactor

during the performance of the project. Experts of the Ministry of the Environment were involved in all stages of the project, including monitoring and testing. They decided to accept the GC results, obtained by Spectrolab, as a better representation of environmental risks. Table 4 indicates that the residual TPH concentration in the small bioreactor, after 12 months of biotreatment, was between 580 to 1870 mg/kg soil, according to GC measurements. Although such results do not comply with the complete remediation goal, the Ministry of the Environment allowed to transfer the treated soil back to the original two sites with no additional treatment. Table 5 indicates that according to the GC measurements, the average residual TPH concentration in the large reactor was between 560 to 6650 mg/kg soil.

Table 3. Residual TPH concentration of the treated soil after 6 months of the bioreactor operation. Random samples were taken from each of the six heaps, from three depths at the same location. The samples were mixed to obtain an average value

Site	Random samples from six different locations (average mg TPH/kg soil)
1	1010
2	199
3	2422
4	1675
5	2198
6	5025

Table 4. Residual TPH determined by GC and IR methods after 12 months of soil treatment in the small reactor. Samples were taken from 1, 2 and 3 m depths in each location, and the average was calculated

Site	IR or GC	Residual TPH concentration (mg TPH/kg soil)			Average mg TPH per kg soil	± Standard deviation
		Sample 1	Sample 2	Sample 3		
1	IR	2698	2582	2577	2619	56
	GC	871	784	678	778	79
2	IR	4956	5185	5018	5053	97
	GC	2037	1695	1866	1866	171
3	IR	3805	3332	3848	3662	234
	GC	579	620	539	579	40
4	IR	6240	6048	5873	6054	150
	GC	1146	1653	1255	1351	218

Treated soil of the large reactor, whose TPH concentration was lower than 1500 mg/kg soil, when determined by the GC method, was allowed to be transferred back to the original small sites, with no further treatment. Treated soil with residual TPH concentration above 1500 mg/kg soil had to be subjected to further treatment.

5 Additional Treatment

The Ministry of the Environment required the soil, which needed further treatment, to be treated by the same bioremediation technology. Therefore, it was decided to construct a bioreactor on the original excavated large site. On top of the exposed soil, low density polyethylene liners were placed. On top of these liners the

Table 5. Residual TPH determined by GC and IR methods after 12 months of soil treatment in the large reactor. Samples were taken from 1, 2 and 3 m depths in each location, and the average was calculated

Site	IR or GC	Residual TPH concentration (mg TPH/kg soil)			Average mg TPH per kg soil	± Standard deviation
		Sample 1	Sample 2	Sample 3		
5	IR	3643	3484	3452	3526	83
	GC	496	652	527	558	67
6	IR	4935	5022	5080	5012	60
	GC	962	1495	1228	1228	266
7	IR	4750	4944	4950	4881	93
	GC	1428	1501	1306	1412	80
8	IR	3466	3050	3931	3482	360
	GC	1165	1090	1270	1175	74
9	IR	5760	5813	5743	5772	30
	GC	3038	3306	3580	3308	221
10	IR	5840	5421	5593	5618	172
	GC	3091	3566	3250	3302	197
11	IR	1577	1398	1462	1479	74
	GC	3093	3001	2820	2971	113
12	IR	2314	2418	2347	2360	43
	GC	4440	4303	3870	4204	243
13	IR	8560	8660	7430	8219	559
	GC	4642	4876	4210	4576	276
14	IR	8807	8100	8944	8617	370
	GC	6726	6408	6820	6651	176
15	IR	8645	8784	9122	8850	200
	GC	6241	6035	6440	6239	165
16	IR	9281	9145	9553	9326	170
	GC	6017	6338	5920	6092	179
17	IR	7779	6956	7211	7325	344
	GC	4317	4937	4050	4435	371
18	IR	7688	7484	7096	7423	254
	GC	2876	2964	2475	2772	213
19	IR	6046	5602	5672	5773	195
	GC	4415	4991	4680	4695	235
20	IR	7465	7589	7140	7398	189
	GC	3292	4021	3742	3685	300

bioreactor was constructed in a similar way to the previous bioreactors. The bioreactor was operated by continuously feeding an aqueous nutrient solution, acclimated microorganisms, and recycled leachates. Aeration was supplied by a vacuum system. It was required that the new bioreactor would continuously operate for an additional 6 months. At the end of this period, no measurement of TPH concentration was required. It was assumed that the isolation of the bioreactor from the soil would prevent leaching of hydrocarbons into the environment.

6 Discussion

This case study of bioremediation is of the largest scale carried out so far in Israel. Due to the complexity of the project, the authorities involved with quality of water and the environment were not fully prepared to anticipate all difficulties entailed in such a project. It was difficult to judge what should be an appropriate criterion for the achievement of satisfactory soil reclamation.

Collective experience in measuring the concentration of various petroleum hydrocarbon constituents in the environment has led scientists to methodologies assessing the potential human health risks of sites affected by petroleum hydrocarbons. To assess these risks, regulatory agencies require information about specific constituents of the petroleum hydrocarbons. Some constitutents are known to be responsible for virtually all of the environmental risks associated with petroleum hydrocarbons (Zemo et al. 1995). Therefore, a risk assessment approach dictates that the analytical suite for reclamation projects should consist of constituent-specific analyses, and no total petroleum hydrocarbon analyses.

The regulatory authorities, in the present reclamation project, adapted a reclamation target, which was not virtually based on risk calculations. Furthermore, it was difficult for them to decide which test protocol should be used for the analysis of soil samples. The diesel oil contamination occurred in the Negev Desert, where the soil is loamy. Review of reclamation studies has not provided sufficient data for the prediction of bioreactor operation under such conditions.

The study indicated that soil permeability is probably one of the crucial factors affecting the efficiency of the bioreactor. In the present case, the very low permeability of the loamy soil was improved by its mixing with hay. But the improvement of the soil permeability was not sufficient to achieve the goals imposed by the Ministry of the Environment. Probably the biodegradation of the hay added organic materials, which were also measured by IR spectrophotometry. It seems that the target and remediation criteria imposed by the Ministry could be met only by improving the permeability of this soil. It seems that mixing with another type of sandy soil could be somehow more effective, and would not increase the organic content in the treated soil. However, the adequate mixing of the loam soil with a permeable agent should be studies prior to carrying out a full-scale reclamation project.

Criteria for safety of the environment should be based on adequate information, and targets of reclamation should be somehow connected with pilot plant studies and gained experience. In a pilot plant, all parameters of bioreactor performance can be studied and evaluated. The pilot plant study should take into account site-specific parameters, as well as criteria of safety and hazards to the environment.

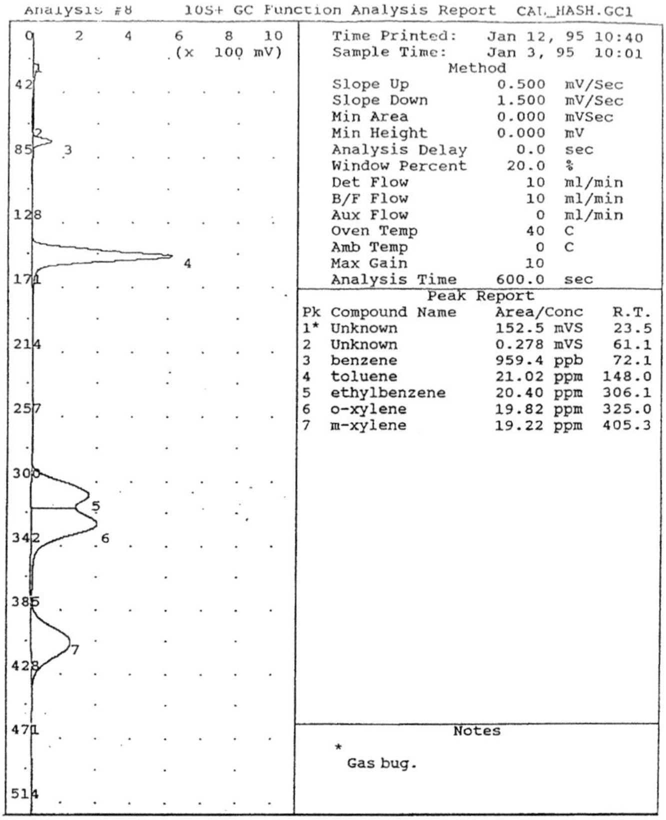

Fig. 5. GC analysis of the BTEX composition in the original diesel

The study showed that targets of bioremediation should be carefully checked by the employment of adequate criteria connected with appropriate risk calculations. There is no point in adopting criteria from other sites that cannot be achieved in the particular site subjected to reclamation.

Figure 5 shows the GC analysis of the raw diesel fuel which contaminated the site. According to this figure, the fuel contained mainly BTEX monoaromatic compounds. According to the monitoring test by GC chromatography, these BTEX compounds were efficiently removed to a residual concentration of less than 1% of the initial value. Such a bioreactor performance is usually considered as very satisfactory. Such facts justified the tendency of the Ministry of the Environment to accept the maximum residual BTEX concentration of 1500 mg/kg soil as sufficient. Furthermore, the lack of an aquifer underneath the contaminated sites, emphasizes the minor risks to the desert environment by such a decision. In various places the reference to TPH does not adequately represent environmental risks. It is suggested that only BTEX compounds are considered as materials which have potential risks to the environment (Zemo et al., 1995), whereas paraffins and olefins, which are also measured by IR spectroscopy, are much less toxic and are not as hazardous as the BTEX compounds.

7 Conclusions

1. In the event of considerable NAPL spill, a careful evaluation of all possible soil remediation methods should be performed.
2. The selected option for soil remediation should be based on guidelines incorporating the negative effect of the NAPL spill on the environment and public health and should identify a possible endpoint target of the selected remediation method.
3. Site-specific conditions like climatic characteristics, soil properties and properties of the spilled NAPL should be taken into account during the planning of the remediation procedure.
4. Remediation methods, not based on off-site soil disposal and incineration, should initially be evaluated by small-scale pilot experiments.
5. Criteria and definitions of successful remediation should be carefully adopted. They should be adequately connected with possible damage to the environment and public health.
6. It is very much recommended that guidelines for the evaluation of remediation targets be defined on national and international levels.
7. A possible target for the reclamation of soil which has been highly contaminated by NAPLs, using bioremediation, should take into account the fact that in many cases, it will not be possible to achieve a 99% cleanup of the contaminant.
8. Bioremediation with the target of almost total removal of contaminating NAPL requires oxygen and moisture penetration to almost all pores of the soil which are contaminated by residual and entrapped NAPL. In many cases, compliance with such a requirement can be achieved only by increasing the permeability of the contaminated soil. Therefore, methods of in situ bioremediation of the contaminated soil cannot be considered in such cases, and the soil should be excavated and treated to increase its permeability. Later it can be subjected to bioremediation. Adequate methods for increasing the permeability of the excavated soil should be evaluated on a small scale prior to performing the full scale treatment, aiming at the increase of the soil permeability. If the soil permeability is increased by mixing the soil with an organic agent, then IR analysis does not represent only the degradation products of NAPL, but also reflects the biodegradation products of the added organic matter.

References

EPA (1986) Test methods for evaluating solid waste. US Environmental Protection Agency, Doc No SW-84b, 3rd edn, Washington, DC

Puri VK, Das BM, Cook EE, Shin EC (1994) Geotechnical properties of crude oil contaminated sand. In: O'Shay TA, Hoddinott KB (eds) Analysis of soil contaminated with petroleum constituents. ASTM STP 1221:61–74

Zemo DA, Graf TE, Embree JW, Bruya JE, Graves KL (1995) Recommended analytical suite for soil and groundwater samples affected by petroleum hydrocarbons. Regional Water Quality Control Board-SF Bay Region, San Francisco

Subject Index

Druck: Strauss Offsetdruck, Mörlenbach
Verarbeitung: Schäffer, Grünstadt

GPSR Compliance

The European Union's (EU) General Product Safety Regulation (GPSR) is a set of rules that requires consumer products to be safe and our obligations to ensure this.

If you have any concerns about our products, you can contact us on ProductSafety@springernature.com

In case Publisher is established outside the EU, the EU authorized representative is:

Springer Nature Customer Service Center GmbH
Europaplatz 3
69115 Heidelberg, Germany

Batch number: 10755284

Printed by Printforce, the Netherlands